Divine Tito F. Wanduku, Gangaram S. Ladde, Jay G. Ladde

Epidemiological Processes in the Biological and Social Sciences

De Gruyter Series in Probability and Stochastics

Volume 6

Divine Tito F. Wanduku, Gangaram S. Ladde,
Jay G. Ladde

Epidemiological Processes in the Biological and Social Sciences

Stochastic Hierarchical Complex Dynamic Network-centric Models

DE GRUYTER

Mathematics Subject Classification 2020
37C75, 34H15, 34H05, 54A05, 60H10, 60H40, 60H35, 60H05, 62E10, 62E15, 62R20, 62R40, 62R10, 92B05, 92B15, 91C20, 91D30, 91D25, 05C90

Authors
Divine Tito F. Wanduku, Ph. D.
Associate Professor of Statistics, & Director, Statistical Consulting Unit (SCU)
Department of Mathematical Sciences
Georgia Southern University
65 Georgia Ave., P.O. Box 8093, Room 3309
Statesboro, GA 30460
USA
dwanduku@georgiasouthern.edu,
wandukudivine@yahoo.com

Jay G. Ladde, MD, FACEP, FAAEM
Senior Associate Program Director of Emergency Medicine, Orlando Health, Orlando & Professor of Emergency Medicine
University of Central Florida
College of Medicine
6850 Lake Nona Blvd.
Orlando, FL 32827
USA
jay.ladde@epcfonline.com

Gangaram S. Ladde, Ph. D.
Professor of Mathematics and Statistics
Founder and Editor: Journal of Stochastic Analysis and Applications
Department of Mathematics and Statistics
4202 East Fowler Avenue, CMC 342
Tampa, FL 33620-5700
USA
gladde@usf.edu

ISBN 978-3-11-161293-5
e-ISBN (PDF) 978-3-11-161321-5
e-ISBN (EPUB) 978-3-11-161395-6
ISSN 2512-9007

Library of Congress Control Number: 2025931474

Bibliographic information published by the Deutsche Nationalbibliothek
The Deutsche Nationalbibliothek lists this publication in the Deutsche Nationalbibliografie; detailed bibliographic data are available on the Internet at http://dnb.dnb.de.

Typesetting: VTeX UAB, Lithuania
Printing and binding: CPI books GmbH, Leck

www.degruyter.com
Questions about General Product Safety Regulation:
productsafety@degruyterbrill.com

MIX
Papier | Fördert gute Waldnutzung
FSC® C083411

To Titomia & Janet Wanduku, whose support made this work possible

This monograph is also dedicated to Professor Gangaram S. Ladde on the occasion of his 84th birthday, in recognition of his profound contributions to mathematics and his inspiring leadership that continues to guide and shape the field

Preface: Foundations and advances in multiscale network population dynamics

This monograph explores the intricate interplay between social, biological, and ecological systems, focusing particularly on hierarchical network structures and stochastic processes within population dynamics. The primary aim is to deepen our understanding of social connectivity and interactions, disease transmission, human mobility, and the emergence of complex behaviors in finite-scale network populations. To this end, the work draws on mathematical modeling, stochastic analysis, and real-world examples, offering a unified framework that can be applied to study these dynamics across various scales and disciplines.

The journey begins in Chapter 2, where we introduce hierarchical network metapopulation models rooted in compact topological spaces. These models are designed to capture the nested layers of human interaction, ranging from local interpersonal connections to global social and mobility networks. This foundational chapter sets the stage for more advanced investigations, emphasizing the dynamic and spatial-temporal evolution of population structures, which is a key feature of complex systems.

In Chapter 3, we dive deeper into the complexities of finite-scale network populations, exploring how human mobility patterns and social structures give rise to heterogeneities within the network. A systematic method for quantifying the nodes within these networks is developed, enabling us to better understand the effects of local interactions on global phenomena, such as disease transmission and the spread of information or rumors.

Building on this foundation, Chapter 4 introduces randomness into the network connectivity model, an essential feature of real-world systems, where connections often emerge unpredictably. In this chapter, random networks, including the Erdős–Rényi model, are employed to study the stochastic nature of connectivity within hierarchical populations. This exploration provides critical insights into the probabilistic aspects of network dynamics.

In Chapter 5, human mobility is incorporated into the network framework, underscoring its critical role in shaping global interconnectedness and influencing the spread of epidemics. This chapter introduces dynamic models for mobility within two-scale networks and extends these findings to large-scale populations, providing valuable insights into the systemic behaviors of modern societies and their vulnerability to external shocks like pandemics.

Chapter 6 transitions into the realm of stochastic epidemic modeling. Expanding upon the framework introduced in Chapter 5, this chapter models the spread of infectious diseases within network populations using stochastic differential equations (SDEs) and Markov chains. By incorporating randomness into epidemic models we achieve a more nuanced understanding of disease trajectories and the impact of human mobility on the dynamics of disease spread.

https://doi.org/10.1515/9783111613215-201

In Chapter 7, we refine the dynamics of disease spread by introducing time delays into the stochastic models. Time delays, such as those related to incubation periods or immunity loss, are pivotal for accurately modeling diseases like measles or malaria. This chapter explores how delays influence epidemic dynamics through the use of delay differential equations (DDEs), accounting for the time lags in disease progression and recovery.

Finally, Chapter 7.5 extends the model by incorporating distributed delays to better capture the variability in immunity periods across individuals. By comparing fixed and distributed delay models this chapter provides deeper insights into immunity dynamics and offers strategies for controlling disease outbreaks, highlighting the critical role of vaccination and other public health interventions.

Together, these chapters provide a comprehensive and evolving framework for understanding population dynamics within the context of stochastic processes, hierarchical networks, and epidemic modeling. The insights gained from these chapters have wide-ranging applications, from epidemiology to network science, and are intended to inform both theoretical development and practical strategies in addressing complex societal challenges.

The content of this monograph is grounded in original research as well as fresh perspectives and generalizations drawn from a collection of scientific, peer-reviewed papers by Wanduku and Ladde [47, 48, 46, 45, 49, 32]. These works have contributed significantly to the development of the theoretical framework presented here, offering both foundational theories and innovative extensions to existing models.

This monograph is written for researchers and practitioners in epidemiology, network science, mathematical modeling, and related fields, with the aim of advancing our understanding of complex, interconnected systems and providing tools to address pressing global challenges.

Acknowledgment

We would like to thank all those who have contributed to this work through discussions, support, and invaluable feedback.

https://doi.org/10.1515/9783111613215-202

Contents

List of Figures

https://doi.org/10.1515/9783111613215-203

List of Tables

https://doi.org/10.1515/9783111613215-204

1 Introduction to stochastic hierarchical dynamic network models and topological elements

1.1 Applications of topology in hierarchical network-centric dynamic modeling

The rapid advancements in technology, scientific development, and educational expansion have brought about significant heterogeneity within the human species. This heterogeneity gives rise to diverse structures in human population dynamics, such as **(i)** spatial structures—citizen, permanent resident, visitor, etc., within human metapopulations, **(ii)** behavioral structures, such as adherence to various preventive measures against infectious diseases, and **(iii)** epidemiological structures, such as different stages of infection progression or varying levels of resistance within a population. As a result, a single human population can be seen as a multipopulation, reflecting this inherent diversity.

As population heterogeneity increases, so does the complexity of interactions between patches and subgroups within this multipopulation structure. For example, increased human mobility between metapopulation patches and interpersonal interactions leads to behavioral changes and further enhances population heterogeneity. Understanding and modeling these interactions provide a foundation for exploring the microscopic factors influencing population dynamics.

A proper understanding of the general topology of human multipopulations is essential for investigating the complex factors affecting these structures. During an epidemic, for instance, disease transmission and contact tracing efforts rely on a network of interpersonal interactions within the population structure. Similarly, the spread of a disease across metapopulations depends on interconnected patches and factors such as human mobility. These dynamic features highlight the critical role of topology in representing and analyzing human population structures.

Moreover, human populations are subject to continuous deformation and transformation over time due to sociocultural and behavioral shifts or physical changes from human mobility. The concepts of *Mathematical Topology* provide tools to understand properties of population structures that remain invariant under continuous deformation. These concepts form the foundation for studying the dynamics of changing populations.

Over the centuries, human societies have progressively developed closer bilateral relationships. In recent years, advancements in communication, transportation, and basic services have facilitated multilateral interactions, making the world resemble a global neighborhood. National and binational challenges have increasingly become global issues, fostering cooperation and understanding of the basic needs of humanity. This process of globalization extends to nearly every aspect of human life.

A key driver of globalization is *human mobility*. Cultural exchanges, the spread of ideas, the emergence of new infectious diseases, and significant world events such as

https://doi.org/10.1515/9783111613215-001

pandemics and outsourcing reflect the impact of human movement. Historical events like the 2009 H1N1 pandemic, the 1918–1919 influenza pandemic, and the recent COVID-19 pandemic underscore the role of mobility in shaping global sociocultural and epidemiological landscapes. This book dedicates a chapter to examining and modeling human mobility and its implications.

The spread of infectious diseases within multipopulation structures influenced by human mobility often displays variable epidemiological parameters. For example, disease incidence rates may vary spatially among subgroups, transmission rates may differ across subpopulations, and basic reproduction numbers may reflect both transmission rates and mobility patterns. Incorporating these factors into a *multipopulation epidemic model* increases its complexity but enhances its realism and representativeness.

Many natural and artificial systems are influenced by *noise* or random variability, which is unavoidable due to internal and external factors. Epidemic dynamics, for instance, may fluctuate due to random variations in poverty rates, sporadic funding for treatments, seasonal changes, or natural disasters. These fluctuations introduce different types of noise, such as *Brownian motion, Markov chain switching, or Lévy jump processes*. Epidemic models incorporating these noises are expressed as systems of *Itô–Doob stochastic differential equations, stochastic differential equations with Markovian switching*, or *stochastic differential equations with Lévy jumps*, respectively.

In dynamical systems theory, an essential aspect is the study of *stochastic differential equations (SDEs)* and *ordinary differential equations (ODEs)*, along with the qualitative analysis of their solutions. Stability theory plays a central role, examining the behavior of system trajectories near equilibria. While deterministic systems follow unique trajectories, stochastic systems exhibit multiple possible paths influenced by random factors. In epidemic models based on SDEs or ODEs, the stability of *disease-free* or *endemic equilibria* determines whether a disease will be eradicated or persist in the population.

This monograph focuses on *stochastic modeling* and *analysis techniques* applied to complex multipopulation epidemic systems. These systems are based on epidemiological processes within topological spaces of interconnected human populations and metapopulations influenced by human mobility and network-centric interactions.

The rest of this chapter focuses on basic definitions and elements of topological and probability spaces used in this book, as well as on the fundamentals of stochastic processes and network/graph theory. Additional references for the concepts in this chapter include [24, 4, 12, 17, 53].

1.2 Basic notations of mathematical topology

In this section, we briefly review fundamental mathematical concepts, terminology, and notations from *Mathematical Topology* that are essential for modeling complex network populations with interconnected levels of human interactions.

The *power set* of a set B, denoted by 2^B, is the set containing all subsets of B, including the set B and the empty set $\emptyset$.

Definition 1.2.1. Given a set X, a *topology* on X is a collection $\mathfrak{T}$ of subsets of X (i. e., $\mathfrak{T} \subset 2^X$) that satisfies the following:
(1) $\emptyset \in \mathfrak{T}$ and $X \in \mathfrak{T}$.
(2) For any $A \in \mathfrak{T}$ and $B \in \mathfrak{T}$, we have $A \cap B \in \mathfrak{T}$, that is, $\mathfrak{T}$ is closed under finite intersection.
(3) For any collection $\{U_i\}_{i\in I} \in \mathfrak{T}$, we have that $\bigcup_{i\in I} U_i \in \mathfrak{T}$, that is, $\mathfrak{T}$ is closed under countable union.

The elements of the topology $\mathfrak{T}$ are commonly called *open sets*, and the elements of the space X are called *points*. The complements $X \setminus A$ of $A \in \mathfrak{T}$ are called *closed sets*. The pair $(X, \mathfrak{T})$ is called a *topological space*.

It is easy to see that in any topological space X, any finite intersection of open sets is open and any countable union of open sets is also an open set.

An example of a topological space is the pair $(\mathbb{R}, \mathfrak{T})$, where $\mathfrak{T}$ is the *natural topology* or *usual topology* on the set of all real numbers $\mathbb{R}$ for which the open sets are open intervals $(a, b) = \{x \mid a < x < b\}$ and their unions. Another example of a topology on any set X is the *discrete topology*, which is the *power set* 2^X of any space X. Also, if the space X is finite, then the natural topology is assumed to be the discrete topology or the power set 2^X on X, unless another topology is selected. In the discrete topology, since all sets and their complements belong to the topology, it is easy to see that all sets are both open and closed.

Some sets are neither open nor closed, e. g., the half-open intervals $[a, b)$ or $(a, b]$ in $\mathbb{R}$, where $a < b$. On the other hand, in any topological space X, the empty set $\emptyset$ and the set X are both open and closed. A topology can also be defined on a *metric space* X by using the notion of open sets in the space and not necessarily the notion of "distance" in the space. Indeed, a metric space is simply a set X with notion of distance between points in the set, called a *metric*. Formally, a metric space is defined as follows.

Definition 1.2.2. Given a set X, a metric for the set X is a positive real-valued function $d : X \times X \to \mathbb{R}_+ = \{x \in \mathbb{R} \mid x \geq 0\}$ that satisfies the following:
(i) For all $x \in X$, $d(x, x) = 0$.
(ii) For all $x, y \in X$, $d(x, y) = d(y, x)$.
(iii) For all $x, y, z \in X$, $d(x, z) \leq d(x, y) + d(y, z)$.
(iv) For all $x, y \in X$, $d(x, y) = 0 \iff x = y$.

The pair (X, d) is called a *metric space* with metric d.

Examples of metric spaces are (1) the set of real numbers $\mathbb{R}$ with the natural metric given by $d(x,y) = |x - y|$. (2) The two-dimensional Euclidean space $\mathbb{R}^2$ with the standard metric $d_2(x, y) = \sqrt{(x_1 - y_1)^2 + (x_2 - y_2)^2}$ for all $x = (x_1, x_2), y = (y_1, y_2) \in \mathbb{R}^2$. In general, the *Euclidean metric* d_2 on the n-dimensional space $\mathbb{R}^n$, where $n > 0, n \in \mathbb{Z}$, is an integer, is defined as follows:

$$d_2(x,y) = \left(\sum_{k=1}^{n} |x_k - y_k|^2 \right)^{\frac{1}{2}}, \tag{1.1}$$

where $x = (x_1, x_2, \ldots, x_n), y = (y_1, y_2, \ldots, y_n) \in \mathbb{R}^n$. In general, various metrics can be defined on $\mathbb{R}^n$ or on subspaces of $\mathbb{R}^n$. However, the Euclidean metric is the natural or standard metric on $\mathbb{R}^n$ and all subspaces of $\mathbb{R}^n$. A generalization of the natural metric on $\mathbb{R}^n$ is given as follows. These metrics are called l^p-metrics (note the small "l"), where $1 \le p \le \infty$.

$$d_p(x,y) = \left(\sum_{k=1}^{n} |x_k - y_k|^p \right)^{\frac{1}{p}}, \quad p \in [1, \infty), \tag{1.2}$$

$$d_\infty(x,y) = \max_{1 \le k \le n} |x_k - y_k|. \tag{1.3}$$

Observe that $\lim_{p \to \infty} d_p(x,y) = d_\infty(x,y)$, and essentially, the metrics in (1.2)–(1.3) are *equivalent* (see [12, 1] for the definition of equivalent metrics).

An important characterization of a metric space (X, d) is its ***completeness*** (see [12]). A sequence $\{x_n\}_{n\ge1} = \{x_1, x_2, \ldots\} \subset X$ in the space X is said to be ***d-Cauchy*** if for every $\epsilon > 0$, there exists an integer $N > 0$ such that $d(x_n, x_m) < \epsilon$ whenever $n, m \ge N$. Suppose that for every d-Cauchy sequence $\{x_n\}_{n\ge1} \subset X$, there exists $x \in X$ such that $\lim_{n\to\infty} x_n = x$, i. e., every d-Cauchy sequence in (X, d) converges to a point in the space. Then the metric space (X, d) is said to be *complete*. Note that in the metric space (X, d), $\lim_{n\to\infty} x_n = x$ is equivalent to $\lim_{n\to\infty} d(x_n, x) = 0$.

Metrics can also be introduced on function spaces, that is, a vector space or a one-dimensional space of functions. A very common example of a function space on the set $\mathbb{R}$ is the space of bounded functions on a set $S \subset \mathbb{R}$, denoted by $B(S)$, that is,

$$B(S) = \{f \mid f : S \to \mathbb{R}; f \text{ is bounded on } S\}. \tag{1.4}$$

A subspace of $B(S)$ in (1.4) is the space of continuous functions on a *compact* subset $K \in \mathbb{R}$ (K is a closed and bounded subset), denoted by $C(K)$, that is,

$$C(K) = \{f \mid f : K \to \mathbb{R}; f \text{ is continuous on } K\}. \tag{1.5}$$

A "usual" metric on $B(S)$ is the *uniform or supremum metric* defined by

$$d_\infty(f,g) = \sup_{x\in S} |f(x) - g(x)|, \quad f,g \in B(S). \tag{1.6}$$

It is easy to see that for the metric space $(C(K), d_\infty)$, (1.6) becomes

$$d_\infty(f,g) = \max_{x\in K} |f(x) - g(x)|, \quad f,g \in C(K). \tag{1.7}$$

Another subspace of $B(S)$ is the function space L^p (note the capital "L") on compact subsets of $\mathbb{R}$, where $1 \le p \le \infty$. The L^p spaces are called *p-integrable spaces* and defined as follows. Let $a < b$, $a, b \in \mathbb{R}$. For $1 \le p < \infty$,

$$L^p([a,b]) = \left\{ f \mid f : [a,b] \to \mathbb{R};\ f \text{ is measurable on } [a,b];\ \int_a^b |f(x)|^p dx < \infty \right\}. \tag{1.8}$$

The metrics (natural) on the L^p spaces are given as follows. For $1 \le p < \infty$, define

$$d_p(f,g) = \left(\int_a^b |f(x)|^p dx \right)^{\frac{1}{p}}, \quad f, g \in L^p([a,b]), \tag{1.9}$$

and for $p = \infty$, define

$$d_\infty(f,g) = \sup_{x \in [a,b]} |f(x) - g(x)|, \quad f, g \in L^\infty([a,b]). \tag{1.10}$$

Note the following about L^p spaces. Clearly, the boundedness of the functions in L^∞ is straightforward. Also, the L^p spaces, for all $1 \le p \le \infty$, are *Banach spaces*, i. e. they are *complete* [12, 1]. However, for the subspace of continuous functions, only L^∞ is a Banach space, and the other L^p spaces, $1 \le p < \infty$ are not complete.

Apart from metrics, a distance on the n-dimensional vector space $\mathbb{R}^n$ can also be defined by a *norm*, denoted by $\|\cdot\|$. In general, norms are introduced on vector spaces, and the norm of a vector is interpreted as its "length". Formally, a norm is defined as follows.

Definition 1.2.3. A norm on $\mathbb{R}^n$ is a nonnegative real-valued function $x \mapsto \|x\|$, where $x = (x_1, x_2, \ldots, x_n) \in \mathbb{R}^n$, that satisfies the following properties:
(i) $\|x\| \ge 0$, and $\|x\| = 0$ if and only if $x = 0$.
(ii) $\|x - y\| = \|y - x\|$.
(iii) $\|ax\| = |a|\|x\|$ for all $a \in \mathbb{R}$.
(iv) $\|x - y\| \le \|x\| + \|y\|$ for all $x = (x_1, x_2, \ldots, x_n), y = (y_1, y_2, \ldots, y_n) \in \mathbb{R}^n$.

The distance between the two points $x = (x_1, x_2, \ldots, x_n), y = (y_1, y_2, \ldots, y_n) \in \mathbb{R}^n$ is defined as $\|x - y\|$. In general, if X is a vector subspace of $\mathbb{R}^n$ endowed with a norm $\|\cdot\|$, then the pair $(X, \|\cdot\|)$ is called a *normed space* or *normed linear space*.

The standard *Euclidean norm* on the vector space $\mathbb{R}^n$ is defined similarly as in (1.1):

$$\|x\|_2 = \left(\sum_{k=1}^{n} |x_k|^2 \right)^{\frac{1}{2}}. \tag{1.11}$$

Other, nonstandard norms on $\mathbb{R}^n$ are defined as follows:

$$\begin{aligned} \|x\|_p &= \left(\sum_{k=1}^{n} |x_k|^p \right)^{\frac{1}{p}}, \quad p \in [1, \infty), \\ \|x\|_\infty &= \max_{1 \le k \le n} |x_k|. \end{aligned} \tag{1.12}$$

Given a normed vector space $(X, \|\cdot\|)$, the norm induces the metric $d : X \times X \to \mathbb{R}_+$ on the space X defined by

$$d(x,y) = \|x - y\|. \tag{1.13}$$

The metric in (1.13) is called the *norm-induced metric*. Thus a normed space can be considered as a particular type of a metric space, where the metric is derived from the norm as in (1.13). In addition, all metric space properties and different types of metrics in (1.3)–(1.10) are easily translated to the normed space. Further, the metric space notions of *completeness, compactness, and connectedness* (see [12, 1]) are also examined in the normed space. Also, it can be shown that the norms in (1.12) are *equivalent*. In fact, a normed space $(X, \|\cdot\|)$ is said to be a *Banach space* if the induced metric d in (1.13) is *complete*. Fundamental to norms and normed linear spaces are *inner products* and *inner product spaces*, where the norm defined in Definition 1.2.3 is induced by the inner product on the space X. The interested readers for inner product spaces, should consult [3].

Given a metric space (X, d) with metric d, a point $a \in X$, and a real number $r > 0$, the set denoted by $B(a;r)$ or $B_X(a;r)$ and defined as

$$B(a;r) = \{x \in X \mid d(x,a) < r\} \tag{1.14}$$

is called the *open ball centered at a with radius r*. The set denoted by $\bar{B}(a;r)$ or $\bar{B}_X(a;r)$ and defined as

$$\bar{B}(a;r) = \{x \in X \mid d(x,a) \leq r\} \tag{1.15}$$

is called the *closed ball centered at a with radius r*.

Using the notions of open sets, the metric space can be characterized as a particular type of topological space by selecting a topology containing open balls in the metric space. For any metric space (X, d), the topology $\mathfrak{T}$ containing all unions of open balls of the form $B(x;r)$ for $x \in X$ and $r > 0$ is called the *metric topology*. This definition implies that in the metric topology, all open sets are unions of open balls in the metric space.

Thus, from (1.13), given a normed space $(X, \|\cdot\|)$, a natural *metric topology* is induced by the norm of the space via the metric (1.13). Hence the normed space X and the induced metric topology on X form a *topological vector space*. Moreover, if the metric $d(x,y) = \|x - y\|$ is *complete*, then the space X is called a *Banach space*.

Given a topological space $(X, \mathfrak{T})$, any open set $V \in \mathfrak{T}$ is completely characterized by the *base* of the topology. The base of a topology is defined as follows (see [12]).

Definition 1.2.4. (1) A base for a topology $\mathfrak{T}$ is any collection $\mathfrak{U} \subset \mathfrak{T}$ such that any open set $V \in \mathfrak{T}$ can be expressed as a union of open sets in $\mathfrak{U}$ contained in V, that is, $U = \bigcup\{U \in \mathfrak{U} \mid U \subset V\}$.

(2) A *neighborhood* of a point $x \in X$ is any set N (open or not) that contains an open set U containing the point x, that is, $x \in U \subset V$ for some open set U. A *neighborhood-base* for a point $x \in X$ is a collection $\mathfrak{N}$ of neighborhoods N of $x \in X$ such that any neighborhood V of $x \in X$ contains some neighborhood $N \in \mathfrak{N}$, that is, $x \in N \subset V$ for some $N \in \mathfrak{N}$.

It is important to note that if $\mathfrak{U}$ is a base for the topology $\mathfrak{T}$ on X, then

$$X = \bigcup \mathfrak{U}, \tag{1.16}$$

that is, every point of X lies in some open set contained in $\mathfrak{U}$.

In any metric space (X, d), the collection $\mathfrak{U}$ containing all open balls $B(x; r)$ for all $x \in X$ and $r > 0$ is a *base* for the metric topology $\mathfrak{T}$ on X. Also, for a fixed $x \in X$, the collection $\mathfrak{N}$ containing all open balls $B(x; r)$ for fixed $r > 0$ is a *neighborhood base* at the point $x \in X$.

For example, in $\mathbb{R}$, the natural distance norm $|x - y|, x, y \in \mathbb{R}$, induces the natural metric $d(x, y) = |x - y|$. The metric topology is the natural topology induced by the metric d, where all open sets are unions of open balls in $\mathbb{R}$; the base of the topology is the open balls, which in this case are all open intervals of the form $(a, b), a < b$. The Euclidean norm in (1.11) also induces a natural topology on $\mathbb{R}^2$, where the base is the collection $\mathfrak{U}$ of all open balls (open circles) $B_{\mathbb{R}^2}(\vec{x}_0; r) = \{(x_1, x_2) \in \mathbb{R}^2 \mid (x_1 - x_{1_0})^2 + (x_2 - x_{2_0})^2 < r^2\}$, $\vec{x}_0 = (x_{1_0}, x_{2_0}) \in \mathbb{R}^2$.

If $(X, \mathfrak{T})$ is a topological space and $Y \subset X$, then the subcollection $\{U \cap Y \mid U \in \mathfrak{T}\}$ is also a topology, called the *relative topology* or *subspace topology*.

Compact topological spaces (see [12, 1]) play an important rule in optimization of functions defined on these spaces. Associated with the term "compactness" is the term "*covering*". Given a topological space $(X, \mathfrak{T})$ (including metric spaces (X, d)), a *covering* of a subset $A \subset X$ is a collection of sets $\mathfrak{U}$ such that $A \subseteq \bigcup\{U \in \mathfrak{U}\}$, that is, the set A is contained in the union of the sets in $\mathfrak{U}$.

Definition 1.2.5. Given a topological space $(X, \mathfrak{T})$, a subset $A \subset X$ is said to be compact if every open cover of A has a finite subcover. That is, $A \subset X$ is compact if for any collection $\mathfrak{U} \subset \mathfrak{T}$ such that $A \subseteq \bigcup\{U \in \mathfrak{U}\}$, there exists a finite subcollection $\mathfrak{V} \subset \mathfrak{U}$ such that $A \subseteq \bigcup\{V \in \mathfrak{V} \subset \mathfrak{U}\}$.

Thus, when $A = X$, we say that the topological space $(X, \mathfrak{T})$ is compact if for any collection $\mathfrak{U} \subset \mathfrak{T}$ such that $X = \bigcup\{U \in \mathfrak{U}\}$, there exists a finite collection $\mathfrak{V} \subset \mathfrak{U}$ such that $X = \bigcup\{V \in \mathfrak{V} \subset \mathfrak{U}\}$.

The reader is directed to other notable results on compactness and coverings such as the *Heine–Borel, Lindelöp, and Bolzano–Weierstrass* results in references like [1, 12]. These results imply and are also implied by the following result.

Theorem 1.2.1. *Let S be a subset of $\mathbb{R}^n$. Then the following three statements are equivalent:*

(a) *S is compact.*
(b) *S is closed and bounded.*
(c) *Every infinite subset of S has an accumulation point in S.*

Proof. See [1]. □

1.3 Basic notations in probability theory and theory of stochastic processes

Probability theory deals with the mathematical analysis of the concept of "chance" in the context of a random experiment. A *random experiment* refers to any activity or process that produces different possible outcomes, with only one outcome occurring at a time. The outcomes or observations of such an experiment are represented by ω, and the set of all possible outcomes is called the *sample space*, denoted by Ω. The subsets of Ω are referred to as *events* of the random experiment.

The concept of σ-algebras is fundamental in probability theory, as it provides a structured way to collect specific subsets or events of Ω that are considered observable or meaningful. These subsets form the foundation for defining measurable spaces and probability measures, which are essential in the rigorous treatment of stochastic processes.

Definition 1.3.1. A collection $\mathfrak{F}$ of subsets of Ω is called a *σ-algebra* if the following properties are satisfied:
(i) $\emptyset \in \mathfrak{F}$.
(ii) If $A \in \mathfrak{F}$, then $\Omega \setminus A \in \mathfrak{F}$.
(iii) If $\{A_i\}_{i=1}^{\infty} \subset \mathfrak{F}$ is a countable collection of events, then $\bigcup_{i=1}^{\infty} A_i \in \mathfrak{F}$.

The pair $(\Omega, \mathfrak{F})$ is called a ***measurable space***.

We define a probability measure on the measurable space $(\Omega, \mathfrak{F})$.

Definition 1.3.2. A *probability measure P* on the measurable space $(\Omega, \mathfrak{F})$ is a function $P : \mathfrak{F} \to [0, 1]$ that satisfies the following properties:
(i) $P(\Omega) = 1$.
(ii) For any countable collection of disjoint events $\{A_i\}_{i=1}^{\infty} \subset \mathfrak{F}$, that is, $A_i \cap A_j = \emptyset$ for $i \neq j$, we have $P(\bigcup_{i=1}^{\infty} A_i) = \sum_{i=1}^{\infty} P(A_i)$.

The triple $(\Omega, \mathfrak{F}, P)$ is called a ***probability space***. If $\mathfrak{F}$ contains all subsets B of Ω that have a zero outer measure P^* defined by

$$P^*(B) = \inf\{P(G) \mid G \in \mathfrak{F}, B \subset G\} = 0, \tag{1.17}$$

then the triple $(\Omega, \mathfrak{F}, P)$ is called a ***complete probability space***.

If $\mathfrak{H}$ is a collection of subsets of Ω, that is, $\mathfrak{H} \subset 2^{\Omega}$, then there exists a smallest σ-algebra, denoted by $\mathfrak{F}_{\mathfrak{H}}$, that contains $\mathfrak{H}$. The σ-algebra $\mathfrak{F}_{\mathfrak{H}}$ is said to be generated by $\mathfrak{H}$.

If $\Omega = \mathbb{R}$ and $\mathfrak{H}$ is the base for the natural topology on $\Omega = \mathbb{R}$, that is, $\mathfrak{H}$ contains all open intervals of the form $(a, b), a < b$, then $\mathfrak{F}_{\mathfrak{H}}$ is called the *Borel σ-algebra*, denoted by $\mathfrak{B}$. The elements of $\mathfrak{B}$ are called *Borel sets*. Note that $\mathfrak{B}$ as a σ-algebra also contains all open sets, closed sets, countable unions of open sets, finite intersections of its sets, etc.

If $(\Omega, \mathfrak{F}, P)$ is a probability space, then a function $Y : \Omega \to \mathbb{R}$ is called an *$\mathfrak{F}$-measurable function*, or an *$\mathfrak{F}$-measurable random variable*, or simply a *random variable* if the inverse images

$$Y^{-1}(U) = \{\omega \in \Omega \mid Y(\omega) \in U\} \in \mathfrak{F} \quad \text{for all Borel sets } U \in \mathfrak{B}. \tag{1.18}$$

If $X : \Omega \to \mathbb{R}$ is an *$\mathfrak{F}$-measurable function*, then the σ-algebra generated by X, denoted by $\mathfrak{F}_X$, is the smallest σ-algebra containing all sets $Y^{-1}(U)$, $U \in \mathfrak{B}$.

Let $(\Omega, \mathfrak{F}, P)$ be a complete probability space, and let $X : \Omega \to \mathbb{R}^n$ be an *$\mathfrak{F}$-measurable multivariate random variable or random vector*. Note for simplicity that *multivariate random variable* and *random variable* are used interchangeable. The random variable X induces the probability measure $\mu_X : \mathfrak{B} \to [0, 1]$ on $\mathbb{R}^n$ defined by

$$\mu_X(B) = P(X^{-1}(B)) = P(\{\omega \in \Omega \mid X(\omega) \in B\}). \tag{1.19}$$

The measure μ_X is called the *distribution of X*, and $(\mathbb{R}^n, \mathfrak{B}, \mu_X)$ is called the probability space induced by the *$\mathfrak{F}$-measurable random variable X*.

The *expected value* of the random variable X is defined as follows. If $X : \Omega \to \mathbb{R}^n$ is an *$\mathfrak{F}$-measurable random variable* and $E[||X||] = \int_\Omega ||X(\omega)||dP(\omega) < \infty$, then the expected value of X is the number

$$E[X] = \int_{\Omega} X(\omega)dP(\omega) = \int_{\mathbb{R}^n} x d\mu_X(x), \tag{1.20}$$

where the first integral is defined with respect to (w. r. t.) the probability space $(\Omega, \mathfrak{F}, P)$, and the second integral is defined w. r. t. the induced probability space $(\mathbb{R}^n, \mathfrak{B}, \mu_X)$.

In general, in $(\mathbb{R}^n, \mathfrak{B}, \mu_X)$, if $f : \mathbb{R}^n \to \mathbb{R}$ is $\mathfrak{B}$-measurable and

$$E[|f(X)|] = \int_{\Omega} |f(X(\omega))|dP(\omega) < \infty,$$

then the expected value of $f(X)$ is the number

$$E[f(X)] = \int_{\Omega} f(X(\omega))dP(\omega) = \int_{\mathbb{R}^n} f(x)d\mu_X(x). \tag{1.21}$$

1.3.1 Norms for L^p spaces

Clearly, real-valued $\mathfrak{F}$-measurable functions $X : \Omega \to \mathbb{R}^n$ in the probability space $(\Omega, \mathfrak{F}, P)$ and the induced space $(\mathbb{R}^n, \mathfrak{B}, \mu_X)$ necessitate the notion of distance between them. Similarly, the notion of norms on the linear multidimensional spaces $\mathbb{R}^n$ in Definition 1.2.3 are extended to the *$\mathfrak{F}$-measurable functions* X on probability spaces $(\Omega, \mathfrak{F}, P)$ and $(\mathbb{R}^n, \mathfrak{B}, \mu_X)$. The so-called *$L^p$-spaces* of *$\mathfrak{F}$-measurable functions* endowed with *L^p norms* are defined similarly to the usual norms in the space of bounded continuous functions (see [12, 1]).

Definition 1.3.3. Let $X : \Omega \to \mathbb{R}^n$ be an $\mathfrak{F}$-multidimensional random variable.
(1) If $p \in [1, \infty)$, then

$$\|X\|_p = \|X\|_{L^p(P)} = \left(\int_\Omega \|X(\omega)\|^p dP(\omega) \right)^{\frac{1}{p}}. \tag{1.22}$$

(2) If $p = \infty$, then

$$\|X\|_\infty = \|X\|_{L^\infty(P)} = \inf\{N \in \mathbb{R} \mid \|X(\omega)\| \le N \text{ almost surely}\}. \tag{1.23}$$

The spaces of measurable functions

$$L^p(P) = L^p(\Omega; \mathbb{R}^n) = \{X : \Omega \to \mathbb{R}^n \mid \|X\|_p < \infty\} \tag{1.24}$$

are called the L^p-spaces.

Note that the L^p-spaces are complete normed linear spaces, i. e., *Banach spaces*. Also, when $p = 2$, the L^2-space is a complete inner product space, i. e., a *Hilbert space*, where the inner product is defined by the expected dot product as shown below.

$$\langle X, Y \rangle_{L^2} = E[X \bullet Y], \quad X, Y \in L^2. \tag{1.25}$$

The following are some useful inequalities that hold in the L^p-spaces. We refer to [5] for more important inequalities in the L^p-spaces.

1. **Holder's inequality:** If $p > 0$, $\frac{1}{p} + \frac{1}{q} = 1$, and $X \in L^p$, $Y \in L^q$, then

$$|E[X \bullet Y]| \le (E|X|^p)^{\frac{1}{p}} (E|Y|^q)^{\frac{1}{q}}. \tag{1.26}$$

   As an implication of this theorem, we have that if $0 < r < p < \infty$ and $X \in L^p$, then $(E\|X\|^r)^{\frac{1}{r}} \le (E\|X\|^p)^{\frac{1}{p}}$.

2. **Minkovski's inequality:** If $p > 0$ and $X, Y \in L^p$, then

$$(E\|X + Y\|^p)^{\frac{1}{p}} \le (E\|X\|^p)^{\frac{1}{p}} + (E\|Y\|^p)^{\frac{1}{p}}. \tag{1.27}$$

3. **Chebyshev's inequality:** If $c, p > 0$ and $X \in L^p$, then for any positive real-valued function g,

$$P(g(X) \geq c) \leq c^{-1} E[g(X)] \tag{1.28}$$

and, in particular,

$$P(\|X\| \geq c) \leq c^{-p} E\|X\|^p. \tag{1.29}$$

4. **Cauchy–Schwarz inequality:** For any random variables X and Y,

$$|E[X \bullet Y]| \leq E[|X \bullet Y|] \leq (E\|X\|^2)^{\frac{1}{2}} (E\|Y\|^2)^{\frac{1}{2}}. \tag{1.30}$$

In the $L^p(P) = L^p(\Omega; \mathbb{R}^n)$-space, there are different notions of convergence of sequences of random variables (see [5, 24, 53]).

Definition 1.3.4. Let $(\Omega, \mathfrak{F}, P)$ be a complete probability space. Also, let $\{X_n\}_{n=1}^{\infty} \subset L^p(\Omega; \mathbb{R}^n)$ be a sequence of random variables, and let $X \in L^p(\Omega; \mathbb{R}^n)$.

1. The sequence $\{X_n\}_{n=1}^{\infty}$ converges *almost surely* to the random variable X if

$$P\left(\left\{\omega \in \Omega \mid \lim_{n \to \infty} \|X_n(\omega) - X(\omega)\|_p = 0\right\}\right) = 1. \tag{1.31}$$

2. The sequence $\{X_n\}_{n=1}^{\infty}$ converges *in probability* to the random variable X if for all $\epsilon > 0$,

$$\lim_{n \to \infty} P(\{\omega \in \Omega \mid \|X_n(\omega) - X(\omega)\|_p > \epsilon\}) = 0. \tag{1.32}$$

3. The sequence $\{X_n\}_{n=1}^{\infty}$ converges *in pth moment* to the random variable X if

$$\lim_{n \to \infty} E[\|X_n(\omega) - X(\omega)\|^p] = 0. \tag{1.33}$$

4. The sequence $\{X_n\}_{n=1}^{\infty}$ converges *in distribution* to the random variable X if the corresponding sequence of distribution functions $\{F_{X_n}\}_{n=1}^{\infty}$ converges to F_X at all continuity points $x \in \mathbb{R}$ of F_X, that is,

$$\lim_{n \to \infty} F_{X_n}(x) = F_X(x). \tag{1.34}$$

1.3.2 Stochastic processes

Let $(\Omega, \mathfrak{F}, P)$ be a probability space. A *filtration* is a family of σ-algebras $\{\mathfrak{F}_t\}_{t \geq 0}$ such that $\mathfrak{F}_t \subset \mathfrak{F}_s \subset \mathfrak{F}$ for all $0 \leq t < s < \infty$, that is, a filtration is a family of increasing sub-σ-algebras. If the filtration satisfies $\mathfrak{F}_t = \bigcap_{s>t} \mathfrak{F}_s$, then it is said to be *right continuous*.

Note that in this book, a special interest is given to *complete* probability spaces $(\Omega, \mathfrak{F}, P)$ that contain a *right continuous* filtration $\{\mathfrak{F}_t\}_{t\geq 0}$ with sub-σ-algebra $\mathfrak{F}_0$ containing all P-null sets (i. e., all sets $A \in \mathfrak{F}$ with $P(A) = 0$).

A *stochastic process* is a collection of random variables $X = \{X_t, t \in I\}$, also denoted $\{X_t\}_{t\in I}$, or $X_t, t \in I$, or $X(t), t \in I$, defined on a probability space $(\Omega, \mathfrak{F}, P)$, that is,

$$X : \Omega \times I \to \mathbb{R}^n, \tag{1.35}$$

where I is called the *index set*, commonly referred to as the time domain. In this book, $I = \mathbb{R}_+ = [0, \infty)$.

For a fixed $t \geq 0$, the function $\Omega \ni \omega \to X(\omega, t) = X_t(\omega) \in \mathbb{R}^n$, is a random variable. Also, for $\omega \in \Omega$, the function $\mathbb{R}_+ \ni t \to X(\omega, t) = X_t(\omega) \in \mathbb{R}^n$ is called a *path or trajectory* of the stochastic process.

In this book, we give special attention to *diffusion processes* with almost surely continuous paths. In particular, we focus on the continuous-time and continuous-state *Wiener process* or *Brownian motion process* (see [23]). The Brownian motion process takes a central place in the study of stochastic processes due to its occurrence in a wide variety of fields and applications. It is characterized by having a *stationary independent increment process*, where each increment follows a *normal distribution*. The formal definition of the Brownian motion process is as follows.

Definition 1.3.5. Let $(\Omega, \mathfrak{F}, P)$ be a probability space, and let $\{\mathfrak{F}_t\}_{t\geq 0}$ be a filtration on this probability space. Also, let $\{W_t\}_{t\geq 0}$ denote a stochastic process that is continuous in state, time and $\mathfrak{F}_t$-*adapted* (i. e., for all $t \geq 0$, W_t is $\mathfrak{F}_t$-measurable). Since $\{\mathfrak{F}_t\}_{t\geq 0}$ is continuous, $\mathfrak{F}_t$ is generated by the collection of random variables $\{W_s : 0 \leq s \leq t\}$. The process $\{W_t\}_{t\geq 0}$ is called a *Wiener process* or *a Brownian motion* if the following properties hold:

1. **Starts at the origin:** $W_0 = 0$ almost surely.
2. **Stationary increments:** For all $0 \leq s < t < \infty$, the increment $W_t - W_s$ is normally distributed with mean 0 and variance $t - s$.
3. **Independent increments:** For all $0 \leq s < t < \infty$, the increment $W_t - W_s$ is independent of $\mathfrak{F}_s$.

1.4 Stability of stochastic differential equations

Let $(\Omega, \mathcal{F}, P)$ be a probability space with filtration $\mathcal{F} = \{\mathfrak{F}_t\}_{t\geq 0}$.

1.4.1 A stochastic integral

Let $\{b(x(t), t), 0 \leq t \leq T\}$ be a stochastic process for some $T > 0$, where $b(x(t), \cdot) : [0, T) \to \mathbb{R}^m$ is an $\mathfrak{F}_t$-*adapted process* (see [23]), i. e., the random variables $b(x(t), t)$ are

$\mathfrak{F}_t$-measurable, almost surely continuously differentiable with respect to $x(t)$, and continuous with respect to t. In addition, let the expectation

$$\mathbb{E}\left[\int_0^T \|b(x(s),s)\|ds\right] < \infty. \tag{1.36}$$

Similarly, let $\{\sigma(x(t),t), 0 \le t \le T\}$ be a stochastic process for some $T > 0$, where $\sigma(x(t),\cdot) : [0,T) \to \mathbb{R}^m$ is $\mathfrak{F}_t$-measurable, almost surely twice continuously differentiable with respect to $x(t)$, and continuous with respect to t. Furthermore, the expectation

$$\mathbb{E}\left[\int_0^T \|\sigma(x(s),s)\|^2 ds\right] < \infty. \tag{1.37}$$

The *stochastic integral* of σ with respect to the Wierner process $\{W_t, t \ge 0\}$ is defined as

$$\int_{t_0}^t \sigma(x(s),s)dW(s) = \lim_{\mu(\mathcal{P})\to 0} \sum_{k=0}^{n-1} \sigma(x(t_k),t_k)[W(t_{k+1}) - W(t_k)], \tag{1.38}$$

where $\mu(\mathcal{P}) = \max_{0\le k\le n}(t_{k+1} - t_k)$ is the mesh of the partition

$$\mathcal{P} = \{t_0 < t_1 < \cdots < t_k < t_{k+1} < \cdots < t_{n-1} < t_n = t\}. \tag{1.39}$$

For any process $x(t), t \ge 0$, an ***Itô–Doob stochastic differential equation*** (see [23]) of $x(t)$ with respect to a *Wiener or Brownian motion process* $\{w(t), t \ge 0\}$ is defined by the expression

$$dx(t) = b(x(t),t)dt + \sigma(x(t),t)dW(t), \tag{1.40}$$

where the functions $b(x(t),t)$ and $\sigma(x(t),t)$ are defined earlier, and over any small time interval $[t, t+\Delta t]$, a new stochastic process with respect to the Wiener process $\{W(t), t \ge 0\}$ emerges given by

$$x(t+\Delta t) = x(t) + \int_t^{t+\Delta t} b(x(s),s)ds + \int_t^{t+\Delta t} \sigma(x(s),s)dW(s), \quad 0 \le t \le t+\Delta t \le T, \tag{1.41}$$

and at any time $t \in [t_0, T), t_0 \ge 0$, the stochastic process $\{x(t), t \in [t_0, T)\}$ given by

$$x(t) = x(t_0) + \int_t^t b(x(s),s)ds + \int_{t_0}^t \sigma(x(s),s)dW(s), \quad t_0 \le t \le T. \tag{1.42}$$

The process $x(t), t \ge 0$, in (1.42) is called an ***Itô–Doob stochastic process***.

Multidimensional stochastic differential equations are defined as follows.

Definition 1.4.1. Let $W_t = (W_1(t), \ldots, W_m(t))$ be an m-dimensional Brownian motion, and let $b : [0, T] \times \mathbb{R}^n \to \mathbb{R}^n$ and $\sigma : [0, T] \times \mathbb{R}^n \to \mathbb{R}^{n\times m}$ be measurable functions. Then the process $X_t = (X_1(t), \ldots, X_m(t))$, $t \in [0, T]$ is the solution of the stochastic differential equation (SDE)

$$dX_t = b(t, X_t)\mathrm{d}t + \sigma(t, X_t)\mathrm{d}W_t, \tag{1.43}$$

where $b(t, X_t) \in \mathbb{R}^n$, $\sigma(t, X_t) \in \mathbb{R}^{n\times m}$, if for all t, X_t satisfies the equation

$$X_t = X_0 + \int_0^t b(s, X_s)ds + \int_0^t \sigma(s, X_s)dW_s. \tag{1.44}$$

Assume that for every initial value $X_t(0) = X_0 \in R^n$, there exists a unique global solution of (1.43) denoted by $X(t; t_0, X_0)$. Suppose that

$$b(t, 0) \equiv \sigma(t, 0) = 0. \tag{1.45}$$

This implies that equation (1.43) has the trivial solution $X_t(t_0, 0) \equiv 0$, which also is an **equilibrium state** for the SDE system.

The stochastic stability of the trivial solution $X_t(t_0, 0) \equiv 0$ is defined subsequently.

1.4.2 Lyapunov's second method

The Lyapunov stability theory, formalized in 1930, has significant applications in control theory, particularly in assessing the stability of dynamical systems. This stability framework evaluates the system robustness to small perturbations, ensuring that minor deviations from the equilibrium state dissipate over time. If a dynamical system possesses a stable equilibrium state, then all trajectories originating near this equilibrium remain in close proximity to it over time.

The *Lyapunov second method*, widely used for analyzing the stability of deterministic dynamical systems, enables stability determination without requiring explicit solutions to the governing differential equations. Before formally stating the theorem for Lyapunov's second method, we introduce some foundational concepts (see [53, 24]).

Let K denote the family of all continuous nondecreasing functions $\mu : \mathbb{R}_+ \to \mathbb{R}_+$ such that $\mu(0) = 0$ and $\mu(r) > 0$ for all $r > 0$. For $h > 0$, let $S_h = \{x \in \mathbb{R}^n : \|x\| < h\}$ denote an open ball in $\mathbb{R}^n$.

A continuous function $V(x, t)$ defined on $S_h \times [t_0, \infty)$ is said to be ***positive-definite*** if

$$V(0, t) \equiv 0 \quad \text{and} \quad V(x, t) \geq \mu(\|x\|)$$

for some $\mu \in K$ and all $(x, t) \in S_h \times [t_0, \infty)$.

A function $V(x,t)$ is to be ***negative-definite*** if $-V$ is positive-definite. A continuous nonnegative function $V(x,t)$ is said to be ***decrescent*** if

$$V(x,t) \leq \mu(\|x\|)$$

for some $\mu \in K$ and all $(x,t) \in S_h \times [t_0,\infty)$.

A function $V(x,t)$ defined on $\mathbb{R}^n \times [t_0,\infty)$ is said to be ***radially unbounded*** if

$$\lim_{|x|\to\infty} \inf_{t\geq t_0} V(x,t) = \infty.$$

Let $C^{1,1}(S_h \times [t_0,\infty), \mathbb{R}_+)$ denote the family of all continuously differentiable functions $V(x,t)$ on $S_h \times [t_0,\infty)$ with first partial derivatives with respect to the components x and t. Define the derivative of $V(x,t)$ along the system trajectory as

$$\dot{V}(x,t) = \frac{\partial V}{\partial t}(t,x) + \sum_{i=1}^{n} \frac{\partial V}{\partial x_i}(t,x) b_i(t,x),$$

where $b(t,x)$ represents the system vector field. If $\dot{V}(x,t) \leq 0$, then $V(t,X_t)$ does not increase, indicating that the distance of X_t from the equilibrium point, as measured by $V(t,X_t)$, does not grow over time. If $\dot{V}(x,t) < 0$, then $V(t,X_t)$ decreases to zero, ensuring that the distance to the equilibrium diminishes over time.

Theorem 1.4.1 (Lyapunov's second method). *Let $V(x,t) \in C^{1,1}(S_h \times [t_0,\infty), \mathbb{R}_+)$ be a positive-definite function. If*

$$\dot{V}(x,t) = \frac{\partial V}{\partial t}(t,X_t) + \sum_{i=1}^{n} \frac{\partial V}{\partial x_i}(t,X_t) b_i(t,X_t) \leq 0$$

for all $(x,t) \in S_h \times [t_0,\infty)$, then the trivial solution $X_t = 0$ is stable.

If, additionally, $V(x,t)$ is positive-definite and decrescent and if $\dot{V}(x,t)$ is negative-definite, then the trivial solution is asymptotically stable.

The Lyapunov second method, as described in Theorem 1.4.1, extends naturally to stochastic systems, enabling stability analysis in the presence of randomness.

1.4.3 Stability in probability

Definition 1.4.2. (i) The trivial solution of equation (1.43) is said to be **stochastically stable (or stable in probability)** if for every pair $\varepsilon \in (0,1)$ and $r > 0$, there exists $\delta = \delta(\varepsilon, r, t_0) > 0$ such that

$$P\{\|x(t,t_0,x_0)\| < r\} \geq 1 - \varepsilon$$

for all $t \geq t_0$ whenever $\|x_0\| < \delta$. Otherwise, the trivial solution is said to be stochastically unstable.

(ii) The trivial solution of equation (1.43) is said to be **stochastically asymptotically stable** if it is stochastically stable and, moreover, for every $\varepsilon \in (0,1)$, there exists $\delta_0 = \delta_0(\varepsilon, t_0) > 0$ such that

$$P\Big\{\lim_{t\to\infty} x(t, t_0, x_0) = 0\Big\} \geq 1 - \varepsilon$$

whenever $|x_0| < \delta_0$.

(iii) The trivial solution of equation (1.43) is said to be **stochastically asymptotically stable in the large** if it is stochastically stable and, moreover, for all $x_0 \in \mathbb{R}^n$,

$$P\Big\{\lim_{t\to\infty} x(t, t_0, x_0) = 0\Big\} = 1.$$

1.4.4 Extension of Lyapunov's second method to stochastic systems

We now extend Theorem 1.4.1 to stochastic systems. Let $0 < h \leq \infty$. Denote by $C^{2,1}(S_h \times \mathbb{R}_+, \mathbb{R}_+)$ the family of all nonnegative functions $V(x,t)$ defined on $S_h \times \mathbb{R}_+$ that are continuously twice differentiable in x and once in t.

The differential operator L associated with equation (1.43) is defined as

$$L = \frac{\partial}{\partial t} + \sum_{i=1}^{n} \frac{\partial V}{\partial x_i}(t,x) b_i(x,t) + \frac{1}{2} \sum_{i,j=1}^{n} \frac{\partial^2 V}{\partial x_i \partial x_j} [\sigma(x,t)\sigma^T(x,t)]_{ij}.$$

In the stochastic case, to establish stability, the inequality $\dot{V}(x,t) \leq 0$ is replaced by $LV(x,t) \leq 0$. The following theorem outlines the stochastic version of Lyapunov's second method (see [53, Theorem 2.4], [53, Theorem 4.4], [24]).

Theorem 1.4.2. (i) *If there exists a positive-definite function $V(x,t) \in C^{2,1}(S_h \times [t_0,\infty), \mathbb{R}_+)$ such that $LV(x,t) \leq 0$ for all $(x,t) \in S_h \times [t_0,\infty)$, then the trivial solution of equation (1.43) is stochastically stable.*

(ii) *If there exists a positive-definite and decrescent function $V(x,t) \in C^{2,1}(S_h \times [t_0,\infty), \mathbb{R}_+)$ such that $LV(x,t)$ is negative-definite, then the trivial solution of equation (1.43) is stochastically asymptotically stable.*

(iii) *If there exists a positive-definite, decrescent, and radially unbounded function $V(x,t) \in C^{2,1}(\mathbb{R}^n \times [t_0,\infty), \mathbb{R}_+)$ such that $LV(x,t)$ is negative-definite, then the trivial solution of equation (1.43) is stochastically asymptotically stable in the large.*

Proof. See [53, p. 111]. □

1.5 Fundamental notations in network science and graph theory

The following fundamental concepts from *graph theory* and *network science* are employed to describe the networks within the complex topological hierarchical structures studied in this book. For an extensive treatment of network theory, we refer to sources such as [4].

Definition 1.5.1. (i) **Network (or Graph):** A network is a mathematical structure consisting of a set of objects, referred to as **nodes or vertices**, and the connections between them, known as **edges or links**. Formally, a network G is defined as a pair $G = (V, E)$, where V represents the set of vertices, and E represents the set of edges. Each edge is an unordered pair of vertices

$$E \subseteq \{\{v_1, v_2\} \mid v_1, v_2 \in V\}.$$

An edge of the form $\{v, v\} \in E$ for some $v \in V$ is called a **loop**. A network that contains no loops is referred to as a **simple graph.**

(ii) **Endvertex:** Given an edge $e = \{u, v\} \in E$ in a graph $G = (V, E)$, the vertices u and v are referred to as the **endvertices** of the edge. Such vertices are said to be **incident** with the edge and are termed **adjacent** (or neighbors), denoted by $u \sim v$.

(iii) **Degree of a Node:** The **degree** (or **valency**) of a vertex $v \in V$ is the number of edges for which v serves as an endvertex. A vertex of degree 0 is referred to as an **isolated vertex**. The degree of a vertex v is denoted by $\deg(v)$ or $d_G(v)$. In a **simple graph**, the degree corresponds to the number of edges incident to v. In graphs with loops, each loop contributes 2 to the degree of its endvertex, as the vertex appears twice in the loop.

1.5.1 Directed graphs

A particular class of networks, known as *directed graphs (digraphs)*, is defined as follows.

Definition 1.5.2. (i) **Directed graph (digraph):** A directed graph, or *digraph*, consists of two sets: $G = (V, A)$, where V is the set of vertices, and A is the set of **arcs**, which are ordered pairs of vertices:

$$A \subseteq \{(v_1, v_2) \mid v_1, v_2 \in V\}.$$

In graphical representations, arcs are depicted as arrows; for example, an arc (u, v) is represented by an arrow pointing from u to v, denoted $u \to v$.

(ii) **Outdegree and indegree:** In a digraph $G = (V, A)$, the **outdegree** (or **outvalency**) of a vertex $v \in V$, denoted $d^+(v)$, is the number of arcs for which v is the origin:

$$d^+(v) = \left|\{w \in V \mid (v, w) \in A\}\right|.$$

The **indegree** (or **invalency**) of a vertex v, denoted $d^-(v)$, is the number of arcs for which v is the destination:

$$d^-(v) = \left|\{w \in V \mid (w, v) \in A\}\right|.$$

2 Finite-scale compact topological spaces: applications in multiscale ecological and social systems

2.1 Introduction

As discussed in Chapter 1, a comprehensive understanding and representation of the general topological structure of human metapopulations form a foundational step toward investigating population dynamics. This framework allows for a mathematical representation of complex microscopic social, biological, and epidemiological factors influencing these dynamics. For example, in the context of disease epidemics, understanding the drivers behind the emergence and spread of new outbreaks within a complex, hierarchical metapopulation necessitates a basic representation of human mobility networks and interpersonal contact networks at various hierarchical levels within the population structure.

Moreover, the structural dynamics of human metapopulations evolve continuously over time and space, analogous to the deformations and stretching of topological spaces. This dynamic nature underscores the relevance of topological concepts in analyzing these systems. Therefore this chapter introduces a novel hierarchical metapopulation model, referred to as the *N-scale structured population* (*for an integer* $N \geq 1$), which naturally emerges from the application of *compact topological spaces*. This model describes a class of metapopulations with nested levels of human interactions—such as **(i)** mobility and **(ii)** social connections— within and between different population patches or individuals. This chapter extends the two-scale network population model studied by Wanduku and Ladde [32, 47, 48, 46, 45, 49, 47, 45], presenting a generalized hierarchical network population structure in a coherent and unified framework.

To aid comprehension of the concepts presented in this chapter, the characteristics of a *two-scale population* are described first. Following this, a generalization to the *multiscale structured population* is provided, offering a robust framework for understanding these hierarchical systems.

2.2 A two-scale compact topological space in $\mathbb{R}^n_+$

Consider the finite-dimensional positive real vector space $\mathbb{R}^n_+$, $n > 0$. The pair $(\mathbb{R}^n_+, d_2)$ is a metric space, where d_2 is the metric induced by the *Euclidean norm* $\|\cdot\|_2$ on the normed linear space $(\mathbb{R}^n_+, \|\cdot\|_2)$, that is, $d_2(x,y) = \|x-y\|_2$ for all $x, y \in \mathbb{R}^n_+$. Let $\mathfrak{T}_{\mathbb{R}^n_+}$ be the *natural topology* on $\mathbb{R}^n_+$ with the *topology base* consisting of all d_2-open balls in $\mathbb{R}^n_+$. The pair $(\mathbb{R}^n_+, \mathfrak{T}_{\mathbb{R}^n_+})$ denotes the superior *topological space* of consideration in this book.

Consider the subspace of consideration for the hierarchical metapopulation given by $C \subset \mathbb{R}^n_+$, equipped with the *relative or subspace topology* $\mathfrak{T}_C$, that is, the pair $(C, \mathfrak{T}_C)$ is the referenced topological space of interest in this book. We assume that the subspace

https://doi.org/10.1515/9783111613215-002

$C \subset \mathbb{R}_+^n$ is *closed and bounded* in $\mathbb{R}_+^n$; in other words, the subspace $C \subset \mathbb{R}_+^n$ is a *compact topological space*. In the rest of this section, we define a hierarchical *finite subcover* for the compact topological space $C \subset \mathbb{R}_+^n$.

Assume that the compact topological space $C \subset \mathbb{R}_+^n$ is composed of two nested structural hierarchical *finite coverings* in the topology $\mathfrak{T}_C$ characterized subsequently. The two hierarchical levels are termed $N = 2$-***scales***. In fact, the following definition is presented for an arbitrary $N > 0$-*scale compact topological space*.

Definition 2.2.1 (A finite-scale compact topological space). A nonempty compact topological space $C \subset \mathbb{R}_+^n$ is called an overall finite-scale compact topological space if (a) it is a union of a finite number of mutually disjoint compact topological subspaces and (b) it has a finite descending sequence of finite-scale subcompact topological spaces.

The following definition will be useful in explaining the scale-structured population defined in Definition 2.2.1.

Definition 2.2.2. A partition of a topological space $(X, \mathfrak{T})$ is a subcollection $\mathfrak{U} \subset \mathfrak{T}$ such that $A \neq \emptyset$ for all $A \in \mathfrak{U}$, $A \cap B = \emptyset$ for all $A, B \in \mathfrak{U}$, and $X = \bigcup\{U \in \mathfrak{U}\}$.

Remark 2.2.1. (i) In other words, Definition 2.2.1 suggests that a nonempty compact topological space $C \subset \mathbb{R}_+^n$ is an overall finite-scale compact topological space if and only if it has a finite descending sequence of *partitions* that cover the space. This definition simplifies the finite covers of the compact topological space C by restricting all finite covers to partitions of C by subsets of smaller scale. Definition 2.2.1 is used to construct a finite-scale compact topological space in $\mathbb{R}_+^n$. In this section, a two-scale compact topology space is presented.

(ii) Denote by $I(a, b)$ the finite set of integers between the integers a and b, inclusive. For example, $I(0, 4) = \{0, 1, 2, 3, 4\}$.

Using Definition 2.2.1 an $N = 2$-scale compact topological space is defined.

Definition 2.2.3. Consider a subspace $C = C_{00}^{00} \subset \mathbb{R}_+^n$ equipped with the relative topology $\mathfrak{T}_C$, i. e., $(C, \mathfrak{T}_C)$ is a topological space. For integers $M_1, M_2 > 0$, the subspace $C = C_{00}^{00}$ is said to have $N = 2$ scales if there exists a finite subcover $\{C_{0r^2}^{r^20}\}_{r^2 \in I(1,M_2)}$ of mutually disjoint sets such that

$$C_{00}^{00} = \bigcup_{r^2=1}^{M_2} C_{0r^2}^{r^20}, \tag{2.1}$$

and for each $r^2 \in I(1, M_2)$, the set $C_{0r^2}^{r^20}$ is also a compact topological subspace with respect to the subspace topology. Furthermore, there exists a finite collection of mutually disjoint sets $\{C_{r^1r^2}^{r^20}\}_{r^1 \in I(1,M_1)}$ that form a subcover for $C_{0r^2}^{r^20}$, that is,

$$C_{0r^2}^{r^20} = \bigcup_{r^1=1}^{M_1} C_{r^1r^2}^{r^20}. \tag{2.2}$$

Therefore the overall $N = 2$-scale compact topological space $C = C_{00}^{00}$ is characterized by two distinct nested partitions, where in the first level, it follows from (2.2) that C is partitioned into $M_2 \times M_1$ sets so that

$$C_{00}^{00} = \bigcup_{r^2=1}^{M_2} \bigcup_{r^1=1}^{M_1} C_{r^1 r^2}^{r^2 0}. \tag{2.3}$$

The second level distinct partition of C consists of M_2 subsets as shown in (2.1).

2.2.1 A two-scale network compact topological space in $\mathbb{R}_+^2$

For a compact topological space $C = C_{00}^{00}$ in the two-dimensional Euclidean space $\mathbb{R}_+^2$, the notations for the finite subcovers in (2.1)–(2.3) are simplified.

Indeed, the subspace $C = C_{00}^{00} \subset \mathbb{R}_+^2$ is a two-scale compact topological space if and only if there exists a second-level partition composed of the finite subcover $\{C_r\}_{r \in I(M_2)}$, $C_r \subset \mathbb{R}_+^2$, such that, by (2.1),

$$C_{00}^{00} = \bigcup_{r^2=1}^{M_2} C_{0r^2}^{r^2 0} = \bigcup_{r=1}^{M_2} C_r, \tag{2.4}$$

and $C_r \cap C_q = \emptyset$ whenever $r \neq q$ and $r, q \in I(1, M_2)$.

The first-level partition of $C = C_{00}^{00}$ is constructed as follows. For each $r \in I(1, M_2)$, there exist M_1 points $s_i^r \in \mathbb{R}_+^2$, $i \in I(1, M_1)$, such that C_r is partitioned by the singletons $\{s_i^r\}_{i \in I(1,M_1)}$ such that, by (2.2),

$$C_r \equiv C_{0r^2}^{r^2 0} = \bigcup_{r^1=1}^{M_1} C_{r^1 r^2}^{r^2 0} = \bigcup_{i=1}^{M_1} \{s_i^r\}, \quad r \in I(1, M_2), \tag{2.5}$$

and, by (2.3),

$$C_{00}^{00} = \bigcup_{r^2=1}^{M_2} C_{0r^2}^{r^2 0} = \bigcup_{r=1}^{M_2} C_r = \bigcup_{r=1}^{M_2} \bigcup_{i=1}^{M_1} \{s_i^r\}. \tag{2.6}$$

In the following, we present a detailed characterization of the compact topological subspace $C = C_{00}^{00}, \mathfrak{T}_C)$ in the metric topological space $(\mathbb{R}_+^2, \mathfrak{T})$. Let $\vec{x} = (x_1, x_2) \in \mathbb{R}_+ \times \mathbb{R}_+ = \mathbb{R}_+^2$. Furthermore, for $r \in I(1, M_2)$, let $y_r = (y_{1_r}, y_{2_r}) \in \mathbb{R}_+^2$ be fixed. Also, let $d_2(\vec{x}, y_r) = \|\vec{x} - y_r\|_2 = \|\vec{x} - y_r\|_{\mathbb{R}^2} = \sqrt{(x_1 - y_{1_r})_1^2 + (x_2 - y_{2_r})^2}$. Let $b_r > 0$. Define the open balls C_r that form a finite subcover for $C = C_{00}^{00}$ as follows:

$$C_r = B(y_r; b_r) = B_{\mathbb{R}_+^2}(y_r; b_r) = \{\vec{x} \in \mathbb{R}_+^2 : \|\vec{x} - y_r\|_{\mathbb{R}^2} < b_r\}, \tag{2.7}$$

where for all $r, q \in I(1, M_2), r \neq q$, the sets C_r and C_q are selected from the metric topology on $C = C^{00}_{00}$ in such as way that the balls are mutually disjoint, i. e., $C_r \cap C_q = \emptyset$, and they form a partition of C, i. e., $C = \bigcup_{r=1}^{M_2} C_r$.

For each $r \in I(1, M_2)$, let $\{s^r_i\}_{i \in I(1,M_1)}$ be distinct points in the interior of the open balls C_r that form a finite subcover. To distinguish between the finite subcovers of two disjoint subsets C_r and C_q, $r \neq q$, C_r has a finite subcover containing n_r singletons $\{s^r_i\}_{i \in I(1,n_r)}$, and C_q has a finite subcover containing n_q singletons $\{s^q_i\}_{i \in I(1,n_q)}$.

For the compact topological subspace C_r, $r \in I(1, M_2)$, it is easy to see that for all points in $\{s^r_i\}_{i \in I(1,n_r)}$, the distance between two distinct points $\|s^r_i - s^r_j\|_{\mathbb{R}^2} > 0, i \neq j$. Also, let

$$C_{rr}(s^r) = \{s^r_i \in C_r : i \in I(1, n_r)\} \tag{2.8}$$

be a finite collection of the n_r distinct points $\{s^r_i\}_{i \in I(1,n_r)}$ that form a subcover of C_r. Define

$$C(s) = \{s^r_i \in C_r : \forall i \in I(1, n_r), \forall r \in I(1, M_2)\} = \bigcup_{r=1}^{M_2} C_{rr}(s^r) \tag{2.9}$$

be the complete collection of all first-level points that cover C. The cardinality of $C(s)$ is

$$n = \sum_{r=1}^{M_2} n_r. \tag{2.10}$$

2.2.2 Application of a two-scale network compact topological space to ecological and social network systems

The human population within a given nation can be represented as a two-scale compact topological space, denoted by $C = C^{00}_{00} \subset \mathbb{R}^2_+$, where C encompasses the entire collection of individuals in the country. For instance, in the case of the USA, C represents the complete population of the nation. This framework allows for the description of different hierarchical levels of partitions within C. As noted in (2.4), the finite subcover $\{C_r\}_{r \in I(M_2)}$ can correspond, for example, to the $M_2 = 50$ states of the USA. Thus the overall population of the USA can be expressed as

$$C^{00}_{00} = \bigcup_{r=1}^{M_2=50} C_r. \tag{2.11}$$

The population C_r, $r \in I(1, M_2 = 50)$, of any state in the USA can be further partitioned into the subpopulations residing in the n_r counties within C_r. These subpopulations are characterized by finite subcovers of singletons $\{s^r_i\}_{i \in I(1,M_1)}$ such that, by (2.5),

$$C_r = \bigcup_{i=1}^{n_r} \{s^r_i\}, \quad r \in I(1, M_2), \tag{2.12}$$

and by (2.6) the complete population $C = C_{00}^{00}$ of the USA is described in two ways by the following union:

$$C_{00}^{00} = \bigcup_{r=1}^{M_2} C_r = \bigcup_{r=1}^{M_2} \bigcup_{i=1}^{n_r} \{s_i^r\}. \tag{2.13}$$

Similarly, a two-scale compact topological space $C = C_{00}^{00}$ can be applied to characterize social networks, such as social media platforms like *Facebook*. Here the complete population of individuals with Facebook accounts corresponds to $C = C_{00}^{00}$. The population can be partitioned into account owners from different continents, as characterized in (2.12), which can be further subdivided into account owners in different countries within these continents, as given in (2.13).

The advantage of using a scale-structured compact topological space to describe the hierarchical levels of partitioning in an ecological or social network lies in its ability to provide a suitable framework for introducing and analyzing complex networks of interactions. Examples include the mobility of species between patches in an ecological space or the social ties between individuals interacting in a social media network like Facebook. The detailed structure and analysis of networks within a finite-scale compact topological space are discussed in subsequent chapters.

2.3 A finite-scale network structured population

In this section, the $N = 2$-scale compact topological space $C = C_{00}^{00} \subset \mathbb{R}_+^n$ equipped with the relative topology $\mathfrak{T}_C$, i. e., $(C, \mathfrak{T}_C)$ in Section 2.2 is generalized to a finite $N > 0$-scale compact topological space $C = C_{00\cdots0}^{00\cdots0} \subset \mathbb{R}_+^n$ with $N > 0$ hierarchical partitions.

Note. Throughout this book, the term *scale* is used to describe nested hierarchical partition levels within a compact topological subspace $C \subset \mathbb{R}_+^n$. Hence, for $N > 0$, a compact topological space with N *scales* is defined as a *closed and bounded* subspace $C \subset \mathbb{R}_+^n$ that contains N nested hierarchical level partitions forming a finite descending sequence of compact subspaces.

In an ecological context, a metapopulation is said to have N scales if there exist $N > 0$ hierarchical decompositions of the population, representing different levels of ecological interactions among sites within various large- and small-scale nested patches of the species' complete population. For example, consider investigating the interactions of a species of ants, such as their aggressiveness. This could involve examining interactions among colonies within a city, interactions between colonies in different cities within the same state in the USA, and interactions among colonies in different states.

Definition 2.2.1 is applied similarly as in Section 2.2 to construct an N-scale compact topological space. Let N and M_k, $k \in I(1, N)$, be positive integers. Here N is referred as an overall "finite-scale", and M_k is the number of mutually disjoint subspaces that form the kth hierarchical level partition of the space $C = C_{00\cdots0}^{00\cdots0}$.

(D1) The Nth-level partition of C:
Let $C = C_{00\cdots0}^{00\cdots0}$ be an N-scale compact topological space. Then there exists a collection of mutually disjoint subspaces $\{C_{00\cdots0r^N}^{00\cdots r^N0}\}_{r^N \in I(1,M_N)}$ that form a finite subcover of the space, that is,

$$C = C_{00\cdots0}^{00\cdots0} = \bigcup_{r^N=1}^{M_N} C_{00\cdots00r^N}^{00\cdots0r^N0}, \tag{2.14}$$

where (2.14) is called the *Nth level partition* of C, and the subsets $C_{00\cdots00r^N}^{00\cdots0r^N0}$ and $C_{00\cdots00q^N}^{00\cdots0q^N0}$ denoted the rth and qth subspaces in the Nth-level partition of C. Moreover, $C_{00\cdots0r^N}^{00\cdots r^N0} \cap C_{00\cdots0q^N}^{00\cdots q^N0} = \emptyset$ whenever $r^N \neq q^N, r^N, q^N \in I(1, M_N)$.

Example 2.3.1. Let C be a three-scale compact topological space, where $N = 3$, and $M_3 = 4$. This implies that C has three nested hierarchical partition levels, $C = C_{000}^{000}$. The third-level partition contains $M_3 = 4$ mutually disjoint sets that form a finite subcover for C. The finite cover is denoted by $\{C_{00r^3}^{0r^30}\}_{r^3 \in I(1,4)} = \{C_{001}^{010}, C_{002}^{020}, C_{003}^{030}, C_{004}^{040}\}$, that is,

$$C = C_{001}^{010} \cup C_{002}^{020} \cup C_{003}^{030} \cup C_{004}^{040}, \tag{2.15}$$

where in this case the index represents $r^3 = 1, 2, 3, 4$.

(D2) The $(N-1)$th-level partition of C:
For each $r^N \in I(1, M_N)$, the compact topological subspace $C_{00\cdots0r^N}^{00\cdots0r^N0}$ has a finite subcover of mutually disjoint compact subspaces $\{C_{00\cdots00r^{N-1}r^N}^{00\cdots0r^{N-1}r^N0}\}_{r^{N-1} \in I(1,M_{N-1,r^N})}$ that forms a partition called the $(N-1)$**th partition** of the subspace $C_{00\cdots0r^N}^{00\cdots0r^N0}$, that is,

$$C_{00\cdots0r^N}^{00\cdots r^N0} = \bigcup_{r^{N-1}=1}^{M_{N-1,r^N}} C_{00\cdots0r^{N-1}r^N}^{00\cdots r^{N-1}r^N0}, \tag{2.16}$$

where $C_{00\cdots00r^{N-1}r^N}^{00\cdots0r^{N-1}r^N0} \cap C_{00\cdots00q^{N-1}r^N}^{00\cdots0q^{N-1}r^N0} = \emptyset$ whenever $r^{N-1}, q^{N-1} \in I(1, M_{N-1,r^N})$, $r^{N-1} \neq q^{N-1}$, and $r^N \in I(1, M_N)$. Thus the $(N-1)$**th partition** of the overall compact topological space C is obtained by combining (2.14) and (2.16) as follows:

$$C = \bigcup_{r^N=1}^{M_N} C_{00\cdots00r^N}^{00\cdots0r^N0} = \bigcup_{r^N=1}^{M_N} \bigcup_{r^{N-1}=1}^{M_{N-1,r^N}} C_{00\cdots00r^{N-1}r^N}^{00\cdots0r^{N-1}r^N0}. \tag{2.17}$$

Example 2.3.2. From Example 2.3.1, the compact topological space $C = C_{000}^{000}$ has the third-level partition composed of the compact topological subspaces $\{C_{00r^3}^{0r^30}\}_{r^3 \in I(1,4)} =$

$\{C_{001}^{010}, C_{002}^{020}, C_{003}^{030}, C_{004}^{040}\}$, where each subspace $C_{00r^3}^{0r^30}$, $r^3 \in I(1,4)$, is further partitioned by the finite subcover $\{C_{0r^2r^3}^{r^2r^30}\}_{r^2 \in I(1,M_{4,r^3}), r^3 \in I(1,4)}$, that is,

$$C_{00r^3}^{0r^30} = \bigcup_{r^2=1}^{M_{4,r^3}} C_{0r^2r^3}^{r^2r^30}, \quad r^3 \in I(1,4). \tag{2.18}$$

Hence from (2.18) we easily see that the compact topological subspaces in the collection $\{C_{001}^{010}, C_{002}^{020}, C_{003}^{030}, C_{004}^{040}\}$ are partitioned by $M_{4,1}, M_{4,2}, M_{4,3}$, and $M_{4,4}$ mutually exclusive subsets, respectively.

Hence the **second-level partition** of the overall compact topological space follows by combining the decompositions in (2.18) and (2.15):

$$C = \bigcup_{r^3=1}^{4} \bigcup_{r^2=1}^{M_{2,r^2}} C_{0r^2r^3}^{r^2r^30}. \tag{2.19}$$

(D3) The $(N-2)$th-level partition of C:
Similarly, for all $r^{N-1} \in I(1, M_{N-1,r^N})$ and $r^N \in I(1, M_N)$, the set $C_{00\cdots00r^{N-1}r^N}^{00\cdots0r^{N-1}r^N0}$ is a compact topological subspace with a finite subcover of mutually exclusive subspaces $\{C_{00\cdots00r^{N-2}r^{N-1}r^N}^{00\cdots0r^{N-2}r^{N-1}r^N0}\}_{r^{N-2} \in I(1,M_{N-2,r^{N-1},r^N})}$, $r^{N-1} \in I(1, M_{N-1,r^N})$, $r^N \in I(1, M_N)$ that form a partition of the space, that is, the $(N-2)$th partition of $C_{00\cdots00r^{N-2}r^{N-1}r^N}^{00\cdots0r^{N-2}r^{N-1}r^N0}$ is

$$C_{00\cdots00r^{N-2}r^{N-1}r^N}^{00\cdots0r^{N-2}r^{N-1}r^N0} = \bigcup_{r^{N-2}=1}^{M_{N-2,r^{N-1},r^N}} C_{00\cdots00r^{N-2}r^{N-1}r^N}^{00\cdots0r^{N-2}r^{N-1}r^N0}, \tag{2.20}$$

where $C_{00\cdots00r^{N-2}r^{N-1}r^N}^{00\cdots0r^{N-2}r^{N-1}r^N0} \cap C_{00\cdots00q^{N-2}r^{N-1}r^N}^{00\cdots0q^{N-2}r^{N-1}r^N0} = \emptyset$ whenever $r^{N-2} \neq q^{N-2}$, $r^{N-2}, q^{N-2} \in I(1, M_{N-2,r^{N-1},r^N})$, and $r^{N-1} \in I(1, M_{N-1,r^N})$, $r^N \in I(1, M_N)$, and the subspace $C_{00\cdots00r^{N-2}r^{N-1}r^N}^{00\cdots0r^{N-2}r^{N-1}r^N0}$ is the rth subspace of the partition of $C_{00\cdots00r^{N-1}r^N}^{00\cdots0r^{N-1}r^N0}$.
Moreover, from (2.14)–(2.20) we easily see that the $(N-2)$th partition of the overall compact topological space C is given by

$$C = \bigcup_{r^N=1}^{M_N} C_{00\cdots00r^N}^{00\cdots0r^N0} = \bigcup_{r^N=1}^{M_N} \bigcup_{r^{N-1}=1}^{M_{N-1,r^N}} C_{00\cdots00r^{N-1}r^N}^{00\cdots0r^{N-1}r^N0} = \bigcup_{r^N=1}^{M_N} \bigcup_{r^{N-1}=1}^{M_{N-1,r^N}} \bigcup_{r^{N-2}=1}^{M_{N-2,r^{N-1},r^N}} C_{00\cdots00r^{N-2}r^{N-1}r^N}^{00\cdots0r^{N-2}r^{N-1}r^N0}. \tag{2.21}$$

Example 2.3.3. Continuing with Example 2.3.2, where $C = C_{000}^{000}$, recall that for each $r^3 \in I(1,4)$, $C_{00r^3}^{0r^30} = \bigcup_{r^2=1}^{M_{2,r^3}} C_{0r^2r^3}^{r^2r^30}$. Also, for each $r^2 \in I(1, M_{2,r^3})$, where $r^3 = 1,2,3,4$, we obtain the following partition: $C_{0r^2r^3}^{r^2r^30} = \bigcup_{r^1=1}^{M_{1,r^2,r^3}} C_{r^1r^2r^3}^{r^2r^30}$. Thus by (2.19) we obtain

$$C = \bigcup_{r^3=1}^{4} \bigcup_{r^2=1}^{M_{2,r^3}} C_{0r^2r^3}^{r^2r^30} = \bigcup_{r^3=1}^{4} \bigcup_{r^2=1}^{M_{2,r^3}} \bigcup_{r^1=1}^{M_{1,r^2,r^3}} C_{r^1r^2r^3}^{r^2r^30}. \tag{2.22}$$

From (2.14)–(2.21) we can introduce the following notations to simplify the complex notations introduced above.

Notation 2.3.1. In view of (2.21), we denote the N-scale compact topological space C and all the subspaces in the $(N-2)$th-level partition of C as follows:

$$C = C^{00\cdots0}_{00\cdots0} \equiv C^{N\times0}_{N\times0}, \quad C^{00\cdots0r^N0}_{00\cdots00r^N} \equiv C^{(N-2)\times0;r^N0}_{(N-1)\times0;r^N}, \quad C^{00\cdots0r^{N-1}r^N0}_{00\cdots00r^{N-1}r^N} \equiv C^{(N-3)\times0;r^{N-1}r^N0}_{(N-4)\times0;r^{N-1}r^N}, \\ C^{00\cdots0r^{N-2}r^{N-1}r^N0}_{00\cdots00r^{N-2}r^{N-1}r^N} \equiv C^{(N-4)\times0;r^{N-2}r^{N-1}r^N0}_{(N-3)\times0;r^{N-2}r^{N-1}r^N}. \tag{2.23}$$

To generalize the notations, for any $k \in I(1,N)$, we denote

$$C^{00\cdots0r^kr^{k+1}r^{k+2}\cdots r^{N-2}r^{N-1}r^N0}_{00\cdots00r^kr^{k+1}\cdots r^{N-3}r^{N-2}r^{N-1}r^N} \equiv C^{(N-k-1)\times0;r^kr^{k+1}r^{k+2}\cdots r^{N-2}r^{N-1}r^N0}_{(N-k)\times0;r^kr^{k+1}\cdots r^{N-3}r^{N-2}r^{N-1}r^N}. \tag{2.24}$$

The notation used in (2.24) can be further simplified:

$$\begin{aligned} C^{00\cdots0r^kr^{k+1}r^{k+2}\cdots r^{N-2}r^{N-1}r^N0}_{00\cdots00r^kr^{k+1}\cdots r^{N-3}r^{N-2}r^{N-1}r^N} &\equiv C^{(N-k-1)\times0;r^kr^{k+1}r^{k+2}\cdots r^{N-2}r^{N-1}r^N0}_{(N-k)\times0;r^kr^{k+1}\cdots r^{N-3}r^{N-2}r^{N-1}r^N} \\ &\equiv C^{(N-k-1)\times0;r^{k:N}0}_{(N-k)\times0;r^{k:N}}. \end{aligned} \tag{2.25}$$

Also, denote

$$C^{(N-k-1)\times0r^kr^{k+1}r^{k+2}\cdots r^{N-2}r^{N-1}r^N0}_{(N-k)\times0q^kr^{k+1}\cdots r^{N-3}r^{N-2}r^{N-1}r^N} \equiv C^{(N-k-1)\times0;r^{k:N}0}_{(N-k)\times0;q^kr^{k+1:N}}, \quad q^k \neq r^k. \tag{2.26}$$

Thus by (2.26) we can simplify the following notation:

$$C^{(N-k-1)\times0q^kq^{k+1}r^{k+2}r^{k+3}\cdots r^{N-2}r^{N-1}r^N0}_{(N-k)\times0q^kq^{k+1}r^{k+2}\cdots r^{N-3}r^{N-2}r^{N-1}r^N} \equiv C^{(N-k-1)\times0;q^{k:k+1}r^{k+2:N}0}_{(N-k)\times0;q^{k:k+1}r^{k+2:N}}, \quad q^{k+1} \neq r^{k+1}. \tag{2.27}$$

Similarly, the cardinality of the finite collection $\{C^{(N-k-1)\times0;r^{k:N}0}_{(N-k)\times0;r^{k:N}}\}_{r^k=M_{k,r^{k+1},r^{k+2},\ldots,r^{N-1},r^N}}$ is denoted by

$$M_{k,r^{k+1},r^{k+2},\ldots,r^{N-1},r^N} = M_{k,r^{k+1:N}}. \tag{2.28}$$

(D4) The kth-level partition of C, $k \in I(1,N)$:
Applying the notations in (2.24), it follows from (2.14)–(2.21) that the partition of any finite-scale compact topological space can be generalized inductively as follows. Suppose the collection of mutually exclusive subspaces $\{C^{[N-(k+1)]\times0;r^{k:N}0}_{(N-k)\times0;r^{k:N}}\}_{r^k=M_{k,r^{k+1:N}}}$ is a partition of the compact topological space $C^{[N-(k+2)]\times0;r^{k+1:N}0}_{[N-(k+1)]\times0;r^{k+1:N}}$ for some $k \in I(1,N)$. It follows that

$$C^{[N-(k+2)]\times0;r^{k+1:N}0}_{[N-(k+1)]\times0;r^{k+1:N}} = \bigcup_{r^k=1}^{M_{k,r^{k+1:N}}} C^{[N-(k+1)]\times0;r^{k:N}0}_{(N-k)\times0;r^{k:N}}. \tag{2.29}$$

It follows from (2.29) that the kth-level partition of C ($k \in I(1,N)$) is given by

$$\begin{aligned} C &= \bigcup_{r^N=1}^{M_N} \bigcup_{r^{N-1}=1}^{M_{N-1,r^N}} \bigcup_{r^{N-2}=1}^{M_{N-2,r^{N-1}r^N}} \cdots \bigcup_{r^{k+1}=1}^{M_{k+1,r^{k+2:N}}} C_{[N-(k+1)]\times 0;r^{k+1:N}}^{[N-(k+2)]\times 0;r^{k+1:N}0} \\ &= \bigcup_{r^N=1}^{M_N} \bigcup_{r^{N-1}=1}^{M_{N-1,r^N}} \bigcup_{r^{N-2}=1}^{M_{N-2,r^{N-1}r^N}} \cdots \bigcup_{r^{K+1}=1}^{M_{K+1,r^{K+2:N}}} \bigcup_{r^k=1}^{M_{k,r^{k+1:N}}} C_{(N-k)\times 0;r^{k:N}}^{[N-(k+1)]\times 0;r^{k:N}0}. \end{aligned} \tag{2.30}$$

(D5) The first-level partition of the compact subspace C:
The above expressions in **(D1)–(D4)** exhibit the induction assumption, and hence by this principle it is easy to see that for each $r^2 \in I(1,M_2)$, the finite collection of mutually disjoint subspaces given by $\{C_{r^{1:N}}^{r^{2:N}0}\}_{r^1 \in I(1,M_{1,r^{2:N}})}$ is the first-level partition of the compact subspace given by

$$C_{0;r^{1:N}}^{r^{2:N}0} = \bigcup_{r^1=1}^{M_{1,r^{2:N}}} C_{r^{1:N}}^{r^{2:N}0}, \tag{2.31}$$

where

$$C_{r^1r^2r^3\ldots r^{N-2}r^{N-1}r^N}^{r^2r^3r^4\ldots r^{N-1}r^N0} \equiv C_{r^{1:N}}^{r^{2:N}0}, \quad C_{0r^2r^3\ldots r^{N-2}r^{N-1}r^N}^{r^2r^3r^4\ldots r^{N-1}r^N0} \equiv C_{0;r^{1:N}}^{r^{2:N}0}, \tag{2.32}$$

and $C_{r^{1:N}}^{r^{2:N}0}$ is the rth subspace in the first-level partition of the hierarchical topological subspace $C_{0;r^{1:N}}^{r^{2:N}0}$. In addition, $C_{r^{1:N}}^{r^{2:N}0} \cap C_{q^1r^{2:N}}^{r^{2:N}0} = \emptyset$ whenever $r^1 \neq q^1$, $r^1, q^1 \in I(1, M_{1,r^{2:N}})$. Therefore note that the **first-level partition** of the overall hierarchical compact topological space is the following:

$$C_{00\cdots 0}^{00\cdots 0} = \bigcup_{r^N=1}^{M_N} \bigcup_{r^{N-1}=1}^{M_{N-1},r^N} \bigcup_{r^{N-2}=1}^{M_{N-2},r^{N-1:N}} \cdots \bigcup_{r^3=1}^{M_3,r^{4:N}} \bigcup_{r^2=1}^{M_2,r^{3:N}} \bigcup_{r^1=1}^{M_1,r^{2:N}} C_{r^{1:N}}^{r^{2:N}0}. \tag{2.33}$$

2.4 Description of networks in finite-scale hierarchical populations

The following basic definitions in *graph theory* and *network science* (see [4]) will be used to describe a finite-scale network population.

Definition 2.4.1.

(i) **A network:** A network or graph is a set of connected objects. The objects in a graph are called **nodes or vertices**, and the connections between the objects are called **links or edges**. Formally, a network G is a pair of sets $G = (V,E)$, where V is the set of vertices or nodes, and E is the set of edges or links. Each edge is a pair of vertices, meaning that each element of E is an unordered pair of elements of the set V. In mathematical terms,

$$E \subseteq \{\{v_1, v_2\} \mid v_1, v_2 \in V\}. \tag{2.34}$$

An edge of the form $\{v, v\} \in E$ for some $v \in V$ is called a **loop**. A network that contains no loops is called a **simple graph**.

(ii) **Endvertex:** Given any edge $e = \{u, v\} \in E$ of a graph $G = (V, E)$ (with or without loops), the vertices u and v are called **endvertices**. Furthermore, u and v are said to be **incident** with the edge e (or vice versa, the edge e is incident with the vertices). Also, u and v are said to be **adjacent** if there is an edge joining them, or equivalently, u is a **neighbor** of v, denoted by $u \sim v$.

(iii) **Degree of a node:** If $v \in V$ is a vertex of a graph (with or without loops), the number of times v appears as the endvertex of some edge is called the **degree or valency** of v in the graph G. A vertex of degree 0 is called an **isolated vertex**. The degree of a node $v \in V$ is denoted by $\deg(v)$ or $d_G(v)$. Note that in a **simple graph**, the degree of any node $v \in V$ is simply the **number of edges incident with** v. Additionally, if the graph is not simple, then each loop contributes 2 to the degree of its endvertex, since a vertex v appears twice as the endvertex of any loop incident with v.

The next definition introduces a special type of network called a *directed network or graph.*

Definition 2.4.2.

(i) **A directed graph:** A directed graph, also called a **digraph**, consists of two sets: $G = (V, A)$, where the elements of V are called the vertices of the digraph, and the elements of A are ordered pairs from V, referred to as the **arcs** of the digraph, that is,

$$A \subseteq \{(v_1, v_2) \mid v_1, v_2 \in V\}. \tag{2.35}$$

When graphing a digraph, arrows are used to represent each arc, e. g., (u, v), where the arrow points from the first vertex u of the ordered pair to the second vertex v, denoted by $u \to v$.

(ii) **Outdegree and indegree of a directed network:** The **outdegree** or **outvalency** of a vertex $v \in V$ in a digraph G, denoted by $d^+(v)$, is the number of arcs where the first entry of the ordered pair is v, i. e.,

$$d^+(v) = \big|\{w \in V \mid (v, w) \in A\}\big|, \tag{2.36}$$

where $|C|$ denotes the **cardinality** of a set C. The **indegree** or **invalency** of a vertex $v \in V$ in a digraph, denoted by $d^-(v)$, is the number of arcs where the second entry of the ordered pair is v, i. e.,

$$d^-(v) = \big|\{w \in V \mid (w, v) \in A\}\big|. \tag{2.37}$$

Definition 2.4.3. The population of individuals in an epidemiological, behavioral, or social dynamic system is called an N-scale network population if the different states of the population can be compartmentalized spatially, behaviorally, socially, or epidemiologically, and there exist hierarchical partition levels between the compartments of the population analogous to an N-scale compact topological space $C = C_{00\cdots 00}^{00\cdots 00}$. Furthermore, there exists a network connecting the compartments in the hierarchical partitions of the space C.

Definitions 2.4.1–2.4.3 are applied in biological and social networks to characterize a finite-scale network metapopulation and a social media network. The simpler case of a two-scale network population is considered first.

2.4.1 Networks in a two-scale network population

We use the descriptions and notations for the two-scale compact topological space $C \subset \mathbb{R}_+^2$ in Section 2.2. Consider a population of individuals (e. g., humans or other organisms) spatially distributed over M mutually exclusive patches called **regions**, denoted by $C_1, C_2, \ldots, C_M$, where by (2.4) the complete population is encompassed by the union of all patches, i. e., $C_{00}^{00} = \bigcup_{r=1}^{M} C_r$. Suppose the population of each region C_r, $r \in I(1, M)$, is further partitioned into n_r distinct sub-patches called **intraregional sites**, denoted by $s_1^r, s_2^r, \ldots, s_{n_r}^r$, where the complete population in the region C_r, $r \in I(1, M)$, is the union of the population of all intraregional patches:

$$C_r = \bigcup_{i=1}^{n_r} \{s_i^r\}.$$

Assume that a network or graph[1] exists between the M regions $C_1, C_2, \ldots, C_M$ and their intraregional sites $s_1^r, s_2^r, \ldots, s_{n_r}^r$, $r \in I(1, M)$.

Example 2.4.1 (Human mobility network). Suppose $C_{00}^{00} = \bigcup_{r=1}^{M} C_r$ is the partition of the population of a country into the distinct states or provinces of the country. In the rth state C_r, the population can be further partitioned into the subpopulations of distinct counties $s_1^r, s_2^r, \ldots, s_{n_r}^r$ of the state C_r, $r \in I(1, M)$. A human mobility network occurs in the two-level partition

$$C = \bigcup_{r=1}^{M} \bigcup_{i=1}^{n_r} \{s_i^r\},$$

where, by Definition 2.4.1 and (2.9),

1 A collection of connected objects, where the objects are called "nodes or vertices", and the connections are the links or edges between the nodes.

$$V = C(s);$$
$$E = \{\{s_i^r, s_j^r\}, \{s_i^r, s_k^q\} \mid s_i^r, s_j^r, s_k^q \in C(s); r \neq q; i \neq j; \quad (2.38)$$
$$i, j \in I(1, n_r); r, q \in I(1, M); k \in I(1, n_q)\}.$$

In other words, the residents of a site s_i^r in a region C_r can either visit other sites s_j^r within the region C_r or visit sites s_l^q in other regions C_q, where $i, j \in I(1, n_r)$, $r \neq q$, and $r, q \in I(1, M)$.

Example 2.4.2 (Social media network). Suppose in a social media network such as **Facebook**, $C_{00}^{00} = \bigcup_{r=1}^{M} C_r$ represents the partition of the complete collection of online account owners globally into distinct continents of the world. In the rth continent C_r, the online account owners can be further partitioned into subcollections of account owners from distinct countries, denoted by $s_1^r, s_2^r, \ldots, s_{n_r}^r$, where $r \in I(1, M)$. A "friendship" network occurs in this two-level partition:

$$C = \bigcup_{r=1}^{M} \bigcup_{i=1}^{n_r} \{s_i^r\},$$

where, by Definition 2.4.1 and (2.9),

$$V = C(s);$$
$$E = \{\{s_i^r, s_j^r\}, \{s_i^r, s_k^q\} \mid s_i^r, s_j^r, s_k^q \in C(s); r \neq q; i \neq j; \quad (2.39)$$
$$i, j \in I(1, n_r); r, q \in I(1, M); k \in I(1, n_q)\},$$

where the pair $\{s_i^r, s_j^r\}$ represents friendship between account owners in the two countries s_i^r and s_j^r within the continent C_r.

Note that in higher-scale network populations, the friendship link $\{s_i^r, s_j^r\}$ can be further broken down into friendships between account owners in the same or different cities. This can be refined further to represent personal-level friendships between individuals socializing within the social media network.

2.4.2 Networks in the *N*-scale network population

Similarly, in this section, we use the notations in Section 2.3 to describe a finite-scale network population. To generalize the networks in a two-scale network population in Section 2.4.1, consider that the complete collection of individuals in the population (e. g., human beings, other organisms, or objects) given by the space C is spatially distributed between M_N mutually exclusive patches called regions. These regions form a partition of C, denoted by the collection $\{C_{(N-1)\times 0;r^N}^{(N-2)\times 0;r^N,0}\}_{r^N \in I(1,M_N)}$, such that the population C is given by

$$C = C_{N\times 0}^{N\times 0} = \bigcup_{r^N=1}^{M_N} C_{(N-1)\times 0;r^N}^{(N-2)\times 0;r^N,0}. \quad (2.40)$$

Suppose each patch of the population $C^{(N-2)\times 0;r^N,0}_{(N-1)\times 0;r^N}$, $r^N \in I(1, M_N)$, can be further decomposed into $(N-1)$ hierarchical partitions similarly to the hierarchical partitions of the compact topological subspaces in Section 2.3, i. e., equations (2.14)–(2.33). Then the complete collection of individuals $C = C^{N\times 0}_{N\times 0}$ will exhibit N hierarchical partitions, where the kth-level partition of C, $k \in I(1, N)$, is given by

$$C = \bigcup_{r^N=1}^{M_N} \bigcup_{r^{N-1}=1}^{M_{N-1,r^N}} \bigcup_{r^{N-2}=1}^{M_{N-2,r^{N-1},r^N}} \cdots \bigcup_{r^{K+1}=1}^{M_{K+1,r^{K+2:N}}} C^{[N-(k+2)]\times 0;r^{k+1:N}0}_{[N-(k+1)]\times 0;r^{k+1:N}}$$
$$= \bigcup_{r^N=1}^{M_N} \bigcup_{r^{N-1}=1}^{M_{N-1,r^N}} \bigcup_{r^{N-2}=1}^{M_{N-2,r^{N-1},r^N}} \cdots \bigcup_{r^{K+1}=1}^{M_{K+1,r^{K+2:N}}} \bigcup_{r^k=1}^{M_{k,r^{k+1:N}}} C^{[N-(k+1)]\times 0;r^{k:N}0}_{(N-k)\times 0;r^{k:N}}. \tag{2.41}$$

The **first hierarchical partition level** of C is given by

$$C = C^{N\times 0}_{N\times 0} = \bigcup_{r^N=1}^{M_N} \bigcup_{r^{N-1}=1}^{M_{N-1},r^N} \bigcup_{r^{N-2}=1}^{M_{N-2},r^{N-1:N}} \cdots \bigcup_{r^3=1}^{M_3,r^{4:N}} \bigcup_{r^2=1}^{M_2,r^{3:N}} \bigcup_{r^1=1}^{M_1,r^{2:N}} C^{r^{2:N}0}_{r^{1:N}}. \tag{2.42}$$

At any hierarchical level $k \in I(1, N)$, $r^k \in I(1, M_{k,r^{k+1:N}})$, assume that a network exists between

$$M^k_{\text{total}} = \sum_{r^N=1}^{M_N} \sum_{r^{N-1}=1}^{M_{N-1,r^N}} \sum_{r^{N-2}=1}^{M_{N-2,r^{N-1},r^N}} \cdots \sum_{r^{K+1}=1}^{M_{K+1,r^{K+2:N}}} M_{k,r^{k+1:N}} \tag{2.43}$$

subregions of C that form the union in equation (2.41). By equation (2.42), at the first hierarchical level partition of C, there are a total of

$$M^N_{\text{total}} = \sum_{r^N=1}^{M_N} \sum_{r^{N-1}=1}^{M_{N-1},r^N} \sum_{r^{N-2}=1}^{M_{N-2},r^{N-1:N}} \cdots \sum_{r^3=1}^{M_3,r^{4:N}} \sum_{r^2=1}^{M_2,r^{3:N}} M_{1,r^{2:N}} \tag{2.44}$$

subregions of C.

Example 2.4.3 (Social media network). Refer to Example 2.4.2. Suppose in the social media network (e. g., **Facebook**), the complete collection of online account owners globally is first partitioned into the M_4 continents given by

$$C = \bigcup_{r^4=1}^{M_4} C^{2\times 0;r^4,0}_{3\times 0;r^4}. \tag{2.45}$$

The account owners from each continent $C^{2\times 0;r^4,0}_{3\times 0;r^4}$, $r^4 \in I(1, M_4)$, are further partitioned according to the M_{3,r^4} countries in the continent:

$$C^{2\times 0;r^4,0}_{3\times 0;r^4} = \bigcup_{r^3=1}^{M_3,r^4} C^{0r^{3:4},0}_{00r^{3:4}}. \tag{2.46}$$

Furthermore, the complete collection of online account owners from each country $C^{0r^{3:4},0}_{00r^{3:4}}$, $r^k \in M_{k,r^{K:4}}$, $k \in I(1,3)$, is partitioned according to the $M_{2,r^{3:4}}$ states in each country:

$$C^{0r^{3:4},0}_{00r^{3:4}} = \bigcup_{r^2=1}^{M_2,r^{4:3}} C^{r^{2:4}0}_{0r^{2:4}}. \tag{2.47}$$

The complete collection of account owners from the state $C^{r^{2:4}0}_{0r^{2:4}}$ is further partitioned into the $M_1, r^{4:1}$ counties of the state:

$$C^{r^{2:4}0}_{0r^{2:4}} = \bigcup_{r^1=1}^{M_1,r^{4:1}} C^{r^{2:4},0}_{r^{1:4}}. \tag{2.48}$$

Thus the complete collection of all online account owners in the social media network globally is given by the union

$$C^{0000}_{0000} = \bigcup_{r^4=1}^{M_4} \bigcup_{r^3=1}^{M_3,r^4} \bigcup_{r^2=1}^{M_2,r^{4:3}} \bigcup_{r^1=1}^{M_1,r^{4:1}} C^{r^{2:4},0}_{r^{1:4}}. \tag{2.49}$$

A friendship network between all online account owners in equation (2.49), socializing in the network according to Definition 2.4.1, will comprise the following:

$$V = C^{0000}_{0000}, \tag{2.50}$$

$$E = \{\{x,y\} \mid x,y \in C^{0000}_{0000};\ x \neq y\}, \tag{2.51}$$

where the pair $\{x,y\}$ represents friendship between account owners from two distinct counties x and y in the global social media network of C.

An example of a three-scale network population is depicted in Figure 2.1, which exhibits the overall finite compact topological space $C = C^{3\times0}_{3\times0} \equiv C^{000}_{000} \subset \mathbb{R}^2_+$ with respect to the metric topology with the usual Euclidean metric d_2. The overall three-scale space has the third partition level decomposition with covering consisting of three subspaces $\{C^{0r^30}_{00r^3}, r^3 \in I(1,3)$, where $C^{0r^30}_{00r^3} \cap C^{0q^30}_{00q^3} = \emptyset$, $r^3 \neq q^3$, and the third hierarchical level partition

$$C^{3\times0}_{3\times0} = \bigcup_{r^3=1}^{3} C^{0r^30}_{00r^3}.$$

For each $r^3 \in I(1,3)$, the subcollection $\{C^{r^2r^30}_{0r^2r^3}, r^2 \in I(1,3)\}$ is a partition of $C^{0r^30}_{00r^3}$, and

$$C^{0r^30}_{00r^3} = \bigcup_{r^2=1}^{3} C^{r^2r^30}_{0r^2r^3},$$

where $C^{r^2r^30}_{0r^2r^3} \cap C^{q^2r^30}_{0q^2r^3} = \emptyset$, $r^2 \neq q^2$, and

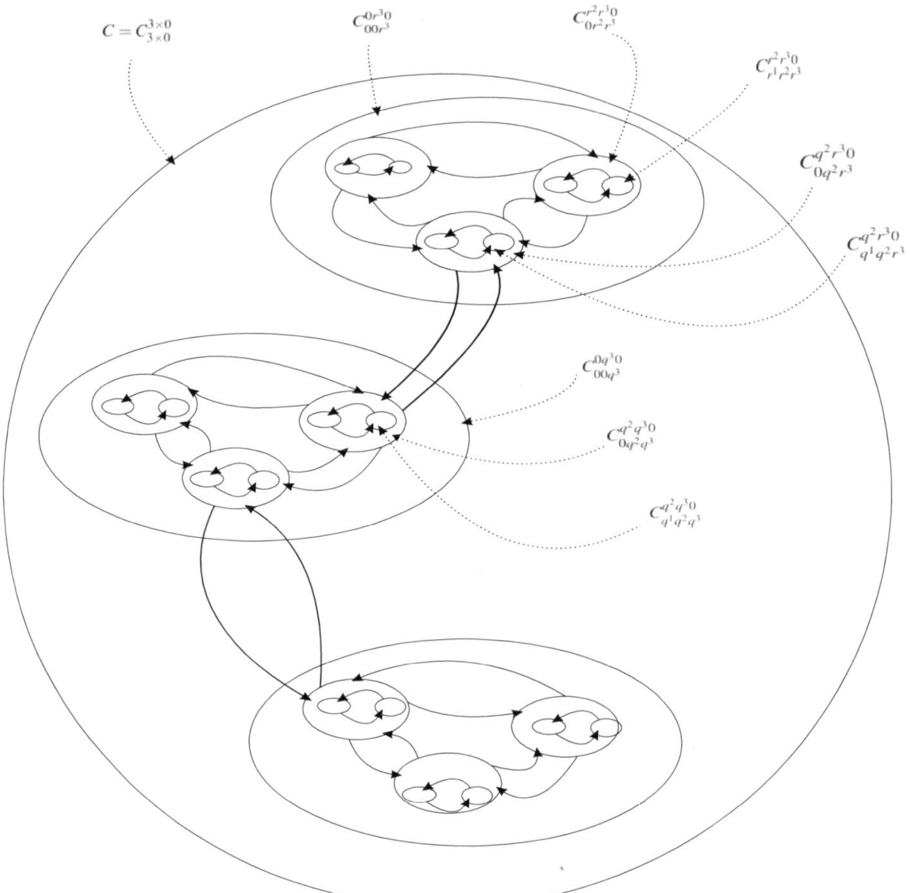

Figure 2.1: An $N = 3$-scale network population on a finite compact topological space $C = C^{3\times 0}_{3\times 0} \equiv C^{000}_{000} \subset \mathbb{R}^2_+$ with the usual Euclidean metric d_2. The overall three-scale space has the third partition level decomposition with covering $\{C^{0r^30}_{00r^3}, r^3 \in I(1,3)$, where $C^{0r^30}_{00r^3} \cap C^{0q^30}_{00q^3} = \emptyset, r^3 \neq q^3$, and $C^{3\times 0}_{3\times 0} = \bigcup_{r^3=1}^{3} C^{0r^30}_{00r^3}$. For each $r^3 \in I(1,3)$, the collection $\{C^{r^2r^30}_{0r^2r^3}, r^2 \in I(1,3)$ is a partition of $C^{0r^30}_{00r^3}$, and $C^{0r^30}_{00r^3} = \bigcup_{r^2=1}^{3} C^{r^2r^30}_{0r^2r^3}$, where $C^{r^2r^30}_{0r^2r^3} \cap C^{q^2r^30}_{0q^2r^3} = \emptyset, r^2 \neq q^2$, and $C^{3\times 0}_{3\times 0} = \bigcup_{r^3=1}^{3}\bigcup_{r^2=1}^{3} C^{r^2r^30}_{0r^2r^3}$ is the second-level partition of C. For each $r^2 \in I(1,3)$, the collection $\{C^{r^2r^30}_{r^1r^2r^3}, r^1 \in I(1,3)$ is a partition of $C^{r^2r^30}_{0r^2r^3}$, and $C^{r^2r^30}_{0r^2r^3} = \bigcup_{r^1=1}^{3} C^{r^2r^30}_{r^1r^2r^3}$, where $C^{r^2r^30}_{r^1r^2r^3} \cap C^{r^2r^30}_{r^1r^2r^3} = \emptyset$, $r^1 \neq q^1$, and $C^{3\times 0}_{3\times 0} = \bigcup_{r^3=1}^{3}\bigcup_{r^2=1}^{3}\bigcup_{r^1=1}^{3} C^{r^2r^30}_{r^1r^2r^3}$ is the first-level partition of C. The solid arrows represent the links between the large- and small-scale coverings in the third hierarchical level partitions of C. For example, the arrows can be the human mobility network or social ties that exist between individuals or groups of individuals in a social media network.

$$C^{3\times 0}_{3\times 0} = \bigcup_{r^3=1}^{3}\bigcup_{r^2=1}^{3} C^{r^2r^30}_{0r^2r^3}$$

is the second hierarchical level partition of C.

For each $r^2 \in I(1,3)$, the subcollection $\{C^{r^2r^30}_{r^1r^2r^3}, r^1 \in I(1,3)\}$ is a partition of $C^{r^2r^30}_{0r^2r^3}$, and

$$C^{r^2r^30}_{0r^2r^3} = \bigcup_{r^1=1}^{3} C^{r^2r^30}_{r^1r^2r^3},$$

where $C^{r^2r^30}_{r^1r^2r^3} \cap C^{r^2r^30}_{r^1r^2r^3} = \emptyset, r^1 \neq q^1$, and

$$C^{3\times0}_{3\times0} = \bigcup_{r^3=1}^{3} \bigcup_{r^2=1}^{3} \bigcup_{r^1=1}^{3} C^{r^2r^30}_{r^1r^2r^3}$$

is the first hierarchical level partition of C.

The solid arrows in Figure 2.1 represent directed links (arcs) connecting the large- and small-scale subspaces, which form coverings in the third hierarchical level partitions of C described above. These links can represent various types of connections depending on the context. For instance:

- **Human mobility networks:** Arrows may signify movement between major and minor geographical regions or subpatches on a landscape, such as a travel between cities and counties within a state.
- **Social networks:** Arrows might depict social ties between individuals or groups, such as friendships, collaborations, or interactions on social media platforms.

This framework emphasizes how directed connections across hierarchical partitions shape the overall structure and dynamics of the finite-scale network population.

3 Node-level decomposition and structural properties of finite-scale networks

3.1 Introduction

The emergence of a finite-scale network structure in a metapopulation, influenced by human mobility networks or social media networks (e. g., friendship networks), introduces heterogeneities in the human population across various patches of the hierarchical-level partitions of the overall finite-scale network population C.

For example, consider a two-scale network population in a USA state. The second-level partition of the state consists of its counties, whereas the first-level partition comprises cities (including towns and unincorporated communities) within these counties. The influence of the human mobility network becomes evident in structures based on the city of residence. Individuals in a given city can be classified as *citizens*, *resident aliens*, *nonresident aliens*, or *visitors* inhabiting the city. Similarly, additional classifications may include human behavior, ethnicity, race, and other demographic factors.

In this chapter, leveraging Definition 2.4.1, we present a method for classifying and quantifying the nodes of the network in the finite-scale network population introduced in Sections 2.4.1 and 2.4.2, based on the links incident with two adjacent nodes. Specifically, we introduce the density of nodes present in a given patch at the kth hierarchical-level partition, $k \in I(1, N)$, of the overall finite-scale network population C (as defined in (2.41)). Furthermore, some structural properties of the network in the finite-scale population are investigated.

3.2 Structural and node-level decomposition in two-scale network populations

Consider the network in the two-scale network population defined in Section 2.4.1 and Examples 2.4.1 and 2.4.2, where the first-level partition is given by $C = \bigcup_{r=1}^{M} \bigcup_{i=1}^{n_r} \{s_i^r\}$, the M regions are denoted by C_r, $r \in I(1, M)$, and the intraregional sites are denoted by s_i^r, $i \in I(1, n_r)$, $r \in I(1, M)$.

For any $r \in I(1, M)$, let the **arc**[1] given by the ordered pair $a_{ij}^{rr} = (s_i^r, s_j^r)$ denote all connections initiated by **residents** of site s_i^r and destined for site s_j^r in the region C_r. Furthermore, let $N_{ij}^{rr}(t)$ be the number of nodes incident with link a_{ij} at any time $t \geq 0$. The arc $a_{ii}^{rr} = (s_i^r, s_i^r)$ denotes the event that the **residents** of site s_i^r initiate contacts only with other nodes in their home site in the region C_r. This implies that $N_{ii}^{rr}(t)$, which is the number of nodes incident with the link $a_{ii}^{rr} = (s_i^r, s_i^r)$ at any time $t \geq 0$, is simply the number of residents of site s_i^r that do not initiate any contacts with other sites at time t.

1 See Definition 2.4.2.

https://doi.org/10.1515/9783111613215-003

Also, denote by the arc $a_{il}^{rq} = (s_i^r, s_l^q)$ all connections initiated by **residents** of site s_i^r in the region C_r and destined for site s_l^q in the region C_q, where $q \neq r$, $r, q \in I(1, M)$, and $l \in I(1, n_q)$. Similarly, $N_{il}^{rq}(t)$ is the number of nodes incident with arc a_{il}^{rq} at any time $t \geq 0$.

In addition, denote by $N_{i0}^{r0}(t)$ the total number of nodes that are residents of site s_i^r in the region C_r incident with either the arcs of type a_{ij}^{rr}, a_{ii}^{rr}, or a_{il}^{rq} at any time $t \geq 0$. It is easy to see that

$$N_{i0}^{r0}(t) = N_{ii}^{rr} + \sum_{j \neq i}^{n_r} N_{ij}^{rr}(t) + \sum_{q \neq r}^{M} \sum_{l=1}^{n_q} N_{il}^{rq}(t), \quad i \in I(1, n_r), r \in I(1, M), \tag{3.1}$$

where $N_{ii}^{rr}(t)$ is the number of residents of site s_i^r in the region C_r that did not initiate any contacts with other sites, $\sum_{j \neq i}^{n_r} N_{ij}^{rr}(t)$ is the number of residents from site s_i^r who initiate contacts with the other $(n_r - 1)$ intraregional sites in C_r, and $\sum_{q \neq r}^{M} \sum_{l=1}^{n_q} N_{il}^{rq}(t)$ is the number of residents of site s_i^r who initiate contacts with other sites in the $M - 1$ subregions C_q of the overall two-scale network population C, where $q \in I(1, M)$, $q \neq r$.

Example 3.2.1 (Human mobility network). Return to the human mobility network in the two-scale network population in Example 2.4.1. The term $N_{ii}^{rr}(t)$ is the number of residents of site s_i^r, $i \in I(1, n_r)$, in the region C_r, $r \in I(1, M)$, who are present in their home site at any time t, $N_{ij}^{rr}(t)$ is the number of residents of s_i^r visiting the site s_j^r within the region C_r where $j \neq i$, and $N_{il}^{rq}(t)$ is the number of residents of site s_i^r in the region C_r that are visiting the site s_l^q in the region C_q, $q \neq r$, $r, q \in I(1, M)$, $l \in I(1, n_q)$. Hence, by (3.1), N_{i0}^{r0} is the total number of residents of site s_i^r within the region C_r who are either present at their home site s_i^r, visiting other sites s_j^r in the region C_r, or visiting sites s_l^q in other regions C_q.

Example 3.2.2 (Social media network). Return to the social media network in Example 2.4.2. In the social media network, the global collection of online account owners is partitioned at the second hierarchical level into distinct continents represented by $C_{00}^{00} = \bigcup_{r=1}^{M_2} C_r$. Within the rth continent C_r, the account owners are further partitioned into subcollections corresponding to the countries $s_1^r, s_2^r, \ldots, s_{n_r}^r$ of that continent, where $r \in I(1, M)$. The arc $a_{ij}^{rr} = (s_i^r, s_j^r)$ represents a friendship connection between two account owners from the countries s_i^r and s_j^r $(j \neq i)$, with the connection initiated by the resident of country s_i^r.

Similarly, the arc $a_{ii}^{rr} = (s_i^r, s_i^r)$ represents a friendship connection between two residents of the country s_i^r. The arc $a_{il}^{rq} = (s_i^r, s_l^q)$, $q \neq r$, represents a friendship connection between two account owners from the distinct countries s_i^r and s_l^q located in the distinct continents C_r and C_q, with the connection initiated by the resident of country s_i^r.

In (3.1), N_{ii}^{rr} is the number of account owners of the country s_i^r who have connections of the type a_{ii}^{rr}, N_{ij}^{rr} is the number of account owners of the country s_i^r who have connections of the type a_{ij}^{rr} with residents of the country s_j^r, and N_{il}^{rq} is the number of account owners of the country s_i^r within the continent C_r who have connections of the type a_{il}^{rq} with residents of the country s_l^q within the continent C_q. This implies that N_{i0}^{r0} is the total number of account owners in the social media network who are residents of the country s_i^r and either have connections with account owners in their home country, represented by the term N_{ii}^{rr}, have connections with account owners from other countries within the continent C_r, represented by $\sum_{j\neq i}^{n_r} N_{ij}^{rr}(t)$, or have connections with account owners from different countries s_l^q within other continents C_q, represented by the sum $\sum_{q\neq r}^{M}\sum_{l=1}^{n_q} N_{il}^{rq}(t)$, where $q \in I(1, M)$, $q \neq r$.

3.3 Structural decomposition and node characterization in finite-scale *N*-network populations

The notations used in Section 3.2 for a two-scale network population are now extended to the overall N-scale network population. We first consider the three-scale network population described in Examples 2.3.1–2.3.3.

Notation 3.3.1. The following general notation used in (3.1) will be extended to any N-scale network population. Given any two nodes in the network, denoted by C_i^m and C_j^n, and the arc denoted by $a_{i;j}^{m;n} = (C_i^m, C_j^n)$ representing the event that a link initiated by an individual residing at site C_i^m is destined for site C_j^n, $N_{i;j}^{m;n}$ denotes the number of nodes of site C_i^m incident with the arc $a_{i;j}^{m;n}$.

3.3.1 Node level decomposition of the thee-scale network population

To obtain a good sense for applying the notations introduced in Notation 3.3.1, we consider a three-scale network population $C = C_{3\times 0}^{3\times 0}$ and decompose the density of the nodes in the three hierarchical level partitions as shown in Example 3.3.1.

Example 3.3.1. Recall the first hierarchical level partition in (2.22) of the three-scale network population denoted by $C = C_{000}^{000}$, that is,

$$C = \bigcup_{r^3=1}^{4}\bigcup_{r^2=1}^{M_{2,r^3}} C_{0r^2r^3}^{r^2r^30} = \bigcup_{r^3=1}^{4}\bigcup_{r^2=1}^{M_{2,r^3}}\bigcup_{r^1=1}^{M_{1,r^2,r^3}} C_{r^1r^2r^3}^{r^2r^30}, \tag{3.2}$$

the **second-level partition** of the overall compact topological space in (2.19) given as

$$C = \bigcup_{r^3=1}^{4}\bigcup_{r^2=1}^{M_{2,r^2}} C_{0r^2r^3}^{r^2r^30}, \tag{3.3}$$

and the third-level partition in (2.15) given as

$$C = C_{001}^{010} \cup C_{002}^{020} \cup C_{003}^{030} \cup C_{004}^{040}. \tag{3.4}$$

A network in the overall three-scale network population C consists of links between nodes that form the partitions in (3.2)–(3.4). The notations in Notation 2.3.1 are used subsequently.

(E.1) Interconnections between sites $C^{r^{2:3}0}_{r^{1:3}}$ and $C^{r^{2:3}0}_{q^1r^{2:3}}$ in the same region $C^{r^{2:3}0}_{0r^{2:3}}$ of the second-level partition of C:
For all $r^k \in I(1, M_{k,r^{k+1:N}})$, $k \in I(1, N)$, consider the part of the network between sites in the region $C^{r^{2:3}0}_{0r^{2:3}}$. Let the arc

$$a^{r^2r^3 0;r^2r^3 0}_{r^1r^2r^3;q^1r^2r^3} \equiv a^{r^{2:3}0;r^{2:3}0}_{r^{1:3};q^1r^{2:3}} = (C^{r^2r^30}_{r^1r^2r^3}, C^{r^2r^30}_{q^1r^2r^3}) \equiv (C^{r^{2:3}0}_{r^{1:3}}, C^{r^{2:3}0}_{q^1r^{2:3}}) \tag{3.5}$$

denote the event that a link initiated by a resident of the site $C^{r^{2:3}0}_{r^{1:3}}$ is destined for the site $C^{r^{2:3}0}_{q^1r^{2:3}}$, where $q^1 \neq r^1, q^1, r^1 \in I(1, M_{1,r^{2:3}})$. Thus the arc $a^{r^{2:3}0;r^{2:3}0}_{r^{1:3};r^{1:3}}$ denotes the event that a link initiated by a resident of the site $C^{r^{2:3}0}_{r^{1:3}}$ is destined for the same site $C^{r^{2:3}0}_{r^{1:3}}$. Recall that both sites $C^{r^{2:3}0}_{q^1r^{2:3}}$ and $C^{r^{2:3}0}_{r^{1:3}}$ are in the partition of the region $C^{r^{2:3}0}_{0r^{2:3}}$. Applying Notation 3.3.1, denote by $N^{r^{2:3}0;r^{2:3}0}_{r^{1:3};r^{1:3}}(t)$ the number of nodes of the site $C^{r^{2:3}0}_{r^{1:3}}$ incident with the arc $a^{r^{2:3}0;r^{2:3}0}_{r^{1:3};r^{1:3}}$ at time t, and by $N^{r^{2:3}0;r^{2:3}0}_{r^{1:3};q^1r^{2:3}}(t)$ denote the number of nodes of the site $C^{r^{2:3}0}_{r^{1:3}}$ incident with the arc $a^{r^{2:3}0;r^{2:3}0}_{r^{1:3};q^1r^{2:3}}$ at time t. We easily see that the sum

$$N^{r^{2:3}0;r^{2:3}0}_{r^{1:3};r^{1:3}}(t) + \sum_{q^1 \neq r^1} N^{r^{2:3}0;r^{2:3}0}_{r^{1:3};q^1r^{2:3}}(t) \tag{3.6}$$

is the total number of nodes of the site $C^{r^{2:3}0}_{r^{1:3}}$ that have initiated links with other nodes either at their home site or at other intraregional sites of $C^{r^{2:3}0}_{0r^{2:3}}$.

(E.2) Interconnections between sites $C^{r^{2:3}0}_{r^{1:3}} \in C^{r^{2:3}0}_{0r^{2:3}}$ and $C^{q^2r^30}_{q^{1:2}r^3} \in C^{q^2r^30}_{0q^2r^3}$, $q^2 \neq r^2$ in the region $C^{0r^30}_{00r^3}$ that are in the third-level partition of C:
Similarly, let the arc

$$a^{r^2r^30;q^2r^30}_{r^1r^2r^3;q^1q^2r^3} \equiv a^{r^{2:3}0;q^2r^30}_{r^{1:3};q^1q^2r^3} = (C^{r^2r^30}_{r^1r^2r^3}, C^{q^2r^30}_{q^1q^2r^3}) \equiv (C^{r^{2:3}0}_{r^{1:3}}, C^{q^2r^30}_{q^{1:2}r^3}) \tag{3.7}$$

represent the event that a link initiated by a resident of the site $C^{r^{2:3}0}_{r^{1:3}}$ is destined at the site $C^{q^2r^30}_{q^{1:2}r^3}$, where $q^2 \neq r^2, q^1, r^1 \in I(1, M_{1,r^{2:3}})$, $q^2, r^2 \in I(1, M_{2,r^3})$. Let $N^{r^{2:3}0;q^2r^30}_{r^{1:3};q^{1:2}r^3}(t)$ denote the number of nodes of the site $C^{r^{2:3}0}_{r^{1:3}}$ incident with the arc $a^{r^{2:3}0;q^2r^30}_{r^{1:3};q^1q^2r^3}$ at time t. We easily see that the sum

$$\sum_{q^2 \neq r^2}^{M_{2,r^3}} \sum_{q^1=1}^{M_{1,q^2r^3}} N^{r^{2:3}0;q^2r^30}_{r^{1:3};q^{1:2}r^3}(t) \tag{3.8}$$

is the total number of nodes of the site $C^{r^{2:3}0}_{r^{1:3}}$ that have initiated links with all other sites $C^{q^2r^30}_{q^{1:2}r^3} \in C^{q^2r^30}_{0q^2r^3}$, $q^2 \neq r^2$, of the region $C^{0r^30}_{00r^3}$ in the third-level partition of C.

(E.3) Interconnections between sites $C_{r^{1:3}}^{r^{2:3}0} \in C_{0r^{2:3}}^{r^{2:3}0}$ and $C_{q^{1:3}}^{q^{2:3}0} \in C_{0q^{2:3}}^{q^{2:3}0}$, $q^3 \neq r^3$, in the overall three-scale network population $C = C_{0\times 3}^{0\times 3}$:
Let the arc

$$a_{r^{1:3};q^{1:3}}^{r^{2:3}0;q^{2:3}0} = (C_{r^1r^2r^3}^{r^2r^30}, C_{q^1q^2q^3}^{q^2q^30}) \equiv (C_{r^{1:3}}^{r^{2:3}0}, C_{q^{1:3}}^{q^{2:3}0}) \tag{3.9}$$

represent the event that a link initiated by a resident of the site $C_{r^{1:3}}^{r^{2:3}0}$ is destined at the site $C_{q^{1:3}}^{q^{2:3}0}$, where $q^3 \neq r^3$, $r^k \in I(1, M_{k,r^{k+1:3}})$, $k = I(1,3)$, $q^k \in I(1, M_{k,r^{k+1:3}})$, $k = I(1,3)$. Let $N_{r^{1:3};q^{1:3}}^{r^{2:3}0;q^{2:3}0}(t)$ denote the number of nodes of the site $C_{r^{1:3}}^{r^{2:3}0}$ incident with the arc $a_{r^{1:3};q^{1:3}}^{r^{2:3}0;q^{2:3}0}$ at time t. We easily see that the sum

$$\sum_{q^3\neq r^3}^{M_3} \sum_{q^2=1}^{M_{2,q^3}} \sum_{q^1=1}^{M_{1,q^2r^3}} N_{r^{1:3};q^{1:2}r^3}^{r^{2:3}0;q^2r^30}(t) \tag{3.10}$$

is the total number of nodes of the site $C_{r^{1:3}}^{r^{2:3}0}$ that have initiated links with all other sites $C_{q^{1:3}}^{q^{2:3}0} \in C_{0q^{2:3}}^{q^{2:3}0}$, $q^3 \neq r^3$, of the region $C_{00q^3}^{0q^30}$ in the third-level partition of C. Hence, by (3.6)–(3.10), the sum

$$N_{r^{1:3};r^{1:3}}^{r^{2:3}0;r^{2:3}0}(t) + \sum_{q^1\neq r^1} N_{r^{1:3};q^1r^{2:3}}^{r^{2:3}0;r^{2:3}0}(t) + \sum_{q^2\neq r^2}^{M_{2,r^3}} \sum_{q^1=1}^{M_{1,q^2r^3}} N_{r^{1:3};q^{1:2}r^3}^{r^{2:3}0;q^2r^30}(t) + \sum_{q^3\neq r^3}^{M_3} \sum_{q^2=1}^{M_{2,q^3}} \sum_{q^1=1}^{M_{1,q^2r^3}} N_{r^{1:3};q^{1:2}r^3}^{r^{2:3}0;q^2r^30}(t) \tag{3.11}$$

is the total number of nodes of the site $C_{r^{1:3}}^{r^{2:3}0}$ that have initiated links with their home site and all other sites in the partition $C = \bigcup_{r^3=1}^{4} \bigcup_{r^2=1}^{M_{2,r^3}} \bigcup_{r^1=1}^{M_{1,r^2,r^3}} C_{r^1r^2r^3}^{r^2r^30}$ of the overall three-scale network population C given in (3.2).

The following remark and additional notations follow from Example 3.3.1.

Remark 3.3.1. (1) The sum (3.6) represents all nodes that initiate connections between the site $C_{r^{1:3}}^{r^{2:3}0}$ and all other sites in the region $C_{0r^{2:3}}^{r^{2:3}0}$. It equals

$$N_{r^{1:3};0r^{2:3}}^{r^{2:3}0;r^{2:3}0}(t) = N_{r^{1:3};r^{1:3}}^{r^{2:3}0;r^{2:3}0}(t) + \sum_{q^1\neq r^1} N_{r^{1:3};q^1r^{2:3}}^{r^{2:3}0;r^{2:3}0}(t). \tag{3.12}$$

(2) Adding the sums (3.6) and (3.8) gives all nodes that initiate connections between the site $C_{r^{1:3}}^{r^{2:3}0}$ and all other sites in the region $C_{00r^3}^{0r^30}$. It equals

$$\begin{aligned} N_{r^{1:3};00r^3}^{r^{2:3}0;0r^30}(t) &= N_{r^{1:3};0r^{2:3}}^{r^{2:3}0;r^{2:3}0}(t) + \sum_{q^2\neq r^2}^{M_{2,r^3}} \sum_{q^1=1}^{M_{1,q^2r^3}} N_{r^{1:3};q^{1:2}r^3}^{r^{2:3}0;q^2r^30}(t) \\ &= N_{r^{1:3};r^{1:3}}^{r^{2:3}0;r^{2:3}0}(t) + \sum_{q^1\neq r^1} N_{r^{1:3};q^1r^{2:3}}^{r^{2:3}0;r^{2:3}0}(t) + \sum_{q^2\neq r^2}^{M_{2,r^3}} \sum_{q^1=1}^{M_{1,q^2r^3}} N_{r^{1:3};q^{1:2}r^3}^{r^{2:3}0;q^2r^30}(t). \end{aligned} \tag{3.13}$$

(3) Adding the sums (3.6), (3.8), and (3.10) gives all nodes that initiate connections between the site $C^{r^{2:3}0}_{r^{1:3}}$ and all other sites in the overall three-scale network population in $C = C^{3\times0}_{3\times0}$. It equals

$$N^{r^{2:3}0;3\times0}_{r^{1:3};3\times0}(t) = N^{r^{2:3}0;0r^30}_{r^{1:3};00r^3}(t) + \sum_{q^3\neq r^3}^{M_3}\sum_{q^2=1}^{M_{2,q^3}}\sum_{q^1=1}^{M_{1,q^2r^3}} N^{r^{2:3}0;q^2r^30}_{r^{1:3};q^{1:2}r^3}(t)$$
$$= N^{r^{2:3}0;r^{2:3}0}_{r^{1:3};r^{1:3}}(t) + \sum_{q^1\neq r^1} N^{r^{2:3}0;r^{2:3}0}_{r^{1:3};q^1r^{2:3}}(t) + \sum_{q^2\neq r^2}^{M_{2,r^3}}\sum_{q^1=1}^{M_{1,q^2r^3}} N^{r^{2:3}0;q^2r^30}_{r^{1:3};q^{1:2}r^3}(t)$$
$$+ \sum_{q^3\neq r^3}^{M_3}\sum_{q^2=1}^{M_{2,q^3}}\sum_{q^1=1}^{M_{1,q^2r^3}} N^{r^{2:3}0;q^2r^30}_{r^{1:3};q^{1:2}r^3}(t). \tag{3.14}$$

3.3.2 Node level decomposition of the general *N*-scale network population

Similar notations used in the decomposition of the three-scale network population in Example 3.3.1 are applied to describe the decomposition of the general N-scale network population. Recall (2.41), the kth-level decomposition of the general N-scale network population $C = C^{N\times0}_{N\times0}$, $k \in I(1, N)$, given by

$$C = \bigcup_{r^N=1}^{M_N}\bigcup_{r^{N-1}=1}^{M_{N-1,r^N}}\bigcup_{r^{N-2}=1}^{M_{N-2,r^{N-1}r^N}}\cdots\bigcup_{r^{K+1}=1}^{M_{K+1,r^{K+2:N}}} C^{[N-(k+2)]\times0;r^{k+1:N}0}_{[N-(k+1)]\times0;r^{k+1:N}}$$
$$= \bigcup_{r^N=1}^{M_N}\bigcup_{r^{N-1}=1}^{M_{N-1,r^N}}\bigcup_{r^{N-2}=1}^{M_{N-2,r^{N-1}r^N}}\cdots\bigcup_{r^{K+1}=1}^{M_{K+1,r^{K+2:N}}}\bigcup_{r^k=1}^{M_{k,r^{k+1:N}}} C^{[N-(k+1)]\times0;r^{k:N}0}_{(N-k)\times0;r^{k:N}}, \tag{3.15}$$

and the **first hierarchical partition level** decomposition of C, given by

$$C = C^{N\times0}_{N\times0} = \bigcup_{r^N=1}^{M_N}\bigcup_{r^{N-1}=1}^{M_{N-1},r^N}\bigcup_{r^{N-2}=1}^{M_{N-2},r^{N-1:N}}\cdots\bigcup_{r^3=1}^{M_3,r^{4:N}}\bigcup_{r^2=1}^{M_2,r^{3:N}}\bigcup_{r^1=1}^{M_1,r^{2:N}} C^{r^{2:N}0}_{r^{1:N}}. \tag{3.16}$$

The following definition characterizes the interconnections between sites $C^{r^{2:N}0}_{r^{1:N}}$ and $C^{r^{2:N}0}_{q^1r^{2:N}}$ of the region $C^{r^{2:N}0}_{0r^{2:N}}$ of the second-level partition of C.

Definition 3.3.1 (Interconnections between nodes in the first hierarchical level partition of $C = C^{N\times0}_{N\times0}$). For all $k \in I(1, M)$ and $r^k \in I(1, M_{k,r^{k+1:N}})$, recall that the subspace $C^{r^{2:N}0}_{0r^{2:N}}$ is part of the second-level partition of C and the sites $C^{r^{2:N}0}_{r^{1:N}}$ and $C^{r^{2:N}0}_{q^1r^{2:N}}$ are part of the partition of the region $C^{r^{2:N}0}_{0r^{2:N}}$, i. e., from (3.15)–(3.16) observe that

$$C_{0r^{2:N}}^{r^{2:N}0} = C_{r^{1:N}}^{r^{2:N}0} \cup \bigcup_{q^1 \neq r^1}^{M_{1,r^{2:N}}} C_{q^1 r^{2:N}}^{r^{2:N}0}, \quad \text{and}$$
$$C = C_{N\times 0}^{N\times 0} = \bigcup_{r^N=1}^{M_N} \bigcup_{r^{N-1}=1}^{M_{N-1,r^N}} \bigcup_{r^{N-2}=1}^{M_{N-2,r^{N-1}r^N}} \cdots \bigcup_{r^{K+1}=1}^{M_{K+1,r^{K+2:N}}} \bigcup_{r^k=1}^{M_{k,r^{k+1:N}}} \cdots \bigcup_{r^2}^{M_{2,r^{3:N}}} C_{0r^{2:N}}^{r^{2:N}0}. \tag{3.17}$$

Let the arc

$$a_{r^{1:N};q^1 r^{2:N}}^{r^{2:N}0;r^{2:N}0} = (C_{r^{1:N}}^{r^{2:N}0}, C_{q^1 r^{2:N}}^{r^{2:N}0}), \quad \forall q^1 \neq r^1, q^1, r^1 \in I(1, M_{1,r^{2:N}}) \tag{3.18}$$

denote the event that a link initiated by a resident of the site $C_{r^{1:N}}^{r^{2:N}0}$ is destined at the site $C_{q^1 r^{2:N}}^{r^{2:N}0}$, where $q^1 \neq r^1$, $q^1, r^1 \in I(1, M_{1,r^{2:N}})$. Thus the arc $a_{r^{1:N};r^{1:N}}^{r^{2:N}0;r^{2:N}0}$ denotes the event that a link initiated by a resident of site $C_{r^{1:N}}^{r^{2:N}0}$ is destined at another resident at the same site. Denote by $N_{r^{1:N};q^1 r^{2:N}}^{r^{2:N}0;r^{2:N}0}(t)$ the number of nodes of the site $C_{r^{1:N}}^{r^{2:N}0}$ incident with the arc $a_{r^{1:N};q^1 r^{2:N}}^{r^{2:N}0;r^{2:N}0}$ at time t. We easily see that

$$N_{r^{1:N};0r^{2:N}}^{r^{2:N}0;r^{2:N}0}(t) = N_{r^{1:N};r^{1:N}}^{r^{2:N}0;r^{2:N}0}(t) + \sum_{q^1 \neq r^1}^{M_{1,r^{2:N}}} N_{r^{1:N};q^1 r^{2:N}}^{r^{2:N}0;r^{2:N}0}(t), \quad t \geq 0, \tag{3.19}$$

is the total number of nodes resident at site $C_{r^{1:N}}^{r^{2:N}0}$ that have initiated links with other nodes either at the home site or with any other intraregional sites of $C_{0r^{2:N}}^{r^{2:N}0}$.

The next definition presents the interconnections between sites $C_{r^{1:N}}^{r^{2:N}0}$ and $C_{q^{1:2} r^{3:N}}^{q^2 r^{3:N}0}$ of the region $C_{00r^{3:N}}^{0r^{3:N}0}$ of the third-level partition of C

Definition 3.3.2 (Interconnections between nodes in the second hierarchical level partition of $C = C_{N\times 0}^{N\times 0}$). For all $k \in I(1, M)$ and $r^k \in I(1, M_{k,r^{k+1:N}})$, recall that the subspace $C_{00r^{3:N}}^{0r^{3:N}0}$ is part of the third-level partition of C, the site $C_{r^{1:N}}^{r^{2:N}0}$ is in the subregion $C_{0r^{2:N}}^{r^{2:N}0} \subset C_{00r^{3:N}}^{0r^{3:N}0}$, and $C_{q^{1:2} r^{3:N}}^{q^2 r^{3:N}0}$ is in the subregion $C_{0q^2 r^{3:N}}^{q^2 r^{3:N}0} \subset C_{00r^{3:N}}^{0r^{3:N}0}$, $q^2 \neq r^2$, i. e., from (3.15)–(3.16) observe that

$$C_{00r^{3:N}}^{0r^{3:N}0} = C_{0r^{2:N}}^{r^{2:N}0} \cup \bigcup_{q^2 \neq r^2}^{M_{2,r^{3:N}}} \bigcup_{q^1=1}^{M_{1,q^2 r^{3:N}}} C_{q^{1:2} r^{3:N}}^{q^2 r^{3:N}0}, \quad \text{and}$$
$$C = C_{N\times 0}^{N\times 0} = \bigcup_{r^N=1}^{M_N} \bigcup_{r^{N-1}=1}^{M_{N-1,r^N}} \bigcup_{r^{N-2}=1}^{M_{N-2,r^{N-1}r^N}} \cdots \bigcup_{r^{K+1}=1}^{M_{K+1,r^{K+2:N}}} \bigcup_{r^k=1}^{M_{k,r^{k+1:N}}} \cdots \bigcup_{r^3=1}^{M_{3,r^{4:N}}} C_{00r^{3:N}}^{0r^{3:N}0}. \tag{3.20}$$

Let the arc

$$a_{r^{1:N};q^{1:2} r^{3:N}}^{r^{2:N}0;q^2 r^{3:N}0} = (C_{r^{1:N}}^{r^{2:N}0}, C_{q^{1:2} r^{3:N}}^{q^2 r^{3:N}0}), \quad q^2 \neq r^2, q^2, r^2 \in I(1 : M_{2,r^{3:N}}),$$
$$r^1 \in I(1, M_{1,r^{2:N}}), \quad q^1 \in I(1, M_{1,q^2 r^{3:N}}), \tag{3.21}$$

denote the event that a link initiated by a resident of the site $C_{r^{1:N}}^{r^{2:N}0}$ is destined at the site $C_{q^{1:2}r^{3:N}}^{q^2r^{3:N}0}$, where $q^2 \neq r^2$, $q^2, r^2 \in I(1 : M_{2,r^{3:N}})$, $r^1 \in I(1, M_{1,r^{2:N}})$, $q^1 \in I(1, M_{1,q^2r^{3:N}})$. Denote by $N_{r^{1:N};q^{1:2}r^{3:N}}^{r^{2:N}0;q^2r^{3:N}0}(t)$ the number of nodes of the site $C_{r^{1:N}}^{r^{2:N}0}$ incident with the arc $a_{r^{1:N};q^{1:2}r^{3:N}}^{r^{2:N}0;q^2r^{3:N}0}$ at time t. We easily to see from (3.19) and (3.20) that

$$N_{r^{1:N};00r^{3:N}}^{r^{2:N}0;0r^{3:N}0}(t) = N_{r^{1:N};0r^{2:N}}^{r^{2:N}0;r^{2:N}0}(t) + \sum_{q^2\neq r^2}^{M_{2,r^{3:N}}} \sum_{q^1=1}^{M_{1,q^2r^{3:N}}} N_{r^{1:N};q^{1:2}r^{3:N}}^{r^{2:N}0;q^2r^{3:N}0}$$
$$= N_{r^{1:N};r^{1:N}}^{r^{2:N}0;r^{2:N}0}(t) + \sum_{q^1\neq r^1}^{M_{1,r^{2:N}}} N_{r^{1:N};q^1r^{2:N}}^{r^{2:N}0;r^{2:N}0}(t) + \sum_{q^2\neq r^2}^{M_{2,r^{3:N}}} \sum_{q^1=1}^{M_{1,q^2r^{3:N}}} N_{r^{1:N};q^{1:2}r^{3:N}}^{r^{2:N}0;q^2r^{3:N}0}, \quad t \geq 0, \tag{3.22}$$

is the total number of nodes resident at site $C_{r^{1:N}}^{r^{2:N}0}$ that have initiated links with other nodes either at the home site or with any other intraregional sites of the region $C_{00r^{3:N}}^{0r^{3:N}0}$.

3.3.3 Application of the principle of mathematical induction for node-level decomposition in the entire *N*-scale network population

We applied the two definitions above to generalize the node-level decomposition for all hierarchical level partitions of the network population C using the *Principle of Mathematical Induction.*

By (3.19) and (3.22) the complete decomposition of the total nodes of the site $C_{r^{1:N}}^{r^{2:N}0}$ that are linked to other nodes of other sites at all levels of the overall N-scale network population C can be obtained by the principle of mathematical induction. For instance, for all $k \in I(1, M)$ and $r^k \in I(1, M_{k,r^{k+1:N}})$, recall that the subspace $C_{(k+1)\times 0r^{k+1:N}}^{k\times 0r^{k+1:N}0}$ is part of the $(k+1)$th-level partition of C and that $C_{r^{1:N}}^{r^{2:N}0} \subset C_{0r^{2:N}}^{r^{2:N}0} \subset C_{00r^{3:N}}^{0r^{3:N}0} \subset \cdots \subset C_{k\times 0r^{k:N}}^{(k-1)\times 0r^{k:N}0} \subset C_{(k+1)\times 0r^{k+1:N}}^{k\times 0r^{k+1:N}0}$ and $C_{q^{1:(k-1)}r^{k:N}}^{q^{2:(k-1)}r^{k:N}0} \subset C_{0q^{2:(k-1)}r^{k:N}}^{q^{2:(k-1)}r^{k:N}0} \subset C_{00q^{3:(k-1)}r^{k:N}}^{0q^{3:(k-1)}r^{k:N}0} \subset \cdots \subset C_{k\times 0q^kr^{k+1:N}}^{(k-1)\times 0q^kr^{k+1:N}0} \subset C_{(k+1)\times 0r^{k+1:N}}^{k\times 0r^{k+1:N}0}$, where $q^k \neq r^k$, $q^k \in I(1, M_{k,r^{k+1:N}})$. From (3.15)–(3.16) and (3.20) observe that for $k = 2$,

$$C_{2\times 0r^{3:N}}^{1\times 0r^{3:N}0} \equiv C_{00r^{3:N}}^{0r^{3:N}0} = C_{0r^{2:N}}^{r^{2:N}0} \cup \bigcup_{q^2\neq r^2}^{M_{2,r^{3:N}}} \bigcup_{q^1=1}^{M_{1,q^2r^{3:N}}} C_{q^{1:2}r^{3:N}}^{q^2r^{3:N}0}, \quad \text{and}$$
$$C = C_{N\times 0}^{N\times 0} = \bigcup_{r^N=1}^{M_N} \bigcup_{r^{N-1}=1}^{M_{N-1,r^N}} \bigcup_{r^{N-2}=1}^{M_{N-2,r^{N-1}r^N}} \cdots \bigcup_{r^{K+1}=1}^{M_{K+1,r^{K+2:N}}} \bigcup_{r^k=1}^{M_{k,r^{k+1:N}}} \cdots \bigcup_{r^3}^{M_{3,r^{4:N}}} C_{2\times 0r^{3:N}}^{1\times 0r^{3:N}0}, \tag{3.23}$$

and we obtain decomposition (3.22).

When $k = 3$, from (3.23) we have

$$C_{3\times 0r^{3:N}}^{2\times 0r^{4:N}0} \equiv C_{000r^{3:N}}^{00r^{4:N}0} = C_{00r^{3:N}}^{0r^{3:N}0} \cup \bigcup_{q^3\neq r^3}^{M_{3,r^{4:N}}} \bigcup_{q^2=1}^{M_{2,q^{2:3}r^{4:N}}} \bigcup_{q^1=1}^{M_{1,q^{2:3}r^{4:N}}} C_{q^{1:3}r^{4:N}}^{q^{2:3}r^{4:N}0}, \quad \text{and}$$

$$C = C_{N\times 0}^{N\times 0} = \bigcup_{r^N=1}^{M_N} \bigcup_{r^{N-1}=1}^{M_{N-1,r^N}} \bigcup_{r^{N-2}=1}^{M_{N-2,r^{N-1}r^N}} \cdots \bigcup_{r^{K+1}=1}^{M_{K+1,r^{K+2:N}}} \bigcup_{r^k=1}^{M_{k,r^{k+1:N}}} \cdots \bigcup_{r^4=1}^{M_{4,r^{5:N}}} C_{3\times 0r^{3:N}}^{2\times 0r^{4:N}0}, \tag{3.24}$$

and similarly to (3.22), the decomposition of the total nodes of site $C_{r^{1:N}}^{r^{2:N}0}$ is given as follows:

$$\begin{aligned} N_{r^{1:N};000r^{4:N}}^{r^{2:N}0;00r^{4:N}0}(t) &= N_{r^{1:N};r^{1:N}}^{r^{2:N}0;r^{2:N}0}(t) + \sum_{q^1\neq r^1}^{M_{1},r^{2:N}} N_{r^{1:N};q^1r^{2:N}}^{r^{2:N}0;r^{2:N}0}(t) \\ &+ \sum_{q^2\neq r^2}^{M_2,r^{2:N}} \sum_{q^1=1}^{M_1,r^{2:N}} N_{r^{1:N};q^1r^{2:N}}^{r^{2:N}0;q^2r^{3:N}0}(t) + \sum_{q^3\neq r^3}^{M_3,r^{4:N}} \sum_{q^2=1}^{M_{2,q^3r^{4:N}}} \sum_{q^1=1}^{M_{1,q^{2:3}r^{4:N}}} N_{r^{1:N};q^{1:3}r^{4:N}}^{r^{2:N}0;q^{2:3}r^{4:N}0}(t), \end{aligned} \tag{3.25}$$

where the new sum $\sum_{q^3\neq r^3}^{M_3,r^{4:N}} \sum_{q^2=1}^{M_2,q^3r^{4:N}} \sum_{q^1=1}^{M_1,q^{2:3}r^{4:N}} N_{r^{1:N};q^{1:3}r^{4:N}}^{r^{2:N}0;q^{2:3}r^{4:N}0}(t)$ is the number of all nodes resident at site $C_{r^{1:N}}^{r^{2:N}0}$ that initiate a link destined at all sites $C_{q^{1:3}r^{4:N}}^{q^{2:3}r^{4:N}0}$, where $q^3 \neq r^3$, $q^1 \in I(1, M_{1,q^{2:3}r^{4:N}})$, $q^2 \in I(1, M_{2,q^3r^{4:N}})$.

The following notation in (3.25) will be used in the rest of this section.

Notation 3.3.2. By (3.25) the term $N_{r^{1:N};000r^{4:N}}^{r^{2:N}0;00r^{4:N}0}(t)$ is denoted by $N_{r^{1:N};3\times 0r^{4:N}}^{r^{2:N}0;2\times 0r^{4:N}0}(t)$ for $k = 3$. In general, for $k \in I(2, N)$, we denote

$$\begin{aligned} N_{r^{1:N};000r^{4:N}}^{r^{2:N}0;00r^{4:N}0}(t) &\equiv N_{r^{1:N};3\times 0r^{4:N}}^{r^{2:N}0;2\times 0r^{4:N}0}(t), \quad k = 3; \\ N_{r^{1:N};k\times 0r^{4:N}}^{r^{2:N}0;(k-1)\times 0r^{4:N}0}(t) &\equiv N_{r^{1:N};000\ldots 0r^{k+1:N}}^{r^{2:N}0;00\ldots 0r^{k+1:N}0}(t), \quad k \geq 2. \end{aligned} \tag{3.26}$$

Step 1. When $k = J \in I(4, N)$, then by (3.23)–(3.24) the overall space N-scale network population can be decomposed explicitly and in a unified manner as follows:

$$C_{J\times 0r^{J:N}}^{(J-1)\times 0r^{J:N}0} = C_{00r^{3:N}}^{0r^{3:N}0} \cup \bigcup_{q^J\neq r^J}^{M_{J,r^{J+1:N}}} \cdots \bigcup_{q^2=1}^{M_{2,q^{3:J}r^{J+1:N}}} \bigcup_{q^1=1}^{M_{1,q^{2:J}r^{J+1:N}}} C_{q^{1:J}r^{J+1:N}}^{q^{J-1:J}r^{J+1:N}0}, \quad \text{and}$$

$$C = C_{N\times 0}^{N\times 0} = \bigcup_{r^N=1}^{M_N} \bigcup_{r^{N-1}=1}^{M_{N-1,r^N}} \bigcup_{r^{N-2}=1}^{M_{N-2,r^{N-1}r^N}} \cdots \bigcup_{r^{K+1}=1}^{M_{K+1,r^{K+2:N}}} \bigcup_{r^k=1}^{M_{k,r^{k+1:N}}} \cdots \bigcup_{r^{(J+1)}=1}^{M_{J+1,r^{J+2:N}}} C_{J\times 0r^{J:N}}^{(J-1)\times 0r^{J+1:N}0}, \tag{3.27}$$

and similarly to (3.22) and (3.25), the decomposition of the total nodes of site $C_{r^{1:N}}^{r^{2:N}0}$ is given as follows:

$$
\begin{aligned}
&N^{r^{2:N}0;(k-1)\times 0 r^{4:N}0}_{r^{1:N};k\times 0 r^{4:N}}(t)\\
&= N^{r^{2:N}0;r^{2:N}0}_{r^{1:N};r^{1:N}}(t) + \sum_{q^1\neq r^1}^{M_1,r^{2:N}} N^{r^{2:N}0;r^{2:N}0}_{r^{1:N};q^1 r^{2:N}}(t)\\
&+ \sum_{q^2\neq r^2}^{M_2,r^{2:N}} \sum_{q^1=1}^{M_1,r^{2:N}} N^{r^{2:N}0;q^2 r^{3:N}0}_{r^{1:N};q^1 r^{2:N}}(t) + \sum_{q^3\neq r^3}^{M_3,r^{4:N}} \sum_{q^2=1}^{M_{2,q^3 r^{4:N}}} \sum_{q^1=1}^{M_{1,q^{2:3} r^{4:N}}} N^{r^{2:N}0;q^{2:3} r^{4:N}0}_{r^{1:N};q^{1:3} r^{4:N}}(t)\\
&+\cdots+\\
&+ \sum_{q^J\neq r^J}^{M_{J,r^{J+1:N}}} \sum_{q^{J-1}=1}^{M_{J-1,r^{J:N}}} \cdots \sum_{q^2=1}^{M_{2,q^{3:J} r^{J+1:N}}} \sum_{q^1=1}^{M_{1,q^{2:J} r^{J=1:N}}} N^{r^{2:N}0;q^{2:J} r^{J+1:N}0}_{r^{1:N};q^{1:J} r^{J+1:N}}(t),
\end{aligned}
\tag{3.28}
$$

where the last sum $\sum_{q^J\neq r^J}^{M_{J,r^{J+1:N}}} \sum_{q^{J-1}=1}^{M_{J-1,r^{J:N}}} \cdots \sum_{q^2=1}^{M_{2,q^{3:J} r^{J+1:N}}} \sum_{q^1=1}^{M_{1,q^{2:J} r^{J=1:N}}} N^{r^{2:N}0;q^{2:J} r^{J+1:N}0}_{r^{1:N};q^{1:J} r^{J+1:N}}(t)$ is the number of all nodes resident at site $C^{r^{2:N}0}_{r^{1:N}}$ that initiate a link destined at all other sites $C^{q^{2:J} r^{J+1:N}0}_{q^{1:J} r^{J+1:N}}$, where $q^J \neq r^J$, $q^J \in I(1, M_{J,r^{J+1:N}})$, $q^{J-1} \in I(1, M_{J-1,q^J r^{J+1:N}})$, $q^{J-2} \in I(1, M_{J-2,q^{J-1:J} r^{J+1:N}})$, $q^{J-3} \in I(1, M_{J-3,q^{J-2:J} r^{J+1:N}}), \ldots, q^2 \in I(1, M_{2,q^{3:J} r^{J+1:N}})$, $q^1 \in I(1, M_{1,q^{2:J} r^{J+1:N}})$.

Step 2. Hence, when $k = N$, by (3.27) the overall space N-scale network population is decomposed explicitly in a unified manner as follows:

$$
\begin{aligned}
&C^{(N-1)\times 0 r^N 0}_{N\times 0 r^N} = C^{(N-2)\times 0 r^N 0}_{(N-1)\times 0 r^N} \cup \bigcup_{q^N\neq r^N}^{M_N} \bigcup_{q^{N-1}=1}^{M_{N-1,q^N}} \bigcup_{q^{N-2}=1}^{M_{N-2,q^{N-1:N}}} \cdots \bigcup_{q^J=1}^{M_{J,q^{J+1:N}}} \cdots \bigcup_{q^2=1}^{M_{2,q^{3:N}}} \bigcup_{q^1=1}^{M_{1,q^{2:N}}} C^{q^{2:N}0}_{q^{1:N}},\\
&J \geq 1, \quad \text{and}\\
&C = C^{N\times 0}_{N\times 0} = \bigcup_{r^N=1}^{M_N} C^{(N-1)\times 0 r^N 0}_{N\times 0 r^N},
\end{aligned}
\tag{3.29}
$$

and similarly to (3.28), the decomposition of the total nodes of site $C^{r^{2:N}0}_{r^{1:N}}$ is given as follows:

$$
\begin{aligned}
&N^{r^{2:N}0;N\times 0}_{r^{1:N};N\times 0}(t)\\
&= N^{r^{2:N}0;r^{2:N}0}_{r^{1:N};r^{1:N}}(t) + \sum_{q^1\neq r^1}^{M_1,r^{2:N}} N^{r^{2:N}0;r^{2:N}0}_{r^{1:N};q^1 r^{2:N}}(t)\\
&+ \sum_{q^2\neq r^2}^{M_2,r^{2:N}} \sum_{q^1=1}^{M_1,r^{2:N}} N^{r^{2:N}0;q^2 r^{3:N}0}_{r^{1:N};q^1 r^{2:N}}(t) + \sum_{q^3\neq r^3}^{M_3,r^{4:N}} \sum_{q^2=1}^{M_{2,q^3 r^{4:N}}} \sum_{q^1=1}^{M_{1,q^{2:3} r^{4:N}}} N^{r^{2:N}0;q^{2:3} r^{4:N}0}_{r^{1:N};q^{1:3} r^{4:N}}(t)\\
&+\cdots\\
&+ \sum_{q^J\neq r^J}^{M_{J,r^{J+1:N}}} \sum_{q^{J-1}=1}^{M_{J-1,r^{J:N}}} \cdots \sum_{q^2=1}^{M_{2,q^{3:J} r^{J+1:N}}} \sum_{q^1=1}^{M_{1,q^{2:J} r^{J=1:N}}} N^{r^{2:N}0;q^{2:J} r^{J+1:N}0}_{r^{1:N};q^{1:J} r^{J+1:N}}(t)
\end{aligned}
$$

$$
\begin{aligned}
&+\cdots\\
&+\sum_{q^{N-1}\neq r^{N-1}}^{M_{N-1,r^N}}\ \sum_{q^{N-2}=1}^{M_{N-2,q^{N-1}r^N}}\cdots\sum_{q^4=1}^{M_{4,q^{5:N-1}r^N}}\ \sum_{q^3=1}^{M_{3,q^{4:N-1}r^N}}\ \sum_{q^2=1}^{M_{2,q^{3:N-1}r^N}}\ \sum_{q^1=1}^{M_{1,q^{2:N-1}r^N}} N_{r^{1:N};q^{1:N-1}r^N}^{r^{2:N}0;q^{2:N-1}r^N0}(t)\\
&+\sum_{q^N\neq r^N}^{M_N}\ \sum_{q^{N-1}=1}^{M_{N-1,r^N}}\ \sum_{q^{N-2}=1}^{M_{N-2,q^{N-1}r^N}}\cdots\sum_{q^4=1}^{M_{4,q^{5:N-1}r^N}}\ \sum_{q^3=1}^{M_{3,q^{4:N-1}r^N}}\ \sum_{q^2=1}^{M_{2,q^{3:N-1}r^N}}\ \sum_{q^1=1}^{M_{1,q^{2:N-1}r^N}} N_{r^{1:N};q^{1:N}}^{r^{2:N}0;q^{2:N}0}(t),
\end{aligned}
\tag{3.30}
$$

where the last sum

$$
\sum_{q^N\neq r^N}^{M_N}\ \sum_{q^{N-1}=1}^{M_{N-1,r^N}}\ \sum_{q^{N-2}=1}^{M_{N-2,q^{N-1}r^N}}\cdots\sum_{q^4=1}^{M_{4,q^{5:N-1}r^N}}\ \sum_{q^3=1}^{M_{3,q^{4:N-1}r^N}}\ \sum_{q^2=1}^{M_{2,q^{3:N-1}r^N}}\ \sum_{q^1=1}^{M_{1,q^{2:N-1}r^N}} N_{r^{1:N};q^{1:N}}^{r^{2:N}0;q^{2:N}0}(t)
$$

represents all nodes that are resident at site $C_{r^{1:N}}^{r^{2:N}0}$ and initiate a link destined at all other sites $C_{q^{1:N}}^{q^{2:N}0}$, where $q^N \neq r^N$, $q^J \in I(1, M_{J,q^{J+1:N}})$, $J \geq 1$.

3.4 Hierarchical network properties in *N*-scale populations

In this section, are investigate some properties of the network characterized in Section 3. We review the following characteristics of networks [4]. From Definitions 2.4.1 and 2.4.2 suppose a network $G = (V, E)$ consists of $M = |V|$ nodes and $L = |E|$ links. Suppose $k_i = deg(i)$ is the *degree* of node $i \in V$. If G is an *undirected network*, then

$$
L = \frac{1}{2}\sum_{i=1}^{M} k_i. \tag{3.31}
$$

The *average degree of the undirected network*, denoted by $\langle k \rangle$, is given by

$$
\langle k \rangle = \frac{1}{M}\sum_{i=1}^{M} k_i = \frac{2L}{M}. \tag{3.32}
$$

Note that the degree $k_i = deg(i)$ of a node $i \in V$ in an undirected network consists of the *incoming degree* $d^-(i)$ and the *outgoing degree* $d^+(i)$, where $deg(i) = d^-(i) + d^+(i)$. If G is a *directed network*, let $k_i^{\text{in}} = d^-(i)$ and $k_i^{\text{out}} = d^+(i)$. Then

$$
L = \sum_{i=1}^{M} k_i^{\text{in}} = \sum_{i=1}^{M} k_i^{\text{out}}. \tag{3.33}
$$

The *average degree of the directed network* is given by

$$
\langle k^{\text{in}} \rangle = \frac{1}{M}\sum_{i=1}^{M} k_i^{\text{in}} = \langle k^{\text{out}} \rangle = \frac{1}{M}\sum_{i=1}^{M} k_i^{\text{out}} = \frac{L}{M}. \tag{3.34}
$$

It is important to note that real life networks, for example, *professional networks, friendship networks, sexual networks, etc.*, are *sparse* with the number of nodes M and links L varying widely (see [4]). The range for L lies between 0 for a network with no connections, and $L_{\max} = \binom{M}{2} = \frac{M(M-1)}{2}$ for a *complete network*, where all M nodes are interconnected with each other, that is,

$$L \in [0, L_{\max}], \quad L_{\max} = \binom{M}{2} = \frac{M(M-1)}{2}. \tag{3.35}$$

An example of real network is the *the scale-free network* (see [4]).

We give the following definition of a *degree distribution* in a network $G = (V, E)$.

Definition 3.4.1. Let $G = (V, E)$ be a network. The probability that a randomly selected node has degree k is called the **degree distribution** of the network. This is denoted by p_k, and $\sum_{k=1}^{\infty} p_k = 1$.

For a simple *fixed* network $G = (V, E)$ with a *finite size* of $M = |V|$ nodes, the degree distribution is approximated by the *relative frequency* approach for probability assuming a *fair sample space*. Let M_k be the number of nodes in the network with degree k. Then

$$p_k = \frac{M_k}{M}. \tag{3.36}$$

This implies that $M_k = M \times p_k$. In general, for any network $G = (V, E)$ with degree distribution for the random variable k denoting the degree of a node in the network, the expected degree is

$$\mathbb{E}[k] = \langle k \rangle = \sum_{k=1}^{\infty} k p_k. \tag{3.37}$$

3.4.1 Network properties of the two-scale population

Consider the network in the two-scale structured population $C = C_{00}^{00}$ in Section 2.4.1 between M_2 regions $C_1, C_2, \ldots, C_{M_2}$ and the intraregional sites $s_1^r, s_2^r, \ldots, s_{n_r}^r$, $r \in I(1, M_2)$, where the first-level partition of the compact topological space representing the complete population is given by (2.13) and (2.6), i. e.,

$$C_{00}^{00} = \bigcup_{r=1}^{M_2} C_r = \bigcup_{r=1}^{M_2} \bigcup_{i=1}^{n_r} \{s_i^r\}. \tag{3.38}$$

Assume that the network $G = (V, E)$ consists of interconnections between nodes s_i^r, $r \in I(1, M_2)$, $i \in I(1, n_r)$, of the set $V = C(s)$ of n nodes in (2.9) and that the links are the pairs $\{s_i^r, s_j^q\}$, $r, q \in I(1, M_2)$, $i, j \in I(1, n_r)$. By (2.10) the cardinality $|C(s)| = n = \sum_{r=1}^{M_2} n_r$. The set of all links that occur in the network is E (see the definitions in Section 2.4). Assume further for simplicity that the network is **undirected**.

For all $r \in I(1, M_2)$ and $i \in I(1, n_r)$, the degree of the node $s_i^r \in V$, denoted by $k_{i0}^{r0} = deg(s_i^r)$, is decomposed into two components:

$$k_{i0}^{r0} = k_{ii}^{rr} + k_{i0}^{rr} + \sum_{q \neq r}^{M_2} k_{i0}^{rq}, \tag{3.39}$$

where k_{ii}^{rr} is the indicator function of the link $\{s_i^r, s_i^r\}$, i. e.,

$$k_{ii}^{rr} = I_{\{s_i^r, s_i^r\}} = \begin{cases} 1 & \text{if the link } \{s_i^r, s_i^r\} \text{ occurs,} \\ 0 & \text{otherwise,} \end{cases} \tag{3.40}$$

and k_{i0}^{rr} is the number of links of the form $\{s_i^r, s_j^r\}, j \neq i$, that occur in the region C_r, i. e.,

$$k_{i0}^{rr} = \sum_{j \neq i}^{n_r} I_{\{s_i^r, s_j^r\}}, \tag{3.41}$$

and the sum

$$\sum_{q \neq r}^{M_2} k_{i0}^{rq}, \quad k_{i0}^{rq} = \sum_{l=1}^{n_q} I_{\{s_i^r, s_l^q\}}, \tag{3.42}$$

is the number of links between the node s_i^r and other nodes s_l^q in other regions $C_q, q \neq r$, $l \in I(1, n_q)$. The following results apply immediately from Section 3.4.

The total number of possible links $L = |E|$ at any time satisfies $0 \leq L \leq L_{\max} = \binom{n}{2}$ and is given by

$$L = \frac{1}{2} \sum_{r=1}^{M_2} \sum_{i=1}^{n_r} k_{i0}^{r0}. \tag{3.43}$$

The *average degree of the undirected network*, denoted by $\langle k \rangle$, is given by

$$\langle k \rangle = \frac{1}{n} \sum_{r=1}^{M_2} \sum_{i=1}^{n_r} k_{i0}^{r0} = \frac{2L}{M}. \tag{3.44}$$

For a simple *fixed* network $G = (V, E)$ with a *finite size* of $n = |V|$ nodes, the degree distribution is approximated by the *relative frequency* approach for probability assuming a *fair sample space*. Suppose M_k^{00} is the number of nodes in $V = C(s)$ in the network with degree k. Then M_k^{00} can be decomposed according the M_2 regions, where M_k^{r0} is the number of nodes of the region $C_r, r \in I(1, M_2)$, that have degree k. Then

$$M_k^{00} = \sum_{r=1}^{M_2} M_k^{r0}, \quad M_k^{r0} = \sum_{i=1}^{n_r} I_{\{k_{i0}^{r0} = k\}}. \tag{3.45}$$

Therefore the probability that a randomly selected node has degree k is given by

$$p_k = \frac{M_k^{00}}{n}, \tag{3.46}$$

which implies that $M_k^{00} = n \times p_k$.

Note that the extension of the results of Section 3.4.1 from the two-scale network population C_{00}^{00} to the N-scale network population $C_{N\times 0}^{N\times 0}$ in Section 2.4.2 is straightforward and is omitted. In the following, we give the properties of the human mobility network in Example 2.4.1.

Example 3.4.1 (Human mobility network). Consider the human mobility network in Examples 2.4.1 and 3.2.1. Given that $C_{00}^{00} = \bigcup_{r=1}^{M_2} C_r$ is the partition of the population of a country into the distinct states of the country C_r, $r \in I(1, M_2)$, the population of the rth state C_r is further partitioned into the distinct counties $s_1^r, s_2^r, \ldots, s_{n_r}^r$ of the state C_r. The human mobility network $G = (V, A)$ is a **directed network** occurring in the two-level partition $C = \bigcup_{r=1}^{M} \bigcup_{i=1}^{n_r} \{s_i^r\}$, where

$$\begin{aligned} &V = C(s), \\ &A = \{(s_i^r, s_j^r), (s_i^r, s_k^q) \mid s_i^r, s_j^r, s_k^q \in C(s); r \neq q; i \neq j; i,j \in I(1, n_r); r, q \in I(1, M); k \in I(1, n_q)\}, \end{aligned} \tag{3.47}$$

and by Definition 2.4.2 the set A signifies the set of all arcs in the network. In other words, the arc (s_i^r, s_j^r) signifies that the residents of site s_i^r in a region C_r initiate a visit to other sites s_j^r within the region C_r, and (s_i^r, s_k^q) signifies that the residents of site s_i^r in a region C_r initiate a visit to other sites s_l^q in other regions C_q, $i, j \in I(1, n_r)$, $r \neq q$, $r, q \in I(1, M)$.

For all $r \in I(1, M_2)$ and $i \in I(1, n_r)$, the indegree and outdegree of the node $s_i^r \in V$, denoted by $k_{i0}^{r0}(\text{in}) = deg^-(s_i^r)$ and $k_{i0}^{r0}(\text{out}) = deg^+(s_i^r)$, respectively, are decomposed into two components

$$\begin{aligned} k_{i0}^{r0}(\text{in}) &= k_{ii}^{rr}(\text{in}) + k_{i0}^{rr}(\text{in}) + \sum_{q \neq r}^{M_2} k_{i0}^{rq}(\text{in}), \\ k_{i0}^{r0}(\text{out}) &= k_{ii}^{rr}(\text{out}) + k_{i0}^{rr}(\text{out}) + \sum_{q \neq r}^{M_2} k_{i0}^{rq}(\text{out}), \end{aligned} \tag{3.48}$$

where $k_{ii}^{rr}(\text{in})$ and $k_{ii}^{rr}(\text{out})$, respectively, are the indicator functions for the link $\{s_i^r, s_i^r\}$, i. e.,

$$k_{ii}^{rr} = I_{\{s_i^r, s_i^r\}} = \begin{cases} 1 & \text{if the link } \{s_i^r, s_i^r\} \text{ occurs}, \\ 0 & \text{otherwise}, \end{cases} \tag{3.49}$$

and $k_{i0}^{rr}(\text{in})$ and $k_{i0}^{rr}(\text{out})$, respectively, are the numbers of arcs of the form (s_j^r, s_i^r) and (s_i^r, s_j^r), $j \neq i$, that occur in the region C_r, i. e.,

$$k_{i0}^{rr}(\text{out}) = \sum_{j \neq i}^{n_r} I_{(s_i^r, s_j^r)}, \quad k_{i0}^{rr}(\text{in}) = \sum_{j \neq i}^{n_r} I_{(s_j^r, s_i^r)}, \tag{3.50}$$

and the sums

$$\begin{aligned} &\sum_{q \neq r}^{M_2} k_{i0}^{rq}(\text{in}), \quad k_{i0}^{rq}(\text{in}) = \sum_{l=1}^{n_q} I_{\{s_l^r, s_i^q\}}, \\ &\sum_{q \neq r}^{M_2} k_{i0}^{rq}(\text{out}), \quad k_{i0}^{rq}(\text{out}) = \sum_{l=1}^{n_q} I_{\{s_i^r, s_l^q\}}, \end{aligned} \tag{3.51}$$

respectively, are the numbers of arcs between the node s_i^r and other nodes s_l^q in other regions C_q, $q \neq r$, $l \in I(1, n_q)$. The following results follow immediately from Section 3.4.

The total number of possible links $L = |A|$ at any time satisfies

$$0 \leq L \leq L_{\max} = \binom{n}{2}$$

and is given by

$$L = \sum_{r=1}^{M_2} \sum_{i=1}^{n_r} k_{i0}^{r0}(\text{in}) = \sum_{r=1}^{M_2} \sum_{i=1}^{n_r} k_{i0}^{r0}(\text{out}). \tag{3.52}$$

The **average degree of the directed network**, denoted by $\langle k \rangle$, is given by

$$\langle k \rangle = \frac{1}{n} \sum_{r=1}^{M_2} \sum_{i=1}^{n_r} k_{i0}^{r0}(\text{in}) = \frac{1}{n} \sum_{r=1}^{M_2} \sum_{i=1}^{n_r} k_{i0}^{r0}(\text{out}) = \frac{L}{M}. \tag{3.53}$$

For a simple *fixed* network $G = (V, A)$ with a *finite size* of $n = |V|$ nodes, the indegree and outdegree distributions can be approximated using the **relative frequency** approach for probability under the assumption of a *fair sample space*. Specifically, let $M_k^{00}(\text{in})$ and $M_k^{00}(\text{out})$ represent the numbers of nodes in $V = C(s)$ in the directed network with indegree and outdegree k, respectively. The quantities $M_k^{00}(\text{in})$ and $M_k^{00}(\text{out})$ can be further decomposed according to the M_2 regions, where $M_k^{r0}(\text{in})$ and $M_k^{r0}(\text{out})$ denote the numbers of nodes in the region C_r with indegree and outdegree k, respectively, for all $r \in I(1, M_2)$. Then

$$\begin{aligned} M_k^{00}(\text{in}) &= \sum_{r=1}^{M_2} M_k^{r0}(\text{in}), \quad M_k^{r0}(\text{in}) = \sum_{i=1}^{n_r} I_{\{k_{i0}^{r0}(\text{in})=k\}}, \\ M_k^{00}(\text{out}) &= \sum_{r=1}^{M_2} M_k^{r0}(\text{out}), \quad M_k^{r0}(\text{out}) = \sum_{i=1}^{n_r} I_{\{k_{i0}^{r0}(\text{out})=k\}}. \end{aligned} \tag{3.54}$$

Therefore the probability p_k^{in} that a randomly selected node has indegree k is given by

$$p_k^{\text{in}} = \frac{M_k^{00}(\text{in})}{n}. \tag{3.55}$$

Similarly, the probability p_k^{out} that a randomly selected node has outdegree k is given by

$$p_k^{\text{out}} = \frac{M_k^{00}(\text{out})}{n}. \tag{3.56}$$

This implies that $M_k^{00}(\text{in}) = n \times p_k^{\text{in}}$ and $M_k^{00}(\text{out}) = n \times p_k^{\text{out}}$.

The network characterized in Chapter 2 for the N-scale network population $C = C_{0\times N}^{0\times N}$ with decomposition in Section 3 exhibits the connectivity between "sites" and "regions" in the N hierarchical level partitions in the overall population C. For most real networks, the connectivity between nodes in a network occurs randomly (refer to [4]). For example, in a social media network, e. g., *Facebook*, where connectivity is based on *friendship*, most online account owners gradually grow their connectivity in both local and global scales based on mutual interests, acquaintances, and other factors that are unpredictable. This is the reason why social networks are random networks. In this chapter the classical random *Erdös–Rényi network* [4] is also introduced in the N-scale network population $C = C_{0\times N}^{0\times N}$ to characterize the *degree distribution* p_k of the network.

4 Random networks in finite-scale structured populations

4.1 Introduction

The network described in Chapter 2 for the N-scale network population $C = C^{0\times N}_{0\times N}$, with its decomposition outlined in Section 3, captures the connectivity between "sites" and "regions" across the N hierarchical partitions of the overall population C. In many real-world networks, the connectivity between nodes occurs randomly (see [4]). For instance, in a social media network such as *Facebook*, where connections are formed through *friendships*, the growth of a user's network happens gradually, based on mutual interests, acquaintances, and other unpredictable factors. Because of this randomness, social networks are often modeled as random networks.

Indeed, the presence of a hierarchical finite-scale structure in a social network becomes apparent whenever the social ties within the network vary in both strength and scope. For example, in a social media platform like *Facebook*, there are distinct hierarchical partitions that reflect the varying degrees of social connections among account owners. The strongest ties, such as those between family members, are placed in the first hierarchical level partition. Ties between friends are categorized in the second level, whereas connections with acquaintances are placed in the third hierarchical level. Lastly, ties with strangers are represented at the fourth level.

This chapter introduces the classic *Erdős–Rényi network* [4], also known as a *random network*, within the context of the N-scale network population $C = C^{0\times N}_{0\times N}$, with the primary objective of characterizing the ***degree distribution*** p_k of the network.

Recall (3.29) in Chapter 3, the first hierarchical level partition of the overall N-scale network population is given by

$$C^{(N-1)\times 0 r^N 0}_{N\times 0 r^N} = C^{(N-2)\times 0 r^N 0}_{(N-1)\times 0 r^N} \cup \bigcup_{q^N \neq r^N}^{M_N} \bigcup_{q^{N-1}=1}^{M_{N-1,q^N}} \bigcup_{q^{N-2}=1}^{M_{N-2,q^{N-1:N}}} \cdots \bigcup_{q^J=1}^{M_{J,q^{J+1:N}}} \cdots \bigcup_{q^2=1}^{M_{2,q^{3:N}}} \bigcup_{q^1=1}^{M_{1,q^{2:N}}} C^{q^{2:N}0}_{q^{1:N}}, \quad J \geq 1,$$
$$C = C^{N\times 0}_{N\times 0} = \bigcup_{r^N=1}^{M_N} C^{(N-1)\times 0 r^N 0}_{N\times 0 r^N}. \tag{4.1}$$

Note that in this chapter, we assume that the compact topological space C (in $\mathbb{R}^n_+$) defined in Chapter 2 is a finite collection of distinct points $C^{q^{2:N}0}_{q^{1:N}}$ in (4.1), consequently equipped with its natural topology, the **discrete topology** $\mathfrak{T}_C = 2^C$. This simplified spatial structure of C allows for easier understanding of the random network between nodes representing individuals or social entities classified in space as explained below. A brief review of random networks is given subsequently. We refer to Barabasi [4] for more applications of random networks.

https://doi.org/10.1515/9783111613215-004

4.2 The Erdős–Rényi network

Definition 4.2.1. A random network (also called Erdős–Rényi network), denoted by $G(M,p)$, consists of M nodes, where every link between a pair of nodes occurs randomly with the same probability p.

Definition 4.2.1 is applied to derive the probability that a given realization of the network has L links, where $L \in [0, L_{\max} = \frac{M(M-1)}{2}]$. Recall (3.35): in a complete network, there are totally $L_{\max} = \frac{M(M-1)}{2}$ links possible. Thus, assuming that links between any pair of nodes in $G(M,p)$ occur independently, the random variable L has the *binomial distribution* with fixed number of trials $L_{\max}$ and *success* probability p. This is denoted $L \sim \mathit{Binomial}(L_{\max}, p)$. Furthermore, the expected number of links is

$$\mathbb{E}[L] \equiv \langle L\rangle = L_{\max} \times p = \frac{M(M-1)}{2}p. \tag{4.2}$$

It follows from (3.32) and (4.2) that the expected degree of each node in the network $G(M,p)$ is

$$\mathbb{E}[k] \equiv \langle k\rangle = \frac{2\langle L\rangle}{M} = (M-1) \times p. \tag{4.3}$$

The average in (4.3) signifies that the random variable k that denotes the degree of any node $i \in I(1,M)$ in the network has the binomial distribution with $M-1$ independent trials representing the maximum number of possible nodes the node i can connect with and success probability p; that is,

$$k \sim \mathit{Binomial}(M-1, p), \quad \mathbb{E}[k] \equiv \langle k\rangle = (M-1) \times p, \quad \mathbb{V}ar[k] = (M-1)p(1-p). \tag{4.4}$$

In addition, when the success probability p is closer to 1, i. e., $p \to 1$, the network $G(M,p)$ is *denser*, the averages satisfy

$$\mathbb{E}[L] \to L_{\max}, \quad \mathbb{E}[k] \to M-1. \tag{4.5}$$

The degree distribution in (4.4) is a approximated by the *Poisson distribution* whenever M is large and the success probability p is small. Indeed, since most real networks are *sparse*, i. e., $\mathbb{E}[k] \equiv \langle k\rangle \ll M$, then from (4.4) we have $p = \frac{\langle k\rangle}{M-1} \ll 1$, and thus

$$p_k = \binom{M-1}{k} p^k (1-p)^{M-1-k} \approx \frac{\lambda^k e^{-\lambda}}{k!}, \quad \lambda = \langle k\rangle. \tag{4.6}$$

The distribution in (4.6) is denoted by

$$k \sim \mathit{Poisson}(\langle k\rangle), \quad \mathbb{E}[k] = \langle k\rangle, \quad \mathbb{V}ar[k] = \langle k\rangle. \tag{4.7}$$

Note that due to the analytical simplicity of the Poisson distribution compared to the binomial distribution, the preferred form for the degree distribution p_k is (4.7). In this book, the distributions (4.7) and (4.4) will be used interchangeably to emphasize different characteristics of the random network. Furthermore, despite the popularity of the random network in many texts of graph theory, the degree distributions in most real networks are not Poisson. We refer to Barabasi [4] for other real network degree distributions.

The rest of this chapter focusses on characterizing a general hierarchical model for the degree distribution of a random network that occurs in the finite-scale network population $C = C_{0\times N}^{0\times N}$ in Chapter 2, with decomposition in Section 3. Since most complex social networks are random, this chapter focusses on complex social networks. According to Wasserman [52], a *social network* is a group of people with some pattern of social ties between them, for example, friendships between people, business relations between companies, and ties between families such as intermarriages and cousins. Examples of real social networks include social media networks on platforms such as *Twitter and Facebook*. Note that complex social networks are the basis for the spread of rumors and ideas in the same manner as the dynamics of communicable diseases in human populations. Thus it is important to investigate random networks as the finite-scale structured population $C = C_{0\times N}^{0\times N}$.

4.3 Application of a random network in a two-scale structured population

Consider the decomposition of the two-scale network population in Section 3.2, where the two-scale network population is defined in Section 2.4.1, and Examples 2.4.1 and 2.4.2. Furthermore, the first-level partition of the overall population C is given by $C = \bigcup_{r=1}^{M}\bigcup_{i=1}^{n_r}\{s_i^r\}$, where the M "regions" are denoted by $C_r, r \in I(1,M)$, and the intraregional "sites" are denoted by s_i^r, $i \in I(1,n_r)$, $r \in I(1,M)$. To define a meaningful random social network in the two-scale network population C, the following assumptions are made.

Assumption 4.3.1. (1) *From* (2.8)–(2.10), *the random network consists of* $n = \sum_{r=1}^{M_2} n_r$ *nodes denoted by* $s_i^r \in C_r$, $i \in I(1,n_r)$, $r \in I(1,M_2)$, *that represent individuals in the social network.*

(2) *In the network, there are* M_2 *social affiliations or circles or groups, denoted by* $C_1, C_2, \ldots, C_{M_2}$, *where for each* $r \in I(1,M_2)$, *the affiliation has* n_r *members given by* $C_r = \bigcup_{i=1}^{n_r} s_{i=1}^r$.

(3) *For any* $r \in I(1,M)$, *the link given by the pair* $a_{ij}^{rr} = \{s_i^r, s_j^r\}$ *denotes a social tie between the nodes* s_i^r *and* s_j^r *in the affiliation* C_r. *Furthermore, let* $N_{ij}^{rr}(t) = \mathbb{I}_{\{a_{ij}^{rr}\}}$, *where* $\mathbb{I}_{\{a_{ij}^{rr}\}}$ *is the indicator of the event that* a_{ij}^{rr} *occurs at time t. In addition, let the probability*

$$p_{i.}^{rr}(t) = \mathbb{P}(N_{ij}^{rr}(t) = 1), \quad j \in I(1,n_r), j \neq i. \tag{4.8}$$

For the Erdős–Rényi network, the chance of the event that $a_{ii}^{rr} = \{s_i^r, s_i^r\}$ *occurs is given by* $\mathbb{P}(N_{ii}^{rr}(t) = 1) = 0$.

Also, denote by $a_{il}^{rq} = \{s_i^r, s_l^q\}$ *a connection between* s_i^r *of the affiliation* C_r *and the node* s_l^q *of the affiliation* C_q*, where* $q \neq r$, $r, q \in I(1, M_3)$*, and* $l \in I(1, n_q)$*. Similarly,* $N_{il}^{rq}(t) = \mathbb{I}_{\{a_{il}^{rq}\}}$ *indicates the occurrence of the link* a_{il}^{rq} *at any time* $t \geq 0$*. Furthermore,*

$$p_{i.}^{rq}(t) = \mathbb{P}(N_{il}^{rq}(t) = 1), \quad l \in I(1, n_q),\ q, r \in I(1, M_2),\ q \neq r. \tag{4.9}$$

Now denote by $N_{i0}^{r0}(t)$ *the* ***degree*** *of the node* s_i^r *given by the total number of distinct nodes of the affiliations* C_r *and* C_q, $q, r \in I(1, M_2)$, $q \neq r$*, incident with either the link* a_{ij}^{rr} *or the link* a_{il}^{rq} *at any time* $t \geq 0$.

It is easy to see from Assumption 4.3.1 that the degree of the node s_i^r is given by the random variable

$$N_{i0}^{r0}(t) = \sum_{j \neq i}^{n_r} N_{ij}^{rr}(t) + \sum_{q \neq r}^{M} \sum_{l=1}^{n_q} N_{il}^{rq}(t), \quad i \in I(1, n_r),\ r \in I(1, M), \tag{4.10}$$

where the sum $\sum_{j \neq i}^{n_r} N_{ij}^{rr}(t)$ is the number of all links that occur between the node site s_i^r and the other $(n_r - 1)$ individuals s_j^r, $j \neq i$, of the affiliation C_r; and the sum $\sum_{q \neq r}^{M} \sum_{l=1}^{n_q} N_{il}^{rq}(t)$ is the number of all possible links that occur between the node s_i^r and all other nodes s_l^q, $l \in I(1, n_q)$, affiliated with the other $M - 1$ social groups C_q of the overall two-scale network population C, where $q \in I(1, M_2)$, $q \neq r$. The degree distribution of the node s_i^r is given in the following theorem.

Theorem 4.3.1. *Let Assumption* 4.3.1 *be satisfied. Denote* $\vec{p^{r\cdot}}_{i.}(t) = (p_{i.}^{rr}, p_{i.}^{rq})$, $q, r \in I(1, M_2)$, $q \neq r$*. Let* $G(n = \sum_{r=1}^{M_2} n_r, \vec{p^{r\cdot}}_{i.}(t))$ *denote the Erdős–Rényi network in the overall two-scale network population* $C = \bigcup_{r=1}^{M} \bigcup_{i=1}^{n_r} \{s_i^r\}$.

(a) *The degree of any node* s_i^r *in the network is given by the random variable* (4.10).
(b) *For mutually independent events* $\{a_{ij}^{rr}\}$ *and* $\{a_{il}^{rq}\}$, $i, j \in I(1, n_r)$, $i \neq j$, $l \in I(1, n_q)$, $q, r \in I(1, M_2)$, $q \neq r$,

$$\sum_{j \neq i}^{n_r} N_{ij}^{rr}(t) \sim \textit{Binomial}(n_r - 1, p_{i.}^{rr}(t)), \quad \textit{and} \tag{4.11}$$

$$\sum_{l=1}^{n_q} N_{il}^{rq}(t) \sim \textit{Binomial}(n_q, p_{i.}^{rq}(t)), \quad q, r \in I(1, M_2),\ q \neq r, \tag{4.12}$$

and at any time $t \geq 0$*, the (cumulative) distribution* $F_{N_{i0}^{r0}(t)}(t)$ *of the random variable* $N_{i0}^{r0}(t)$ *is a convolution of multiple binomial distributions in* (4.11) *given by*

$$F_{N_{i0}^{r0}(t)}(x) = \left(F_{\textit{Binomial}(n_r - 1, p_{i.}^{rr}(t))} * \sum_{q \neq r}^{M} F_{\textit{Binomial}(n_q, p_{i.}^{rq}(t))} \right)(x) \quad \textit{for all } x. \tag{4.13}$$

Proof. The proof of this result follows from the fact that the binomial distribution is a sum of independent identically distributed Bernoulli random variables. Furthermore, the distribution of the sum of independent random variables in (4.10) is a convolution of the cumulative distribution functions of the independent random variables. □

The random network $G(n = \sum_{r=1}^{M_2} n_r, p^{\vec{r}}{}_{i.}(t))$ in the two-scale structured population in Theorem 4.3.1 is extended in a natural way by mathematical induction to the N-scale network population. The results are stated, and the proofs follow immediately.

4.4 Application of a random network in the N-scale structured population

Drawing analogy with Section 4.3, reconsider the decomposition of the N-scale network population in Section 3.3, where (3.29) gives the total number of individuals (nodes) of the “site” $C_{r^{1:N}}^{r^{2:N}0}$, $r^k \in I(1, M_{k,r^{k+1:N}})$, $k \in I(1, N)$, that initiate a connection with other “sites” in the overall N-scale network population with the first hierarchical level partition given by (3.29) and (3.30). For a clearer understanding of the random social network in the N-scale structured population C, the case of $N = 3$ in Example 3.3.1 will be used. The following assumptions are made for the random network in the N-scale network population C in (4.1).

4.4.1 Assumptions for the random network in the N-scale network population

(Assumption 1) From Section 2.3, assume that in the network, all N hierarchical level partitions of C given in (2.14)–(2.33) represent nest social affiliations or social groups, where the kth-level partition of C, $k \in I(1, N)$, from (2.30) is given by

$$C = \bigcup_{r^N=1}^{M_N} \bigcup_{r^{N-1}=1}^{M_{N-1,r^N}} \bigcup_{r^{N-2}=1}^{M_{N-2,r^{N-1}r^N}} \cdots \bigcup_{r^{K+1}=1}^{M_{K+1,r^{K+2:N}}} C_{[N-(k+1)]\times 0;r^{k+1:N}}^{[N-(k+2)]\times 0;r^{k+1:N}0}$$
$$= \bigcup_{r^N=1}^{M_N} \bigcup_{r^{N-1}=1}^{M_{N-1,r^N}} \bigcup_{r^{N-2}=1}^{M_{N-2,r^{N-1}r^N}} \cdots \bigcup_{r^{K+1}=1}^{M_{K+1,r^{K+2:N}}} \bigcup_{r^k=1}^{M_{k,r^{k+1:N}}} C_{(N-k)\times 0;r^{k:N}}^{[N-(k+1)]\times 0;r^{k:N}0}. \tag{4.14}$$

Furthermore, from (2.42) and (2.10), assume that at any hierarchical level $k \in I(1, N)$, $r^k \in I(1, M_{k,r^{k+1:N}})$, a social network exists between the

$$M_{\text{total}}^k = \sum_{r^N=1}^{M_N} \sum_{r^{N-1}=1}^{M_{N-1,r^N}} \sum_{r^{N-2}=1}^{M_{N-2,r^{N-1}r^N}} \cdots \sum_{r^{K+1}=1}^{M_{K+1,r^{K+2:N}}} M_{k,r^{k+1:N}} \tag{4.15}$$

subregions of C that form the union (2.41), and for $k \geq 2$, each subregion represents a **social affiliation or social class** in the network. Also, from (2.44), in the first hierarchical level partition of C, there are a total of

$$M^1_{\text{total}} = \sum_{r^N=1}^{M_N} \sum_{r^{N-1}=1}^{M_{N-1},r^N} \sum_{r^{N-2}=1}^{M_{N-2},r^{N-1:N}} \cdots \sum_{r^3=1}^{M_3,r^{4:N}} \sum_{r^2=1}^{M_2,r^{3:N}} M_1, r^{2:N} \tag{4.16}$$

subregions of C that form the union in (2.42) given by

$$C = C^{N\times 0}_{N\times 0} = \bigcup_{r^N=1}^{M_N} \bigcup_{r^{N-1}=1}^{M_{N-1},r^N} \bigcup_{r^{N-2}=1}^{M_{N-2},r^{N-1:N}} \cdots \bigcup_{r^3=1}^{M_3,r^{4:N}} \bigcup_{r^2=1}^{M_2,r^{3:N}} \bigcup_{r^1=1}^{M_1,r^{2:N}} C^{r^{2:N}0}_{r^{1:N}}, \tag{4.17}$$

and each subregion here represents a **social entity or individual** in the network.

Example 4.4.1. Refer to Example 3.3.1 and recall (3.2), i. e.,

$$C = \bigcup_{r^3=1}^{4} C^{0r^30}_{00r^3} = \bigcup_{r^3=1}^{4} \bigcup_{r^2=1}^{M_{2,r^3}} C^{r^2r^30}_{0r^2r^3} = \bigcup_{r^3=1}^{4} \bigcup_{r^2=1}^{M_{2,r^3}} \bigcup_{r^1=1}^{M_{1,r^2,r^3}} C^{r^2r^30}_{r^1r^2r^3}. \tag{4.18}$$

It is assumed that C is a finite collection of individuals of distinct nested social affiliations, where the cardinality $M^1_{\text{total}} = |C| = \sum_{r^3=1}^{4} \sum_{r^2=1}^{M_{2,r^3}} M_{1,r^2,r^3}$. In (4.18), it is assumed that the points $C^{r^2r^30}_{r^1r^2r^3}$, $r^k \in I(1, M_{k,r^{k+1:4}})$, $k \in I(2,3)$, represent the nodes of the network, which are individuals socializing, for instance, in a social media network such as Facebook. For each $r^3 \in I(1,4)$, the M_{2,r^3} subsets

$$\{C^{r^2r^30}_{0r^2r^3}\}_{r^2=1}^{M_{2,r^3}} = C^{1r^30}_{01r^3}, C^{2r^30}_{02r^3}, \ldots, C^{M_{2,r^3}r^30}_{0M_{2,r^3}r^3} \tag{4.19}$$

represent distinct social affiliations for all nodes in the set $C^{0r^30}_{00r^3}$, $r^3 \in I(1,4)$, that is, (4.19) may represent distinct national representations of all individuals socializing in the network who are from a given continent $C^{0r^30}_{00r^3}$, $r^3 \in I(1,4)$. Thus, for $r^3 \in I(1,4)$, the total number of individuals in the social network form a specific continent

$$C^{0r^30}_{00r^3} = \bigcup_{r^2=1}^{M_{2,r^3}} \bigcup_{r^1=1}^{M_{1,r^2,r^3}} C^{r^2r^30}_{r^1r^2r^3}. \tag{4.20}$$

Note that in (4.18) it is assumed there are four continents in the social network given by

$$\{C^{0r^30}_{00r^3}\}_{r^3=1}^{4} = C^{010}_{001}, C^{020}_{002}, C^{030}_{003}, C^{040}_{004}. \tag{4.21}$$

Hence the decomposition of C in (4.18) implies that every individual in the social network has a given nationality from a given continent.

(Assumption 2) From (3.30) assume that the decomposition given by

$$
\begin{aligned}
&N_{r^{1:N};N\times 0}^{r^{2:N}0;N\times 0}(t) \\
&= N_{r^{1:N};r^{1:N}}^{r^{2:N}0;r^{2:N}0}(t) + \sum_{q^1\neq r^1}^{M_1,r^{2:N}} N_{r^{1:N};q^1r^{2:N}}^{r^{2:N}0;r^{2:N}0}(t) \\
&+ \sum_{q^2\neq r^2}^{M_2,r^{2:N}} \sum_{q^1=1}^{M_1,r^{2:N}} N_{r^{1:N};q^1r^{2:N}}^{r^{2:N}0;q^2r^{3:N}0}(t) + \sum_{q^3\neq r^3}^{M_3,r^{4:N}} \sum_{q^2=1}^{M_{2,q^3r^{4:N}}} \sum_{q^1=1}^{M_{1,q^{2:3}r^{4:N}}} N_{r^{1:N};q^{1:3}r^{4:N}}^{r^{2:N}0;q^{2:3}r^{4:N}0}(t) \\
&+ \cdots \\
&+ \sum_{q^J\neq r^J}^{M_{J,r^{J+1:N}}} \sum_{q^{J-1}=1}^{M_{J-1,r^{J:N}}} \cdots \sum_{q^2=1}^{M_{2,q^{3:J}r^{J+1:N}}} \sum_{q^1=1}^{M_{1,q^{2:J}r^{J=1:N}}} N_{r^{1:N};q^{1:J}r^{J+1:N}}^{r^{2:N}0;q^{2:J}r^{J+1:N}0}(t) \\
&+ \cdots \\
&+ \sum_{q^{N-1}\neq r^{N-1}}^{M_{N-1,r^N}} \sum_{q^{N-2}=1}^{M_{N-2,q^{N-1}r^N}} \cdots \sum_{q^4=1}^{M_{4,q^{5:N-1}r^N}} \sum_{q^3=1}^{M_{3,q^{4:N-1}r^N}} \sum_{q^2=1}^{M_{2,q^{3:N-1}r^N}} \sum_{q^1=1}^{M_{1,q^{2:N-1}r^N}} N_{r^{1:N};q^{1:N-1}r^N}^{r^{2:N}0;q^{2:N-1}r^N0}(t) \\
&+ \sum_{q^N\neq r^N}^{M_N} \sum_{q^{N-1}=1}^{M_{N-1,r^N}} \sum_{q^{N-2}=1}^{M_{N-2,q^{N-1}r^N}} \cdots \sum_{q^4=1}^{M_{4,q^{5:N-1}r^N}} \sum_{q^3=1}^{M_{3,q^{4:N-1}r^N}} \sum_{q^2=1}^{M_{2,q^{3:N-1}r^N}} \sum_{q^1=1}^{M_{1,q^{2:N-1}r^N}} N_{r^{1:N};q^{1:N}}^{r^{2:N}0;q^{2:N}0}(t)
\end{aligned}
\tag{4.22}
$$

represents the **degree** of the node $C_{r^{1:N}}^{r^{2:N}0}$ in the first hierarchical level partition of C in (2.42), i. e., in the partition

$$
C = C_{N\times 0}^{N\times 0} = \bigcup_{r^N=1}^{M_N} \bigcup_{r^{N-1}=1}^{M_{N-1},r^N} \bigcup_{r^{N-2}=1}^{M_{N-2},r^{N-1:N}} \cdots \bigcup_{r^3=1}^{M_3,r^{4:N}} \bigcup_{r^2=1}^{M_2,r^{3:N}} \bigcup_{r^1=1}^{M_1,r^{2:N}} C_{r^{1:N}}^{r^{2:N}0}, \tag{4.23}
$$

where the term $N_{r^{1:N};r^{1:N}}^{r^{2:N}0;r^{2:N}0}(t) = I_{a_{r^{1:N};r^{1:N}}^{r^{2:N}0;r^{2:N}0}}$ indicates the event that the link $a_{r^{1:N};r^{1:N}}^{r^{2:N}0;r^{2:N}0} = \{C_{r^{1:N}}^{r^{2:N}0}, C_{r^{1:N}}^{r^{2:N}0}\}$ occurs.

The sum

$$
\sum_{q^1\neq r^1}^{M_1,r^{2:N}} N_{r^{1:N};q^1r^{2:N}}^{r^{2:N}0;r^{2:N}0}(t) = \sum_{q^1\neq r^1}^{M_1,r^{2:N}} I_{a_{r^{1:N};q^1r^{2:N}}^{r^{2:N}0;r^{2:N}0}} \tag{4.24}
$$

is the total number of possible links of the form $a_{r^{1:N};q^1r^{2:N}}^{r^{2:N}0;r^{2:N}0} = \{C_{r^{1:N}}^{r^{2:N}0}, C_{q^1r^{2:N}}^{r^{2:N}0}\}$ between the node $C_{r^{1:N}}^{r^{2:N}0}$ and nodes $C_{q^1r^{2:N}}^{r^{2:N}0}$, $q^1 \neq r^1$, in the overall N-scale network C.

For any $J \geq 2$, the sum

$$\sum_{q^J \neq r^J}^{M_{J,r^{J+1:N}}} \sum_{q^{J-1}=1}^{M_{J-1,r^{J:N}}} \cdots \sum_{q^2=1}^{M_{2,q^{3:J}r^{J+1:N}}} \sum_{q^1=1}^{M_{1,q^{2:J}r^{J=1:N}}} N_{r^{1:N};q^{1:J}r^{J+1:N}}^{r^{2:N}0;q^{2:J}r^{J+1:N}0}(t)$$
$$= \sum_{q^J \neq r^J}^{M_{J,r^{J+1:N}}} \sum_{q^{J-1}=1}^{M_{J-1,r^{J:N}}} \cdots \sum_{q^2=1}^{M_{2,q^{3:J}r^{J+1:N}}} \sum_{q^1=1}^{M_{1,q^{2:J}r^{J=1:N}}} I_{a_{r^{1:N};q^{1:J}r^{J+1:N}}^{r^{2:N}0;q^{2:J}r^{J+1:N}0}} \tag{4.25}$$

is the total number of possible links (see (3.5)–(3.21)) of the form $a_{r^{1:N};q^{1:J}r^{J+1:N}}^{r^{2:N}0;q^{2:J}r^{J+1:N}0} = \{C_{r^{1:N}}^{r^{2:N}0}, C_{q^{1:J}r^{J+1:N}}^{q^{2:J}r^{J+1:N}0}\}$ between the node $C_{r^{1:N}}^{r^{2:N}0}$ and all other nodes $C_{q^{1:J}r^{J+1:N}}^{q^{2:J}r^{J+1:N}0}$, $q^J \neq r^J$, in the overall N-scale network C.

Example 4.4.2. In continuation with Example 4.4.1, for any node representing a social entity given by the points $C_{r^1r^2r^3}^{r^2r^30}$, $\forall r^k \in I(1, M_{k,r^{k+1:4}})$, $k \in I(2,3)$, the degree of each node from (3.14) and (4.22) is decomposed as follows:

$$N_{r^{1:3};3\times0}^{r^{2:3}0;3\times0}(t) = N_{r^{1:3};00r^3}^{r^{2:3}0;0r^30}(t) + \sum_{q^3\neq r^3}^{M_3} \sum_{q^2=1}^{M_{2,q^3}} \sum_{q^1=1}^{M_{1,q^2r^3}} N_{r^{1:3};q^{1:2}r^3}^{r^{2:3}0;q^2r^30}(t)$$
$$= N_{r^{1:3};r^{1:3}}^{r^{2:3}0;r^{2:3}0}(t) + \sum_{q^1\neq r^1} N_{r^{1:3};q^1r^{2:3}}^{r^{2:3}0;r^{2:3}0}(t) + \sum_{q^2\neq r^2}^{M_{2,r^3}} \sum_{q^1=1}^{M_{1,q^2r^3}} N_{r^{1:3};q^{1:2}r^3}^{r^{2:3}0;q^2r^30}(t)$$
$$+ \sum_{q^3\neq r^3}^{M_3} \sum_{q^2=1}^{M_{2,q^3}} \sum_{q^1=1}^{M_{1,q^2r^3}} N_{r^{1:3};q^{1:3}}^{r^{2:3}0;q^{2:3}0}(t), \tag{4.26}$$

where the first term $N_{r^{1:3};00r^3}^{r^{2:3}0;0r^30}(t) = I_{\{a_{r^{1:3};00r^3}^{r^{2:3}0;0r^30}\}}$ is the indicator random variable for the event that the link $\{a_{r^{1:3};00r^3}^{r^{2:3}0;0r^30}\}$ occurs at time t, where the link is defined as $a_{r^{1:3};00r^3}^{r^{2:3}0;0r^30} = \{C_{r^{1:3}}^{r^{2:3}0}, C_{r^{1:3}}^{r^{2:3}0}\}$.
Also, the second sum

$$\sum_{q^1\neq r^1} N_{r^{1:3};q^1r^{2:3}}^{r^{2:3}0;r^{2:3}0}(t) = \sum_{q^1\neq r^1} I_{a_{r^{1:3};q^1r^{2:3}}^{r^{2:3}0;r^{2:3}0}} \tag{4.27}$$

is the total number of possible links of the form $a_{r^{1:3};q^1r^{2:3}}^{r^{2:3}0;r^{2:3}0} = \{C_{r^{1:3}}^{r^{2:3}0}, C_{q^1r^{2:3}}^{r^{2:3}0}\}$, $q^1 \neq r^1$, that occur in the population C. That is, (4.27) is the total number of possible connections that occur in the social network by chance between the individual $C_{r^{1:3}}^{r^{2:3}0}$ of nationality $C_{0r^{2:3}}^{r^{2:3}0}$ in the continent $C_{00r^3}^{0r^30}$ with all other individuals $C_{q^1r^{2:3}}^{r^{2:3}0}$, $q^1 \neq r^1$, of the same nationality $C_{0r^{2:3}}^{r^{2:3}0}$.

The third sum

$$\sum_{q^2 \neq r^2}^{M_{2,r^3}} \sum_{q^1=1}^{M_{1,q^2r^3}} N_{r^{1:3};q^{1:2}r^3}^{r^{2:3}0;q^2r^30}(t) = \sum_{q^2 \neq r^2}^{M_{2,r^3}} \sum_{q^1=1}^{M_{1,q^2r^3}} I_{a_{r^{1:3};q^{1:2}r^3}^{r^{2:3}0;q^2r^30}} \tag{4.28}$$

is the total number of possible links of the form $a_{r^{1:3};q^{1:2}r^3}^{r^{2:3}0;q^2r^30} = \{C_{r^{1:3}}^{r^{2:3}0}, C_{q^{1:2}r^3}^{q^2r^30}\}$, $q^2 \neq r^2$, that occur in the population C. That is, (4.28) is the total number of possible connections that occur in the social network by chance between the individual $C_{r^{1:3}}^{r^{2:3}0}$ of nationality $C_{0r^{2:3}}^{r^{2:3}0}$ in the continent $C_{00r^3}^{0r^30}$ with all other individuals $C_{q^{1:2}r^3}^{q^2r^30}$, $q^2 \neq r^2$, of different nationality $C_{0q^2r^3}^{q^2r^30}$ but of the same continent $C_{00r^3}^{0r^30}$.
The last sum

$$\sum_{q^3 \neq r^3}^{M_3} \sum_{q^2=1}^{M_{2,q^3}} \sum_{q^1=1}^{M_{1,q^2r^3}} N_{r^{1:3};q^{1:3}}^{r^{2:3}0;q^{2:3}0}(t) = \sum_{q^3 \neq r^3}^{M_3} \sum_{q^2=1}^{M_{2,q^3}} \sum_{q^1=1}^{M_{1,q^{2:3}}} I_{a_{r^{1:3};q^{1:3}}^{r^{2:3}0;q^{2:3}0}} \tag{4.29}$$

is the total number of possible links of the form $a_{r^{1:3};q^{1:3}}^{r^{2:3}0;q^2r^30} = \{C_{r^{1:3}}^{r^{2:3}0}, C_{q^{1:3}}^{q^{2:3}0}\}$, $q^3 \neq r^3$, that occur in the population C. That is, (4.29) is the total number of possible connections that occur in the social network by chance between the individual $C_{r^{1:3}}^{r^{2:3}0}$ of nationality $C_{0r^{2:3}}^{r^{2:3}0}$ in the continent $C_{00r^3}^{0r^30}$ with all other individuals $C_{q^{1:3}}^{q^{2:3}0}$, $q^2 \neq r^2$, of different nationality $C_{0q^{2:3}}^{q^{2:3}0}$ and different continent $C_{00q^3}^{0q^30}$, $q^3 \neq r^3$.

(Assumption 3) From (4.24) and (4.25), assume that for each $J \geq 2$, the probabilities of the links in the social network occurring are as defined below. For the sum

$$\begin{aligned} &\sum_{q^1 \neq r^1}^{M_1,r^{2:N}} N_{r^{1:N};q^1r^{2:N}}^{r^{2:N}0;r^{2:N}0}(t) = \sum_{q^1 \neq r^1}^{M_1,r^{2:N}} I_{a_{r^{1:N};q^1r^{2:N}}^{r^{2:N}0;r^{2:N}0}}, \\ &p_{r^{1:N};0r^{2:N}}^{r^{2:N}0;r^{2:N}0} = \mathbb{P}(I_{a_{r^{1:N};q^1r^{2:N}}^{r^{2:N}0;r^{2:N}0}} = 1), \quad q^1 \in I(1, M_{1,r^{2:N}}),\ q^1 \neq r^1. \end{aligned} \tag{4.30}$$

Furthermore, it is assumed that all possible links of the form $a_{r^{1:N};q^1r^{2:N}}^{r^{2:N}0;r^{2:N}0} = \{C_{r^{1:N}}^{r^{2:N}0}, C_{q^1r^{2:N}}^{r^{2:N}0}\}$, $q^1 \neq r^1$, between the node $C_{r^{1:N}}^{r^{2:N}0}$ and nodes $C_{q^1r^{2:N}}^{r^{2:N}0}$, $q^1 \neq r^1$, that occur in the overall N-scale network C are *mutually independent*.
Similarly, for any $J \geq 2$ and the sum

$$\begin{aligned} &\sum_{q^J \neq r^J}^{M_{J,r^{J+1:N}}} \sum_{q^{J-1}=1}^{M_{J-1,r^{J:N}}} \cdots \sum_{q^2=1}^{M_{2,q^{3:J}r^{J+1:N}}} \sum_{q^1=1}^{M_{1,q^{2:J}r^{J=1:N}}} N_{r^{1:N};q^{1:J}r^{J+1:N}}^{r^{2:N}0;q^{2:J}r^{J+1:N}0}(t) \\ &= \sum_{q^J \neq r^J}^{M_{J,r^{J+1:N}}} \sum_{q^{J-1}=1}^{M_{J-1,r^{J:N}}} \cdots \sum_{q^2=1}^{M_{2,q^{3:J}r^{J+1:N}}} \sum_{q^1=1}^{M_{1,q^{2:J}r^{J=1:N}}} I_{a_{r^{1:N};q^{1:J}r^{J+1:N}}^{r^{2:N}0;q^{2:J}r^{J+1:N}0}}, \\ &p_{r^{1:N};0q^{2:J}r^{J+1:N}}^{r^{2:N}0;q^{2:J}r^{J+1:N}0} = \mathbb{P}(I_{a_{r^{1:N};q^{1:J}r^{J+1:N}}^{r^{2:N}0;q^{2:J}r^{J+1:N}0}} = 1), \quad q^1 \in I(1, M_{1,q^{2:J}r^{J+1:N}}). \end{aligned} \tag{4.31}$$

Also, it is assumed that all possible links of the form $a^{r^{2:N}0;q^{2:J}r^{J+1:N}0}_{r^{1:N};q^{1:J}r^{J+1:N}} = \{C^{r^{2:N}0}_{r^{1:N}}, C^{q^{2:J}r^{J+1:N}0}_{q^{1:J}r^{J+1:N}}\}$, $q^J \neq r^J$, between the node $C^{r^{2:N}0}_{r^{1:N}}$ and nodes of the form $C^{q^{2:J}r^{J+1:N}0}_{q^{1:J}r^{J+1:N}}$, $q^J \neq r^J$, that occur in the overall N-scale network C are *mutually independent*. All other terms in expression (4.22) are similarly defined as in (4.30)–(4.31).

Example 4.4.3. Continuing with Example 4.4.2, from (4.26), the assumption for the first term $N^{r^{2:3}0;0r^3 0}_{r^{1:3};00r^3}(t) = I_{\{a^{r^{2:3}0;0r^3 0}_{r^{1:3};00r^3}\}}$ is that the event $\{a^{r^{2:3}0;0r^3 0}_{r^{1:3};00r^3}\}$ does not occur in the social media network, since individuals do not randomly connect with themselves. This implies that $\mathbb{P}(I_{\{a^{r^{2:3}0;0r^3 0}_{r^{1:3};00r^3}\}} = 1) = 0$.

For the second term

$$\sum_{q^1 \neq r^1} N^{r^{2:3}0;r^{2:3}0}_{r^{1:3};q^1 r^{2:3}}(t) = \sum_{q^1 \neq r^1} I_{\{a^{r^{2:3}0;r^{2:3}0}_{r^{1:3};q^1 r^{2:3}}\}}, \quad p^{r^{2:3}0;r^{2:3}0}_{r^{1:3};0r^{2:3}} = \mathbb{P}(I_{\{a^{r^{2:3}0;r^{2:3}0}_{r^{1:3};q^1 r^{2:3}}\}} = 1), \tag{4.32}$$

where for any $q^1 \in I(1, M_{1,q^1 r^{2:3}})$, $p^{r^{2:3}0;r^{2:3}0}_{r^{1:3};0r^{2:3}}$ is the "success" probability that a link occurs between the individual $C^{r^{2:3}0}_{r^{1:3}}$ of nationality $C^{r^{2:3}0}_{0r^{2:3}}$ in the continent $C^{0r^3 0}_{00r^3}$ with all other individuals $C^{r^{2:3}0}_{q^1 r^{2:3}}$, $q^1 \neq r^1$, of the same nationality $C^{r^{2:3}0}_{0r^{2:3}}$.

The third term

$$\begin{aligned} \sum_{q^2 \neq r^2}^{M_{2,r^3}} \sum_{q^1=1}^{M_{1,q^2 r^3}} N^{r^{2:3}0;q^2 r^3 0}_{r^{1:3};q^{1:2}r^3}(t) &= \sum_{q^2 \neq r^2}^{M_{2,r^3}} \sum_{q^1=1}^{M_{1,q^2 r^3}} I_{a^{r^{2:3}0;q^2 r^3 0}_{r^{1:3};q^{1:2}r^3}}, \\ p^{r^{2:3}0;q^2 r^3 0}_{r^{1:3};0q^2 r^3} &= \mathbb{P}(I_{\{a^{r^{2:3}0;q^2 r^3 0}_{r^{1:3};q^{1:2}r^3}\}} = 1), \end{aligned} \tag{4.33}$$

where for $q^2 \neq r^2$ and all $q^1 \in I(1, M_{1,q^{1:2}r^3})$, $p^{r^{2:3}0;q^2 r^3 0}_{r^{1:3};0q^2 r^3}$ is the "success" probability that a link occurs between the individual $C^{r^{2:3}0}_{r^{1:3}}$ of nationality $C^{r^{2:3}0}_{0r^{2:3}}$ in the continent $C^{0r^3 0}_{00r^3}$ and all other individuals $C^{q^2 r^3 0}_{q^{1:2}r^3}$, $q^2 \neq r^2$, of different nationality $C^{q^2 r^3 0}_{0q^2 r^3}$ but of the same continent $C^{0r^3 0}_{00r^3}$.

For the last term

$$\begin{aligned} \sum_{q^3 \neq r^3}^{M_3} \sum_{q^2=1}^{M_{2,q^3}} \sum_{q^1=1}^{M_{1,q^2 r^3}} N^{r^{2:3}0;q^{2:3}0}_{r^{1:3};q^{1:3}}(t) &= \sum_{q^3 \neq r^3}^{M_3} \sum_{q^2=1}^{M_{2,q^3}} \sum_{q^1=1}^{M_{1,q^{2:3}}} I_{a^{r^{2:3}0;q^{2:3}0}_{r^{1:3};q^{1:3}}}, \\ p^{r^{2:3}0;q^{2:3}0}_{r^{1:3};0q^{2:3}} &= \mathbb{P}(I_{\{a^{r^{2:3}0;q^{2:3}0}_{r^{1:3};q^{1:3}}\}} = 1), \end{aligned} \tag{4.34}$$

where for any $q^1 \in I(1, M_{1,q^{2:3}})$, $p^{r^{2:3}0;q^{2:3}0}_{r^{1:3};0q^{2:3}}$ is the "success" probability that a link occurs between the individual $C^{r^{2:3}0}_{r^{1:3}}$ of nationality $C^{r^{2:3}0}_{0r^{2:3}}$ in the continent $C^{0r^3 0}_{00r^3}$ with all other individuals $C^{q^{2:3}0}_{q^{1:3}}$, $q^2 \neq r^2$, of different nationality $C^{q^{2:3}0}_{0q^{2:3}}$ and of a different continent $C^{0q^3 0}_{00q^3}$, $q^3 \neq r^3$. It is also assumed that all links in (4.32)–(4.34) are mutually independent.

The following result is based on assumptions **(Assumption 1)–(Assumption 3)** in Section 4.4.1.

Theorem 4.4.1. *Let **(Assumption 1)–(Assumption 3)** in Section* 4.4.1 *be satisfied. Let $G(M^1_{\text{total}}, \vec{p}(t))$ denote the Erdős–Rényi network in the overall N-scale network population $C = C^{N\times 0}_{N\times 0} = \bigcup_{r^N=1}^{M_N} \bigcup_{r^{N-1}=1}^{M_{N-1},r^N} \bigcup_{r^{N-2}=1}^{M_{N-2},r^{N-1:N}} \cdots \bigcup_{r^3=1}^{M_3,r^{4:N}} \bigcup_{r^2=1}^{M_2,r^{3:N}} \bigcup_{r^1=1}^{M_1,r^{2:N}} C^{r^{2:N}0}_{r^{1:N}}$. For $r^k, q^k \in I(1, M_{k,r^{k+1:N}})$, $k \in I(1,N)$, the following are true.*

(a) *The degree of any node $C^{r^{2:N}0}_{r^{1:N}}$ in all the N-hierarchical level partitions of the overall network population C is given by the random variable* (4.22)–(4.31).

(b) *For $J \geq 2$, assume that all possible links of the form $a^{r^{2:N}0;q^{2:J}r^{J+1:N}0}_{r^{1:N};q^{1:J}r^{J+1:N}} = \{C^{r^{2:N}0}_{r^{1:N}}, C^{q^{2:J}r^{J+1:N}0}_{q^{1:J}r^{J+1:N}}\}$, $q^J \neq r^J$, between the node $C^{r^{2:N}0}_{r^{1:N}}$ and all other nodes of the form $C^{q^{2:J}r^{J+1:N}0}_{q^{1:J}r^{J+1:N}}$, $q^J \neq r^J$, that occur at any time t in the overall N-scale network C are **mutually independent**. Then the following distributions hold. For the terms*

$$N^{r^{2:N}0;r^{2:N}0}_{r^{1:N};r^{1:N}}(t) = I_{a^{r^{2:N}0;r^{2:N}0}_{r^{1:N};r^{1:N}}(t)}, \quad \mathbb{P}(I_{a^{r^{2:N}0;r^{2:N}0}_{r^{1:N};r^{1:N}}(t)} = 1) = 0; \tag{4.35}$$

$$\begin{aligned}
&\sum_{q^1\neq r^1}^{M_1,r^{2:N}} N^{r^{2:N}0;r^{2:N}0}_{r^{1:N};q^1r^{2:N}}(t) = \sum_{q^1\neq r^1}^{M_1,r^{2:N}} I_{a^{r^{2:N}0;r^{2:N}0}_{r^{1:N};q^1r^{2:N}}} \sim Binomial(M_1, r^{2:N} - 1, p^{r^{2:N}0;r^{2:N}0}_{r^{1:N};0r^{2:N}}),\\
&p^{r^{2:N}0;r^{2:N}0}_{r^{1:N};0r^{2:N}} = \mathbb{P}(I_{a^{r^{2:N}0;r^{2:N}0}_{r^{1:N};q^1r^{2:N}}} = 1), \quad q^1 \neq r^1,\ \forall q^1 \in I(1, M_{1,r^{2:N}});
\end{aligned} \tag{4.36}$$

and the term

$$\begin{aligned}
&\sum_{q^J\neq r^J}^{M_{J,r^{J+1:N}}} \sum_{q^{J-1}=1}^{M_{J-1,r^{J:N}}} \cdots \sum_{q^2=1}^{M_{2,q^{3:J}r^{J+1:N}}} \sum_{q^1=1}^{M_{1,q^{2:J}r^{J=1:N}}} N^{r^{2:N}0;q^{2:J}r^{J+1:N}0}_{r^{1:N};q^{1:J}r^{J+1:N}}(t)\\
&\quad = \sum_{q^J\neq r^J}^{M_{J,r^{J+1:N}}} \sum_{q^{J-1}=1}^{M_{J-1,r^{J:N}}} \cdots \sum_{q^2=1}^{M_{2,q^{3:J}r^{J+1:N}}} \sum_{q^1=1}^{M_{1,q^{2:J}r^{J=1:N}}} I_{a^{r^{2:N}0;q^{2:J}r^{J+1:N}0}_{r^{1:N};q^{1:J}r^{J+1:N}}}, \quad \&\\
&\sum_{q^1=1}^{M_{1,q^{2:J}r^{J=1:N}}} I_{a^{r^{2:N}0;q^{2:J}r^{J+1:N}0}_{r^{1:N};q^{1:J}r^{J+1:N}}} \sim Binomial(M_{1,q^{2:J}r^{J=1:N}}, p^{r^{2:N}0;q^{2:J}r^{J+1:N}0}_{r^{1:N};0q^{2:J}r^{J+1:N}}), \quad \&\\
&p^{r^{2:N}0;q^{2:J}r^{J+1:N}0}_{r^{1:N};0q^{2:J}r^{J+1:N}} = \mathbb{P}(I_{a^{r^{2:N}0;q^{2:J}r^{J+1:N}0}_{r^{1:N};q^{1:J}r^{J+1:N}}} = 1), \quad q^1 \in I(1, M_{1,q^{2:J}r^{J+1:N}}).
\end{aligned} \tag{4.37}$$

Furthermore, it follows from (4.36)–(4.37) *that at any time $t \geq 0$, the (cumulative) distribution $F_{N^{r^{2:N}0;N\times 0}_{r^{1:N};N\times 0}(t)}$ of the random variable $N^{r^{2:N}0;N\times 0}_{r^{1:N};N\times 0}(t)$ is a convolution of multiple binomial distributions in* (4.36)–(4.37)*) given by*

$$\begin{aligned}
&F_{N^{r^{2:N}0;N\times 0}_{r^{1:N};N\times 0}(t)}(x)\\
&= \Bigg(F_{Binomial(M_1, r^{2:N}-1, p^{r^{2:N}0;r^{2:N}0}_{r^{1:N};0r^{2:N}})}
\end{aligned}$$

$$
\begin{aligned}
&* \sum_{q^2 \neq r^2}^{M_2, r^{2:N}} F_{binomial(M_1, r^{2:N}, p_{r^{1:N};0r^{2:N}}^{r^{2:N}0;q^2 r^{3:N}0}(t))} * \sum_{q^3 \neq r^3}^{M_3, r^{4:N}} \sum_{q^2=1}^{M_{2,q^3 r^{4:N}}} F_{Binomial(M_{1,q^{2:3} r^{4:N}}, p_{r^{1:N};0q^{2:3} r^{4:N}}^{r^{2:N}0;q^{2:3} r^{4:N}0}(t))} \\
&* \cdots \\
&* \sum_{q^J \neq r^J}^{M_{J, r^{J+1:N}}} \sum_{q^{J-1}=1}^{M_{J-1, r^{J:N}}} \cdots \sum_{q^2=1}^{M_{2,q^{3:J} r^{J+1:N}}} F_{Binomial(M_{1,q^{2:J} r^{J=1:N}}, p_{r^{1:N};0q^{2:J} r^{J+1:N}}^{r^{2:N}0;q^{2:J} r^{J+1:N}0}(t))} \\
&* \cdots \\
&* \sum_{q^{N-1} \neq r^{N-1}}^{M_{N-1, r^N}} \sum_{q^{N-2}=1}^{M_{N-2,q^{N-1} r^N}} \cdots \sum_{q^4=1}^{M_{4,q^{5:N-1} r^N}} \sum_{q^3=1}^{M_{3,q^{4:N-1} r^N}} \sum_{q^2=1}^{M_{2,q^{3:N-1} r^N}} F_{Binomial(M_{1,q^{2:N-1} r^N}, p_{r^{1:N};0q^{2:N-1} r^N}^{r^{2:N}0;q^{2:N-1} r^N 0}(t))} \\
&* \sum_{q^N \neq r^N}^{M_N} \sum_{q^{N-1}=1}^{M_{N-1, r^N}} \sum_{q^{N-2}=1}^{M_{N-2,q^{N-1} r^N}} \cdots \sum_{q^4=1}^{M_{4,q^{5:N-1} r^N}} \sum_{q^3=1}^{M_{3,q^{4:N-1} r^N}} \sum_{q^2=1}^{M_{2,q^{3:N-1} r^N}} F_{Binomial(M_{1,q^{2:N-1} r^N}, p_{r^{1:N};q^{1:N}}^{r^{2:N}0;q^{2:N}0}(t))} \Bigg)(x) \\
&\forall x.
\end{aligned} \tag{4.38}
$$

Proof. For the Erdős–Rényi network, the result in (4.35) is implied. The rest of the proof of the result follows from the fact that the binomial distribution is the sum of independent identically distributed *Bernoulli* random variables. Furthermore, the distribution of the sum of independent random variables in (4.22)–(4.31) is a convolution of the cumulative distribution functions of the independent random variables. □

Example 4.4.4. Continuing with Example 4.4.3, it is easy to see from (4.26) that the distribution of the following random variable is a convolution of multiple independent binomial random variables as shown below:

$$
\begin{aligned}
N_{r^{1:3};3\times 0}^{r^{2:3}0;3\times 0}(t) = {} & N_{r^{1:3};r^{1:3}}^{r^{2:3}0;r^{2:3}0}(t) + \sum_{q^1 \neq r^1} N_{r^{1:3};q^1 r^{2:3}}^{r^{2:3}0;r^{2:3}0}(t) + \sum_{q^2 \neq r^2}^{M_{2,r^3}} \sum_{q^1=1}^{M_{1,q^2 r^3}} N_{r^{1:3};q^{1:2} r^3}^{r^{2:3}0;q^2 r^3 0}(t) \\
& + \sum_{q^3 \neq r^3}^{M_3} \sum_{q^2=1}^{M_{2,q^3}} \sum_{q^1=1}^{M_{1,q^2 r^3}} N_{r^{1:3};q^{1:3}}^{r^{2:3}0;q^{2:3}0}(t) \\
\sim \Bigg(& F_{Binomial(M_{1,r^{2:3}}-1, p_{r^{1:3};0r^{2:3}}^{r^{2:3}0;r^{2:3}0}(t))} * \sum_{q^2 \neq r^2}^{M_{2,r^3}} F_{Binomial(M_{1,q^2 r^3}, p_{r^{1:3};0q^2 r^3}^{r^{2:3}0;q^2 r^3 0}(t))} \\
& * \sum_{q^3 \neq r^3}^{M_3} \sum_{q^2=1}^{M_{2,q^3}} F_{Binomial(M_{1,q^2 r^3}, p_{r^{1:3};0q^{2:3}}^{r^{2:3}0;q^{2:3}0}(t))} \Bigg)(t).
\end{aligned} \tag{4.39}
$$

4.5 Other applications of the finite-scale network population framework

The successful application of the Erdős–Rényi network within the finite-scale network population framework in Section 3.3, utilizing the node-level decomposition and structural properties of the network in the scale-structured population, as well as the results for the degree distribution provided in Theorems 4.3.1 and 4.4.1, underscores the value of the N-scale network compact topological space framework introduced in Chapter 2. This framework is particularly relevant for investigating a wide range of real-world complex network systems with hierarchical structures in ecological and social dynamical systems.

The remainder of this monograph explores applications to epidemic systems for infectious diseases that operate within network-centric epidemiological processes in N-scale network population frameworks, where the prevalence and persistence of epidemics are driven by the *human mobility network*. Additionally, as many social media networks propagate *rumors* over time in a manner analogous to infectious disease dynamics, the dynamic modeling of rumor epidemics is also discussed.

Although the intricate framework of the N-hierarchical level partitions in the finite-scale compact topological space $C_{N\times 0}^{N\times 0}$, introduced in Chapter 2, is essential for elucidating the underlying network properties in infectious disease models, the associated complex notations for partitions and node decompositions in Section 3.3 present significant challenges that may overshadow the dynamic modeling and analysis of infectious diseases and rumors.

Therefore the subsequent chapters of this book will primarily address the dynamics of infectious diseases and rumors within the two-scale network population framework. Where necessary, results will be generalized to the N-scale network population by employing the principle of mathematical induction.

5 The dynamic human mobility process in a finite-scale network population

5.1 Introduction

In Chapter 2, the N-scale network population was defined, and the human mobility network within the population was introduced, as exemplified in Examples 2.4.1 and 3.2.1. This chapter focuses on the dynamic modeling and analysis of the human mobility process within a metapopulation influenced by a finite-scale human mobility network.

Modern advancements in communication, transportation, and basic services have significantly facilitated national, bilateral, and multilateral interactions, creating a highly interconnected world that feels as small and cohesive as a neighborhood. This interconnectedness has fostered cooperation and understanding of the basic needs of the human species within the global community, making the concept of globalization pervasive across almost all aspects of human life.

A key driving force behind globalization is human mobility. Cultural exchange and understanding, the spread of ideas and rumors, the emergence of new infectious disease epidemics and pandemics (e. g., COVID-19 and H1N1), the endemicity of infectious diseases in new areas, outsourcing of jobs and resources, and changes in economic development and environmental conditions are among the significant byproducts of human mobility. For instance, the recent COVID-19 pandemic, caused by *SARS-CoV-2*, originated in Wuhan, China, in 2019 and rapidly spread worldwide due to the highly effective and efficient global human mobility systems [26].

From a theoretical standpoint, human mobility networks serve as an essential framework for understanding not only movement patterns but also the systemic spread of interconnected phenomena, such as economic shifts, sociopolitical dynamics, and biological processes like disease transmission. These networks encapsulate the inherent complexities of modern populations, which are often influenced by multiscale and hierarchical structures. The investigation of mobility processes within such frameworks provides a powerful lens for capturing both local and global interactions across populations.

Numerous compartmental theoretical and ecological dynamic models for human mobility in metapopulations and multipopulation structures have been developed in the context of infectious disease epidemic modeling. Many of these studies investigate the impacts of human mobility on the spread of infectious diseases between communities, driven by the movement of people, goods, vectors, and animals across the globe. Common approaches for modeling human mobility and its effects on infectious disease dynamics include discrete-time difference equations, deterministic and stochastic ordinary and partial differential equations (ODEs and PDEs), and delay and nondelay ODEs (see [45, 46, 48, 32, 49]).

https://doi.org/10.1515/9783111613215-005

This chapter builds on these foundational approaches to human mobility by incorporating a finite-scale hierarchical perspective. The two-scale human mobility network introduced in Examples 2.4.1 and 3.2.1 is revisited. The node-level decomposition of the two-scale network population, discussed in Section 3.2 and Example 3.2.1, is employed to construct an ODE-based mean-field dynamic model for human mobility within the population. This model captures the intricate interplay between intraregional and interregional mobility patterns and the hierarchical structure of the network.

Furthermore, the dynamic ODE model is analyzed to elucidate various real-world spatial patterns within the population as influenced by the network properties in the two-scale population, particularly when the dynamic system reaches a steady state. These steady states represent equilibrium conditions that highlight the long-term trends and distributions of human movement within the structured network. Generalizations of these results to the N-scale network population are obtained through the principle of mathematical induction, providing a robust framework for extending the analysis to larger and more complex hierarchical populations.

The results presented in this chapter hold significant implications for multiple domains, such as urban planning, public health, and network optimization. By understanding the steady-state dynamics and transient behavior of human mobility, researchers can better predict the spread of infectious diseases, optimize resource allocation, and inform policy decisions in interconnected societies.

This chapter builds upon and generalizes the studies conducted by Wanduku and Ladde (see [45, 46, 48, 32, 49]) and the mean-field approaches developed by Wanduku [44]. It sets the stage for subsequent chapters, where the dynamic interaction between human mobility and communicable disease dynamics in finite-scale network populations is explored in depth.

5.2 The two-level hierarchic human mobility process

In this section, we analyze the human mobility process within a two-level hierarchic structure of the network population. The framework considers the decomposition of the population into hierarchical partitions, where mobility dynamics occur both within and across levels of the hierarchy. This analysis builds on the two-scale network structure introduced in Chapter 2, leveraging the network structural and node-level decomposition properties.

The two-level hierarchic mobility process models how individuals interact and move between different "sites" within the same hierarchical level and across higher-level partitions, capturing the layered complexity inherent in real-world mobility networks. This structured approach provides a foundation for examining dynamic patterns in human mobility that drive critical phenomena, such as epidemic spreading and resource allocation.

The two-scale compact topological space $C = C_{00}^{00}$ in $\mathbb{R}_+^n$ characterized in Sections 2.2, the two-scale network in Section 2.4.1, and the decomposition of the network population in Section 3.2 are used to describe the dynamic human mobility process in the two-scale network population. The assumptions for the human mobility dynamic model are summarized in the following.

5.2.1 Assumptions for the human mobility model in the two-scale network population

Assumption 1. Decomposition of hierarchic process. Consider a human population distributed into M distinct spatial regions $C_1, C_2, \dots, C_M$. Each region $C_r, r \in I(1, M)$, consists of n_r distinct sites $s_1^r, s_2^r, \dots, s_{n_r}^r$ spatially distributed within the region. Residents of sites in a region can either visit other sites within the region or visit sites in other regions.

Indeed, recall (2.8)–(2.10), the mathematical decomposition of the two-scale network population $C = \bigcup_{r=1}^{M} C_r$, where the compact topological space $(C, \mathfrak{T}_C)$ consists of the subspace $C \subset \mathbb{R}_+^2$ equipped with the relative topology $\mathfrak{T}_C$ with respect to the metric d_2 induced by the Euclidean norm $\|\cdot\|_2$ on $\mathbb{R}_+^2$ defined subsequently.

Suppose $\vec{x} = (x_1, x_2) \in \mathbb{R}_+ \times \mathbb{R}_+ = \mathbb{R}_+^2$ is arbitrary, and for $r \in I(1, M_2)$, let $y_r = (y_{1_r}, y_{2_r}) \in \mathbb{R}_+^2$ be fixed. The metric space $(\mathbb{R}_+^2, d_2)$ designed with the Euclidean metric induced by the Euclidean norm defined as $d_2(\vec{x}, y_r) = \|\vec{x} - y_r\|_2 = \|\vec{x} - y_r\|_{\mathbb{R}^2} = \sqrt{(x_1 - y_{1_r})_1^2 + (x_2 - y_{2_r})^2}$. For some real number $b_r > 0, r \in I(1, M)$, consider the finite subcover for $C = \bigcup_{r=1}^{M_2} C_r$, where the collection $\{C_r\}_{r \in I(1,M)}$ consists of open balls

$$C_r = B(y_r; b_r) = B_{\mathbb{R}_+^2}(y_r; b_r) = \{\vec{x} \in \mathbb{R}_+^2 : \|\vec{x} - y_r\|_{\mathbb{R}^2} < b_r\}. \tag{5.1}$$

Furthermore, for all $r, q \in I(1, M), r \neq q$, the sets C_r and C_q are selected from the metric topology on C in such as way that the balls are mutually disjoint, i. e., $C_r \cap C_q = \emptyset$, and they form a partition of C, i. e., $C = \bigcup_{r=1}^{M_2} C_r$. Also, as in Section 2.3, the second hierarchical level partition of C is given by $C = \bigcup_{r=1}^{M_2} C_r$, and the first hierarchical level partition is given by $C = \bigcup_{r=1}^{M_2} \bigcup_{i=1}^{n_r} \{s_i^r\}$, where for each $r \in I(1, M)$, it is assumed without loss of generality (w.l.o.g) that the collection $\{s_i^r\}_{i \in I(1,n_r)}$ consists of distinct points in the interior of the open balls C_r that form a finite subcover $C_r = \bigcup_{i=1}^{n_r} \{s_i^r\}$. For the compact topological subspace C_r, $r \in I(1, M)$, it is easy to see that for all points in $\{s_i^r\}_{i \in I(1,n_r)}$, the distance between two distinct points $\|s_i^r - s_j^r\|_{\mathbb{R}^2} > 0$ for $i \neq j$.

Now define

$$C_{rr}(s^r) = \{s_i^r \in C_r : i \in I(1, n_r)\} \tag{5.2}$$

as a finite collection of n_r distinct points $\{s_i^r\}_{i \in I(1,n_r)}$ that form a subcover of C_r. Also, let

$$C(s) = \{s_i^r \in C_r : \forall i \in I(1, n_r), \forall r \in I(1, M_2)\} = \bigcup_{r=1}^{M_2} C_{rr}(s^r) \tag{5.3}$$

be the complete collection of all first-level points that cover C. The cardinality of $C(s)$ is given by

$$n = \sum_{r=1}^{M_2} n_r. \tag{5.4}$$

Assumption 2. Population decomposition and aggregation process. Let N_{ii}^{rr} be the number of residents of site s_i^r, $i \in I(1, n_r)$, in the region $C_r, r \in I(1, M)$, residing at their home site at time t. Let N_{ij}^{rr} be the number of residents of s_i^r, $i, j \in I(1, n_r)$, in the region $C_r, r \in I(1, M)$, visiting a distinct site s_j^r within the region C_r, where $j \neq i$. Let N_{il}^{rq} be the number of residents of s_i^r, $i \in I(1, n_r)$, in the region C_r, $r, q \in I(1, M)$, visiting site s_l^q, $l \in I(1, n_q)$, in the region C_q, where $q \neq r$. Let N_{i0}^{r0} be the total number of residents of site s_i^r that are either residing at their home site, or visiting other sites within their home region C_r, or visiting sites in all other regions C_q, $r, q \in I(1, M)$, $q \neq r$. It follows from Subsection 3.2 that

$$N_{i0}^{r0} = N_{ii}^{rr} + \sum_{j \neq i}^{n_r} N_{ij}^{rr} + \sum_{q \neq r}^{M} \sum_{l=1}^{n_q} N_{il}^{rq}, \quad i \in I(1, n_r),\ r \in I(1, M). \tag{5.5}$$

Assumption 3. Intra- and interregional probabilistic visiting rates. For each $r \in I(1, M)$, residents of site s_i^r in the region C_r leave on trips to other sites within the region at the per capita rate σ_i^r. The visitors then distribute themselves among the $n_r - 1$ sites $s_j^r, j \neq i$, with probabilistic rate v_{ij}^{rr}. Also, residents of site s_i^r in the region C_r leave on trips to other regions at the per capita rate γ_i^r. The residents of site s_j^r that leave on trips to other regions distribute themselves among $M - 1$ destinations with probability γ_{i0}^{rq} to region $C_q, q \neq r$. Collectively, the residents of sites in the region C_r leave their region to visit other sites in the region C_q with grand total rate $\gamma^{rq} = \sum_{i=1}^{n_r} \gamma_{i0}^{rq}$. The residents that leave the region C_r to visit sites in the region C_q distribute themselves among n_q destinations with probability γ_{0l}^{rq} to site s_l^q, $l \in I(1, n_q)$.

Assumption 4. Interregional probabilistic return rates. For each $r \in I(1, M)$, persons traveling from site s_i^r to s_j^r in the region C_r have the per capita probabilistic return rate ρ_{ij}^{rr}. Also, for each $q \in I(1, M)$, $q \neq r$, residents from all other regions that came to site s_l^q, $l \in I(1, n_q)$, in the region C_q, $q \neq r$, leave the site s_l^q to return to their home region with rate ρ_l^q. This rate ρ_l^q further distributes among the $M - 1$ regions C_r, $r \neq q$, with probabilities ρ_{0l}^{rq}. Hence the grand total per capita return rate of the residents of the region C_r that came to the n_q sites in the region C^q, $q \neq r$, is $\rho^{rq} = \sum_{l=1}^{n_q} \rho_{0l}^{rq}$. This return rate ρ^{rq} of residents of the region C_r from the region C_q then distribute among the n_r sites s_i^r, $i \in I(1, n_r)$, in the region C_r, $r \neq q$, with probability rate ρ_{i0}^{rq}.

Assumption 3 and **Assumption 4** are reformulated mathematically using *continuous-time Markov chains* (*CTMCs*) for clearer understanding of the underlying random jump processes and rates assumed. Indeed, the underlying counting processes $\{A_{ij}^{rr}(t), t \geq 0\}$ and $\{A_{il}^{rq}(t), t \geq 0\}$ are **birth-and-death processes (in fact, pure-birth processes)**, where $\{A_{ij}^{rr}(t), t \geq 0\}$ counts the transitions over continuous time $[0, t)$ for residents of site s_i^r visiting other sites in their home region C_r, and $\{A_{il}^{rq}(t), t \geq 0\}$ counts the transitions over continuous time $[0, t)$ for residents of site s_i^r visiting other sites in other regions C_q, $q \in I(1, M)$, $q \neq r$. Furthermore, over any small time interval $[t, t + \Delta t)$, only single transitions in the stochastic processes occur; $A_{ij}^{rr}(t)$ and $A_{il}^{rq}(t)$ are mutually independent random variables for all $t \geq 0$, $i, j \in I(1, n_r)$, and $q, r \in I(1, M)$, and the times between transitions are independent and identically **exponentially distributed**. That is, the **infinitesimal generator rate matrix** for the jump processes counting transitions over the interval $[t, t + \Delta t)$ are as follows:

$$p_{ij;ab}^{rr}(\Delta t) = \mathbb{P}(A_{ij}^{rr}(t + \Delta t) = b | A_{ij}^{rr}(t) = a) = \begin{cases} \lambda_{ij;a,a+1}^{rr} \Delta t + o(\Delta t), & b - a = 1, \\ 1 - \lambda_{ij;a,a+1}^{rr} \Delta t + o(\Delta t), & b - a = 0, \\ 0, & |b - a| > 1, \end{cases} \quad a, b = 0, 1, 2, \ldots, \tag{5.6}$$

where

$$\lambda_{ij;a,a+1}^{rr} = \sigma_{ij}^{rr} = \sigma_i^r v_{ij}^{rr}, \quad a \in I(0, \infty), \tag{5.7}$$

is the **homogeneous birth-rate** at every state a for all transitions from state s_i^r to s_j^r, $j \neq i$. Similarly,

$$p_{il;ab}^{rq}(\Delta t) = \mathbb{P}(A_{il}^{rq}(t + \Delta t) = b | A_{il}^{rq}(t) = a) = \begin{cases} \lambda_{il;a,a+1}^{rq} \Delta t + o(\Delta t), & b - a = 1, \\ 1 - \lambda_{il;a,a+1}^{rq} \Delta t + o(\Delta t), & b - a = 0, \\ 0, & |b - a| > 1, \end{cases} \quad a, b = 0, 1, 2, \ldots, \tag{5.8}$$

where

$$\lambda_{il;a,a+1}^{rq} = \gamma_{il}^{rq} = \gamma_i^r \gamma_{0l}^{rq}, \quad a \in I(0, \infty), \tag{5.9}$$

is the **homogeneous birth-rate** at every state a for all transitions from state s_i^r to s_l^q, $q \neq r$.

We have a similar CTMC expression for the stochastic process underlying the return rates in **Assumption 4.** Indeed, the underlying counting processes $\{B_{ij}^{rr}(t), t \geq 0\}$ and $\{B_{il}^{rq}(t), t \geq 0\}$ are **pure-birth processes,** where $\{B_{ij}^{rr}(t), t \geq 0\}$ counts the transitions over continuous time $[0, t)$ for the return of residents of site s_i^r visiting other sites s_j^r, $j \neq i$, in their home region C_r, and $\{B_{il}^{rq}(t), t \geq 0\}$ counts the transitions over continuous time $[0, t)$ for the return of residents of site s_i^r visiting other sites s_l^q in other regions C_q, $q \in I(1, M)$, $q \neq r$. Furthermore, over any small time interval $[t, t+\Delta t)$, only single transitions in the stochastic processes occur; $B_{ij}^{rr}(t)$ and $B_{il}^{rq}(t)$ are mutually independent random variables for all $t \geq 0$, $i, j \in I(1, n_r)$, and $q, r \in I(1, M)$; and the times between transitions are independent and identically **exponentially distributed**. That is, the **infinitesimal generator rate matrix** for the jump processes counting transitions over the interval $[t, t + \Delta t)$ are as follows:

$$p_{ij;ab}^{rr}(\Delta t) = \mathbb{P}(B_{ij}^{rr}(t + \Delta t) = b | B_{ij}^{rr}(t) = a)$$
$$= \begin{cases} \mu_{ij;a,a+1}^{rr} \Delta t + o(\Delta t), & b - a = 1, \\ 1 - \mu_{ij;a,a+1}^{rr} \Delta t + o(\Delta t), & b - a = 0, \\ 0, & |b - a| > 1, \end{cases} \quad a, b = 0, 1, 2, \ldots, \tag{5.10}$$

where

$$\mu_{ij;a,a+1}^{rr} = \rho_{ij}^{rr}, \quad a \in I(0, \infty), \tag{5.11}$$

is the **homogeneous birth-rate** at every state a for all *return* transitions of residents from site s_i^r visiting site s_j^r, $j \neq i$. Similarly,

$$p_{il;ab}^{rq}(\Delta t) = \mathbb{P}(B_{il}^{rq}(t + \Delta t) = b | B_{il}^{rq}(t) = a)$$
$$= \begin{cases} \mu_{il;a,a+1}^{rq} \Delta t + o(\Delta t), & b - a = 1, \\ 1 - \mu_{il;a,a+1}^{rq} \Delta t + o(\Delta t), & b - a = 0, \\ 0, & |b - a| > 1, \end{cases} \quad a, b = 0, 1, 2, \ldots, \tag{5.12}$$

where

$$\mu_{il;a,a+1}^{rq} = \rho_{il}^{rq} = \rho_{i0}^{rq} \rho^{rq} \rho_{0l}^{rq} \rho_l^q, \quad a \in I(0, \infty), \tag{5.13}$$

is the **homogeneous birth-rate** at every state a for all return transitions for residents of site s_i^r visiting s_l^q, $q \neq r$.

5.3 Probability modeling and estimation of the human mobility network

In this section, we develop a probabilistic model for the human mobility network, capturing the exponential transition rates in the infinitesimal generators given by (5.10)–(5.13). These rates represent the dynamics at which residents leave and return to their home sites and regions within the two-scale network population.

Due to the inherent symmetry in travel patterns, the probabilistic formulation of the return rates parallels that of the visiting rates. Therefore we further present a detailed derivation of the visiting rates, which can also serve as a framework for deriving the return rates.

In the following, we define a new term called the *accessible domain* of a node in the human mobility network. Recall Examples 2.4.1 and 3.2.1 that the nodes of the human mobility network are the sites s_i^r, $r \in I(1, M)$, $i \in I(1, n_r)$. The term called the *accessible domain* of residents of site s_i^r in the region C_r refers to all sites within the region C_r and also in other regions C_q accessible to residents of site s_i^r. For this purpose, we introduce a few notions and definitions.

For $r \in I(1, M)$, $i \in I(1, n_r)$, define

$$\begin{aligned} I^r(1, M) &= \{q \in I(1, M) : r \neq q, \gamma^{rq} > 0\}, \quad I^r(1, M) \subseteq I(1, M), \\ I_i^r(1, n_r) &= \{j \in I(1, n_r) : j \neq i, v_{ij}^{rr} > 0\}, \quad I_i^r(1, n_r) \subseteq I(1, n_r), \end{aligned} \tag{5.14}$$

and for $q \in I(1, M)$,

$$I_i^r(1, n_q) = \{l \in I(1, n_q) : \gamma^{rq} > 0 \;\; \gamma_{0l}^{rq} > 0\} \quad \text{and} \quad I_i^r(1, n_q) \subseteq I(1, n_q). \tag{5.15}$$

Definition 5.3.1 (Inter- and intraregional accessible domains). For all $i \in I(1, n_r)$ and $r \in I(1, M)$, $C_{rr}(s_i^r) = \{s_j^r \in C_{rr}(s^r) : j \in I_i^r(1, n_r)\}$ is the intraregional accessible domain for residents of site s_i^r in the region C_r.

For all $r \in I(1, M)$, $i \in I(1, n_r)$, and $q \in I^r(1, M)$, $C_{rq}(s_i^r) = \{s_l^q \in C_{qq}(s^q) : l \in I_i^r(1, n_q)\}$ is the interregional accessible domain for residents of site s_i^r in the region C_r.

For $i \in I(1, n_r)$,

$$\begin{aligned} C(s_i^r) &= \{s_j^q \in C(s) : j \in I_i^r(1, n_r), q \in I^r(1, M)\} \\ &= \bigcup_{q \in I^r(1,M)} \bigcup_{i \in I_i^r(1,n_q)} C_{rq}(s_i^r) = [C_{rr}(s_i^r)] \cup \left[\bigcup_{q \in I^r(1,M)} C_{rq}(s_i^r) \right] \end{aligned} \tag{5.16}$$

is the aggregate inter- and intraregional accessible domain of residents for site s_i^r in the region C_r.

5.3.1 Estimation of the probabilistic visiting and return rates

The decomposition of the total population of site s_i^r in (5.5) in Section 5.2.1, **Assumption 2** will be used to estimate the visiting and return probabilistic rates within the finite accessible domain in Definition 5.3.1.

(1) **Intraregional visiting rates.** Residents of site s_i^r leave on trips to other sites s_j^r within the region at the per capita rate $\sigma_{ij}^{rr} = \sigma_i^r v_{ij}^{rr}$, $i, j \in I(1, n_r)$, $j \neq i$, where σ_i^r and v_{ij}^{rr} are defined in the previous section.

Indeed, let $N_{i0}^{rr}(C_{rr}(s_i^r))$ be the total number of residents of site s_i^r that leave the site to visit other sites in $C_{rr}(s_i^r)$. Furthermore, let $T_{i,\text{total}}^{rr}$ be the total time during which the visiting to sites in $C_{rr}(s_i^r)$ takes place. Then

$$\sigma_i^r \approx \frac{N_{i0}^{rr}(C_{rr}(s_i^r))}{N_{i0}^{rr} \cdot T_{i,\text{total}}^{rr}}. \tag{5.17}$$

Also, let E_i^r be the event that residents leave their site s_i^r to visit other sites in $C_{rr}(s_i^r)$, and let E_{ij}^{rr} be the event that residents leave their site s_i^r and visit site s_j^r. Then the intraregional probability visiting rates are given by

$$\begin{aligned} &\mathbb{P}(E_i^r) = \sigma_i^r, \quad v_{ij}^{rr} = \mathbb{P}(E_{ij}^{rr} \mid E_i^r) = \frac{N_{ij}^{rr}}{N_{i0}^{rr}(C_{rr}(s_i^r))}, \\ &\sigma_{ij}^{rr} = \mathbb{P}(E_{ij}^{rr}) = \mathbb{P}(E_{ij}^{rr} \mid E_i^r)\mathbb{P}(E_i^r) = \sigma_i^r v_{ij}^{rr}. \end{aligned} \tag{5.18}$$

(2) Interregional visiting rates. Residents of site s_i^r in the region C_r, $r \in I(1,M)$, leave on trips to other sites s_l^q in other regions C_q, $q \neq r$, at the per capita rate $\gamma_{il}^{rq} = \gamma_i^r \gamma_{i0}^{rq} \gamma^{rq} \gamma_{0l}^{rq}$, $i \in I(1,n_r)$, $l \in I(1,n_q)$, $r \neq q$.
This can be justified as follows: for all $q \in I^r(1,M)$, let $N_{i0}^{rq}(C_{rq}(s_i^r))$ be the total number of residents of site s_i^r that leave the site to visit other sites in $C_{rq}(s_i^r)$, and let F_i^r be the event representing this movement. In addition, let $T_{i,\text{total}}^{rq}$ be the total time during which the visiting to sites in $C_{rq}(s_i^r)$ takes place. Then

$$\gamma_i^r = \mathbb{P}(F_i^r) \approx \frac{\sum_{z \in I^r(1,M)} N_{i0}^{rz}(C_{rz}(s_i^r))}{N_{i0}^{rr} \cdot \sum_{z \in I^r(1,M)} T_{i,\text{total}}^{rz}}. \tag{5.19}$$

Now for $q \in I(1,M)$ and $l \in I_i^r(1,n_q)$, let F_{i0}^{rq} be the event that residents of site s_i^r leave to visit the region C_q (the specific destination site s_l^q in C_q is not taken into account at this point). Furthermore, let F^{rq} be the event that residents of the region C_r visit the region C_q, and let F_{0i}^{rq} be the event that the residents of the region C_r visiting region the region C_q, go directly to visit the site s_l^q in C_q (the specific site s_i^r of origin in C_r is not taken into account at this point). Applying basic conditional probability rules, the rates are estimated as follows:

$$\begin{aligned} \gamma_{i0}^{rq} &= \mathbb{P}(F_{i0}^{rq} \mid F_i^r) \approx \frac{N_{i0}^{rq}(C_{rq}(s_{i0}^r))}{\sum_{z \in I^r(1,M)} N_{i0}^{rz}(C_{rz}(s_i^r))}, \\ \gamma^{rq} &= \mathbb{P}(F^{rq}) = \sum_{i=1}^{n_r} P(F_{i0}^{rq} \mid F_i^r) = \sum_{i=1}^{n_r} \gamma_{i0}^{rq}, \\ \gamma_{0l}^{rq} &= \mathbb{P}(F_{0l}^{rq} \mid F^{rq}) \approx \frac{N_{il}^{rq}}{N_{i0}^{rq}(C_{rq}(s_i^r))}. \end{aligned} \tag{5.20}$$

Therefore, if F_{il}^{rq} is the event that residents of site s_i^r in the region C_r travel to site s_l^q in the region C_q, then $F_{il}^{rq} = F_{i0}^{rq} \cap F_{0l}^{rq}$, where F_{i0}^{rq} and F_{0l}^{rq} are independent events.

This is because the events of leaving a site of residence and arriving at another site of visitation in other regions are independent, i. e., the leaving any sites s_i^r of origin for residents of the region C_r that visit the region C_q is independent of selecting a site s_i^q to visit in the region C_q, $q \neq r$. This also implies that the event F_{0l}^{rq} is independent of the site s_i^r, and the destination for residents of region C_r visiting any region C_q, $q \neq r$, is independent of the event F_{i0}^{rq}. Thus we have the following estimate:

$$\begin{aligned} \gamma_{il}^{rq} &= \mathbb{P}(F_{il}^{rq}) = P(F_{i0}^{rq} \cap F_{0l}^{rq}) = \mathbb{P}(F_{i0}^{rq}) \cdot \mathbb{P}(F_{0l}^{rq}) \\ &= \mathbb{P}(F_{i0}^{rq} \mid F_i^r)\mathbb{P}(F_i^r) \cdot \mathbb{P}(F_{0l}^{rq} \mid F^{rq})\mathbb{P}(F^{rq}) = \gamma_i^r \gamma_{i0}^{rq} \gamma^{rq} \gamma_{0l}^{rq}. \end{aligned} \tag{5.21}$$

(3) **Intraregional return rates.** Residents traveling from home site s_i^r to s_j^r within the region C_r, $r \in I(1, M)$, have the per capita return rate ρ_{ij}^{rr}, $i, j \in I(1, n_r)$, $i \neq j$. The estimation of the return rate is similar to that of the visitation rate in (5.18).

(4) **Interregional return rates.** Residents traveling from their home site s_i^r in the region C_r to visit the site s_l^q in the region C_q, $q \neq r$, have the per capita return rate

$$\rho_{il}^{rq} = \rho_{0l}^{rq} \rho_l^q \rho_{i0}^{rq} \rho^{rq}, \quad i \in I(1, n_r),\ l \in I(1, n_q),\ r \neq q. \tag{5.22}$$

The estimation of the rate in (5.22) is similar to that of (5.21).

The large two-scale hierarchic mobility network structure is illustrated in Figures 5.2 and 5.1.

5.4 Stochastic modeling and derivation of the mean-field dynamic model for the human mobility process

The jump processes in **Assumption 4** in are used to derive a stochastic model and mean-field dynamic model for the human mobility process in the two-scale network population C in **Assumption 1**, which characterizes the stochastic dynamics of decomposition (5.5) of the number of residents of site s_i^r at the three different locations in C, namely: (1) the dynamics of the population N_{ii}^{rr}, (2) the dynamics of N_{ij}^{rr}, and (3) the dynamics of N_{il}^{rq}. First, consider the following vector representations of the state of the system in **Assumption 2**.

Definition 5.4.1. For $i \in I(1, n_r)$, $a \in I_i^r(1, n_u)$, $r \in I(1, M)$, and $u \in I^r(1, M)$, in view of **Assumption 2**, applying the random variables $N_{ii}^{rr}, N_{ij}^{rr}, N_{ia}^{rq}$, $u = q \neq r$, $a = l$, the matrix $N_{00}^{00} \in \mathbb{R}^{n^2}$ is defined as follows:

$$\begin{aligned} N_{ia}^{ru} &\in \mathbb{R}_+ = [0, \infty), \\ N_{i0}^{ru} &= (N_{i1}^{ru}, N_{i2}^{ru}, \ldots, N_{i,nu}^{ru})^T \in \mathbb{R}_+^{n_u}, \\ N_{00}^{ru} &= (N_{10}^{ruT}, N_{20}^{ruT}, \ldots, N_{n_r0}^{ruT})^T \in \mathbb{R}_+^{n_r n_u}, \end{aligned} \tag{5.23}$$

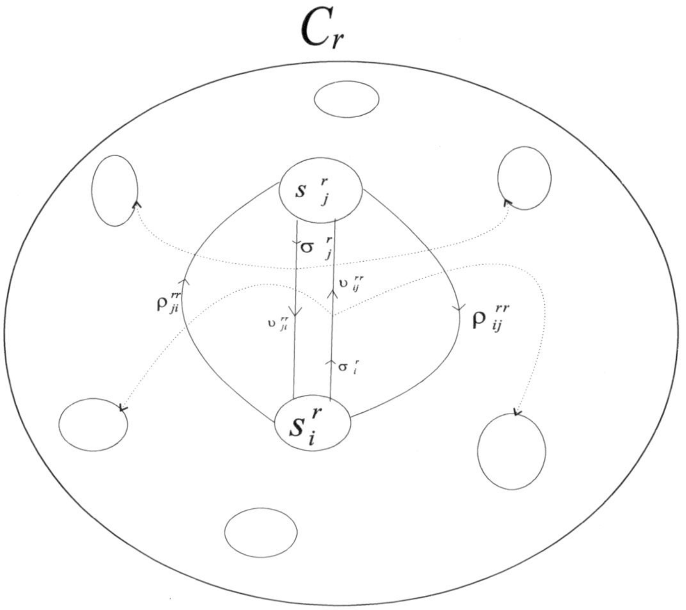

Figure 5.1: Shows the intraregional mobility network between n_r sites s_i^r in the region C_r, $r \in I(1, M)$, $i \in I(1, n_r)$. The dotted lines and curves represent connections with other sites in the region C_r. Furthermore, the parameters in the diagram are defined in Subsection 5.2.1.

$$N_{00}^{r0} = (N_{00}^{r1T}, N_{00}^{r2T}, \ldots, N_{00}^{rMT})^T \in \mathbb{R}_+^{n_r \sum_{u=1}^{M} n_u},$$
$$N_{00}^{00} = (N_{00}^{10T}, N_{00}^{20T}, \ldots, N_{00}^{M0T})^T \in \mathbb{R}_+^{(\sum_{r=1}^{M} n_r)(\sum_{u=1}^{M} n_u)}.$$

Let $(\Omega, \{\mathfrak{F}_t\}_{t\geq 0}, \mathbb{P})$ be a complete probability space, where $\mathfrak{F}_t = \sigma(N_{ii}^{rr}(t), N_{ij}^{rr}(t), N_{il}^{rq}(t))$ is the σ-algebra generated by the random variables in (5.23), $i, j \in I(1, n_r)$, $r, q \in I(1, M)$. Define the stochastic process

$$\mathfrak{N} = \{(N_{ii}^{rr}(t), N_{ij}^{rr}(t), N_{il}^{rq}(t)) \mid i, j \in I(1, n_r), r, q \in I(1, M)\}. \tag{5.24}$$

From the assumptions outlined in Subsection 5.2.1, particularly **Assumption 4**, and utilizing the counting processes that represent transitions to visit intraregional and interregional sites, we consider the following discrete time intervals:

$$t, t + \Delta t, t + 2\Delta t, t + 3\Delta t, \ldots, t + k\Delta t, \ldots, \quad k \in I(0, \infty). \tag{5.25}$$

Based on these intervals, a *discrete-time approximation of the Markov jump process (or continuous-time Markov chain, CTMC)* is formulated. This model describes the time evolution of the states $N_{ii}^{rr}(t)$, $N_{ij}^{rr}(t)$, and $N_{il}^{rq}(t)$ over the time interval $[t, t + \Delta t)$ for $i, j \in I(1, n_r)$ and $r, q \in I(1, M)$:

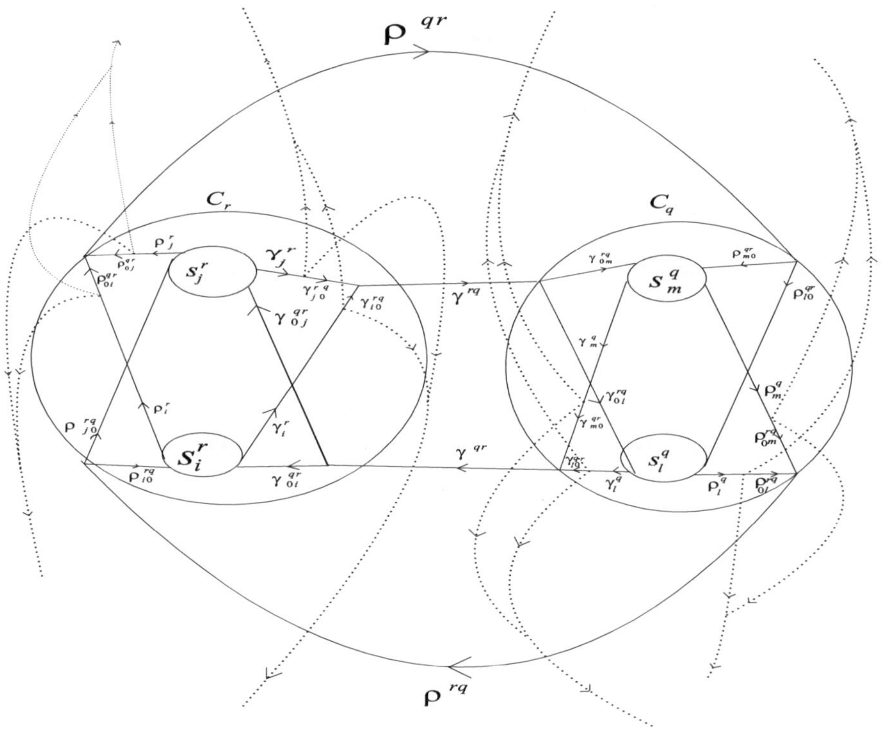

Figure 5.2: The interregional mobility network between M regions $C_r, r \in I(1, M)$, and n_r sites s_i^r that are present in each region C_r. The dotted lines and curves represent connections with other sites in other regions. The parameters are defined in Subsection 5.2.1.

$$
\begin{aligned}
N_{ii}^{rr}(t+\Delta t) &= N_{ii}^{rr}(t) - \sum_{j\neq i}^{n_r} A_{ij}^{rr} - \sum_{q\neq r}^{M}\sum_{l=1}^{n_q} A_{il}^{rq} + \sum_{j\neq i}^{n_r} B_{ij}^{rr} + \sum_{q\neq r}^{M}\sum_{l=1}^{n_q} B_{il}^{rq},\\
N_{ij}^{rr}(t+\Delta t) &= N_{ij}^{rr}(t) + A_{ij}^{rr} - B_{ij}^{rr},\\
N_{il}^{rq}(t+\Delta t) &= N_{il}^{rq}(t) + A_{il}^{rq} - B_{il}^{rq}.
\end{aligned}
\tag{5.26}
$$

Furthermore, the stochastic process $\mathfrak{N}$ in (5.24) satisfies the *Markov property*, that is, for $i, j \in I(1, n_r)$ and $r, q \in I(1, M)$,

$$
\begin{aligned}
&\mathbb{P}\big((N_{ii}^{rr}(t+\Delta t), N_{ij}^{rr}(t+\Delta t), N_{il}^{rq}(t+\Delta t)) \mid \mathfrak{F}_t\big)\\
&\quad = \mathbb{P}\big((N_{ii}^{rr}(t+\Delta t), N_{ij}^{rr}(t+\Delta t), N_{il}^{rq}(t+\Delta t)) \mid N_{00}^{00}(t)\big).
\end{aligned}
\tag{5.27}
$$

In view of (5.26), denote

$$
\Delta N_{ia}^{ru}(t) = \begin{cases} N_{ii}^{rr}(t+\Delta t) - N_{ii}^{rr}(t), & u = r, a = i,\\ N_{ij}^{rr}(t+\Delta t) - N_{ij}^{rr}(t), & u = r, a = j \neq i,\\ N_{il}^{rq}(t+\Delta t) - N_{il}^{ru}(t), & u = q \neq r, a = l. \end{cases}
\tag{5.28}
$$

It is easy to see from (5.27) that the infinitesimal transition probabilities for the Markov chain model $\mathfrak{N}$ are given as follows: for $i,j \in I(1,n_r)$ and $r,q \in I(1,M)$,

$$\begin{aligned}
&P(\Delta N_{ii}^{rr}(t)=1,\Delta N_{ij}^{rr}(t)=-1,j\neq i,\Delta N_{ia}^{ru}(t)=0,u\neq r,a\in I(1,n_u)|N_{00}^{00}(t))\\
&\quad=\rho_{ij}^{rr}N_{ij}^{rr}(t)\Delta t+\mathbf{o}(\Delta),\\
&P(\Delta N_{ii}^{rr}(t)=1,\Delta N_{ij}^{rr}(t)=0,j\neq i,\Delta N_{ia}^{ru}(t)=-1,u\neq r,a\in I(1,n_u)|N_{00}^{00}(t))\\
&\quad=\rho_{ia}^{ru}N_{ia}^{ru}\Delta t+\mathbf{o}(\Delta),\\
&P(\Delta N_{ii}^{rr}(t)=-1,\Delta N_{ij}^{rr}(t)=1,j\neq i,\Delta N_{ia}^{ru}(t)=1,u\neq r,a\in I(1,n_u)|N_{00}^{00}(t))\\
&\quad=(\gamma_i^r+\sigma_i^r)N_{ii}^{rr}(t)\Delta t+\mathbf{o}(\Delta),\\
&P(\Delta N_{ii}^{rr}(t)=-1,\Delta N_{ij}^{rr}(t)=1,j\neq i,\Delta N_{ia}^{ru}(t)=0,u\neq r,a\in I(1,n_u)|N_{00}^{00}(t))\\
&\quad=\sigma_{ij}^{rr}N_{ij}^{rr}(t)\Delta t+\mathbf{o}(\Delta),\\
&P(\Delta N_{ii}^{rr}(t)=-1,\Delta N_{ij}^{rr}(t)=0,j\neq i,\Delta N_{ia}^{ru}(t)=1,u\neq r,a\in I(1,n_u)|N_{00}^{00}(t))\\
&\quad=\gamma_{ia}^{ru}N_{ii}^{rr}(t)\Delta t+\mathbf{o}(\Delta).
\end{aligned} \tag{5.29}$$

From (5.28)–(5.23) it follows that the expectation operator $\mathbb{E}[\cdot]$ applied to the vector

$$\Delta N_{00}^{00}(t)=(\Delta N_{00}^{10T}(t),\Delta N_{00}^{20T}(t),\dots,\Delta N_{00}^{M0T}(t))^T\in\mathbb{R}_+^{(\sum_{r=1}^M n_r)(\sum_{u=1}^M n_u)} \tag{5.30}$$

leads to the expression

$$\mathbb{E}[\Delta N_{00}^{00}(t)]=g(\Delta N_{00}^{00}(t))\Delta t+\mathbf{o}(\Delta t), \tag{5.31}$$

where

$$g(\Delta N_{00}^{00}(t))=g(\Delta N_{ia}^{ru}(t)),\quad \forall i\in I(1,n_r),\ a\in I(1,n_u),\ r,u\in I(1,M),$$
$$g(\Delta N_{ia}^{ru}(t))=\begin{cases}\mathbb{E}(\mathbb{E}(\Delta N_{ii}^{rr}(t)\mid N_{00}^{00}(t)))=\sum_{k=1}^{n_r}\rho_{ik}^{rr}N_{ik}^{rr}+\sum_{q\neq r}^{M}\sum_{l=1}^{n_q}\rho_{il}^{rq}N_{il}^{rq}-(\gamma_i^r+\sigma_i^r)N_{ii}^{rr},\\ \quad u=r,a=i,\\ \mathbb{E}(\mathbb{E}(\Delta N_{ij}^{rr}(t)\mid N_{00}^{00}(t)))=\sigma_{ij}^{rr}N_{ii}^{rr}-\rho_{ij}^{rr}N_{ij}^{rr},\quad u=r,a=j\neq i,\\ \mathbb{E}(\mathbb{E}(\Delta N_{il}^{rq}(t)\mid N_{00}^{00}(t)))=\gamma_{il}^{rq}N_{ii}^{rr}-\rho_{il}^{rq}N_{il}^{rq},\quad u=q\neq r,a=l.\end{cases} \tag{5.32}$$

Thus from (5.31)–(5.32), dividing by Δt and taking the limit as $\Delta t\to 0$ lead to the following approximation:

$$\lim_{\Delta t\to 0}\frac{\mathbb{E}[\Delta N_{00}^{00}(t)]}{\Delta t}=g(\Delta N_{00}^{00}(t)). \tag{5.33}$$

The expression in (5.33) is the **mean-field dynamics approximation** for the rate of change of the average population represented by the random vector in (5.23), that is, from (5.32) the expression in (5.33) can be rewritten explicitly as follows:

$$\lim_{\Delta t \to 0} \frac{\mathbb{E}[\Delta N_{00}^{00}(t)]}{\Delta t} = g(\Delta N_{00}^{00}(t)) = g(\Delta N_{ia}^{ru}(t)), \quad i \in I(1,n_r),\ a \in I(1,n_u),\ r,u \in I(1,M),$$

$$g(\Delta N_{ia}^{ru}(t)) = \begin{cases} \lim_{\Delta t \to 0} \frac{\mathbb{E}(\Delta N_{ii}^{rr}(t))}{\Delta t} = \sum_{k=1}^{n_r} \rho_{ik}^{rr} N_{ik}^{rr}(t) + \sum_{q \neq r}^{M} \sum_{l=1}^{n_q} \rho_{il}^{rq} N_{il}^{rq}(t) - (\gamma_i^r + \sigma_i^r) N_{ii}^{rr}(t), \\ \quad u = r,\ a = i \\ \lim_{\Delta t \to 0} \frac{\mathbb{E}(\Delta N_{ij}^{rr}(t))}{\Delta t} = \sigma_{ij}^{rr} N_{ii}^{rr}(t) - \rho_{ij}^{rr} N_{ij}^{rr}(t), \quad u = r,\ a = j \neq i, \\ \lim_{\Delta t \to 0} \frac{\mathbb{E}(\Delta N_{il}^{rq}(t))}{\Delta t} = \gamma_{il}^{rq} N_{ii}^{rr}(t) - \rho_{il}^{rq} N_{il}^{rq}(t), \quad u = q \neq r,\ a = l. \end{cases} \tag{5.34}$$

Notation 5.4.1. To simplify and minimize the number of new notations, denote for $i,j \in I(1,n_r), j \neq i$ and $r,q \in I(1,M), q \neq r$, the rates of change of the following averages by

$$\begin{aligned} \frac{dN_{ii}^{rr}(t)}{dt} &\equiv \frac{d\mathbb{E}[N_{ii}^{rr}(t)]}{dt} = \lim_{\Delta t \to 0} \frac{\mathbb{E}(\Delta N_{ii}^{rr}(t))}{\Delta t}, \\ \frac{dN_{ij}^{rr}(t)}{dt} &\equiv \frac{d\mathbb{E}[N_{ij}^{rr}(t)]}{dt} = \lim_{\Delta t \to 0} \frac{\mathbb{E}(\Delta N_{ij}^{rr}(t))}{\Delta t}, \quad \text{and} \\ \frac{dN_{il}^{rq}(t)}{dt} &\equiv \frac{d\mathbb{E}[N_{il}^{rq}(t)]}{dt} = \lim_{\Delta t \to 0} \frac{\mathbb{E}(\Delta N_{il}^{rq}(t))}{\Delta t}. \end{aligned} \tag{5.35}$$

By applying notation (5.35) to the average dynamics in (5.32)–(5.33) leads to the following large-scale interconnected linear system of differential equations describing the *mean-field dynamics* of the human mobility process in the two-scale network C:

$$\frac{dN_{ii}^{rr}}{dt} = \sum_{k=1}^{n_r} \rho_{ik}^{rr} N_{ik}^{rr} + \sum_{q \neq r}^{M} \sum_{l=1}^{n_q} \rho_{il}^{rq} N_{il}^{rq} - (\gamma_i^r + \sigma_i^r) N_{ii}^{rr}, \tag{5.36}$$

$$\frac{dN_{ij}^{rr}}{dt} = \sigma_{ij}^{rr} N_{ii}^{rr} - \rho_{ij}^{rr} N_{ij}^{rr}, \quad i \neq j, \tag{5.37}$$

$$\frac{dN_{il}^{rq}}{dt} = \gamma_{il}^{rq} N_{ii}^{rr} - \rho_{il}^{rq} N_{il}^{rq}, \quad r \neq q, \tag{5.38}$$

$$i \in I(1,n_r),\ l \in I_i^r(1,n_q),\ r,q \in I^r(1,M), \tag{5.39}$$

where all the parameters in (5.36)–(5.39), are nonnegative and at time $t = 0$,

$$N_{ii}^{rr}(0) = N_{i0}^{rr}, \quad N_{ij}^{rr}(0) = 0, \quad N_{il}^{rq}(0) = 0, \tag{5.40}$$

and $N_{ii}^{rr}, N_{ij}^{rr}, N_{il}^{rq}, i,j \in I_i^r(1,n_r), l \in I_i^r(1,n_q), r,q \in I(1,M)$, present the mean values of the random variables in (5.23) defined as before.

Remark 5.4.1. It is important to note that residents of every site $s_i^r, i \in I(1,n_r)$, can only reach out to other sites in their accessible domain $C(s_i^r)$. Thus the sums in (5.36) reduce to the sum over all $q \in I^r(1,M)$, $k \in I_i^r(1,n_r)$, and $l \in I_i^r(1,n_q)$.

5.5 Steady states of the two-scale mean-field dynamic model for the human mobility process

In this section, we derive the steady states of the average dynamics described in (5.36)–(5.39). These steady states represent the expected constant states for the stochastic process defined in (5.24) and (5.23), which models human mobility in the two-level hierarchical population structure $C = C_{00}^{00}$ outlined in Subsection 5.2.1, particularly under **Assumption 1**.

It is important to note that the steady states obtained characterize the system behavior when *the average dynamics exhibits a zero growth rate*. This condition is formally defined below.

The concept of steady states derived from mean-field differential equations plays a pivotal role in understanding the long-term behavior of complex stochastic systems. By linking the steady states to the constant paths of the stochastic process we can analyze how the inherent variability in individual movements translates into stable macroscopic patterns within the population. Such insights are essential for interpreting the equilibrium conditions of human mobility networks, especially in finite-scale structured populations.

Moreover, identifying steady states provides a foundation for evaluating the stability of the system. Perturbations around these steady states can reveal the resilience of the network to disruptions such as changes in travel patterns or regional connectivity. This is particularly relevant in contexts where the mobility process is influenced by external factors like policy interventions, environmental changes, or socioeconomic dynamics.

Beyond mobility itself, the steady-state analysis also offers a framework for investigating related phenomena. For instance, the steady states derived here can serve as baseline conditions for exploring how infectious diseases spread through the mobility network. By comparing the disease dynamics under steady-state mobility conditions to those under transient dynamics we can better understand the role of human mobility in driving epidemic outbreaks and persistence.

The results presented in this section thus have broader implications, providing a stepping stone for interdisciplinary applications in public health, urban planning, and network science. The steady states not only reflect equilibrium configurations of mobility but also highlight the interplay between individual-level stochastic behavior and population-level systemic properties.

Definition 5.5.1. Denote the equilibrium states of N_{ii}^{rr}, N_{ij}^{rr}, and N_{il}^{rq} by N_{ii}^{*rr}, N_{ij}^{*rr}, and N_{il}^{*rq}, respectively. It follows from (5.36)–(5.39) that at the equilibrium,

$$\frac{dN_{ii}^{rr}}{dt} \mid (N_{ii}^{*rr}, N_{ij}^{*rr}, N_{il}^{*rq}) = 0, \quad \frac{dN_{ij}^{rr}}{dt} \mid (N_{ii}^{*rr}, N_{ij}^{*rr}, N_{il}^{*rq}) = 0,$$
$$\frac{dN_{il}^{rq}}{dt} \mid (N_{ii}^{*rr}, N_{ij}^{*rr}, N_{il}^{*rq}) = 0. \tag{5.41}$$

Clearly, the steady state of (5.36)–(5.39) characterizes the path for the underlying *birth-and-death processes* exhibited in system (5.26)–(5.29) with the average dynamics of the stochastic process in (5.24) and (5.23) obtained when the **total birth rate into each state** N_{ii}^{rr}, N_{ij}^{rr}, and N_{il}^{rq}, is equal to the **total death rate in the state.**

Solving system (5.41) for $(N_{ii}^{*rr}, N_{ij}^{*rr}, N_{il}^{*rq})$ leads to the following:

$$N_{ij}^{*rr} = \frac{\sigma_{ij}^{rr} N_{ii}^{*rr}}{\rho_{ij}^{rr}}, \quad i \neq j, \tag{5.42}$$

$$N_{il}^{*rq} = \frac{\gamma_{il}^{rq} N_{ii}^{*rr}}{\rho_{il}^{rq}}, \quad r \neq q, \tag{5.43}$$

$$i, j \in I(1, n_r), \ l \in I(1, n_q); \ r, q \in I(1, M).$$

Rewriting (5.5) in terms of the steady states leads to

$$N_{i0}^{*rr} = N_{ii}^{*rr} + \sum_{k \neq i}^{n_r} N_{ik}^{*rr} + \sum_{q \neq r}^{M} \sum_{l=1}^{n_q} N_{il}^{*rq}, \quad i \in I(1, n_r). \tag{5.44}$$

Now substituting (5.42) and (5.43) into (5.44) and factorizing N_{ii}^{*rr}, we have

$$N_{i0}^{*rr} = N_{ii}^{*rr} \left(1 + \sum_{k \neq i}^{n_r} \frac{\sigma_{ik}^{rr}}{\rho_{ik}^{rr}} + \sum_{q \neq r}^{M} \sum_{l=1}^{n_q} \frac{\gamma_{il}^{rq}}{\rho_{il}^{rq}} \right), \quad i \in I(1, n_r). \tag{5.45}$$

From (5.45) it follows that

$$N_{ii}^{*rr} = N_{i0}^{*rr} \left(1 + \sum_{k \neq i}^{n_r} \frac{\sigma_{ik}^{rr}}{\rho_{ik}^{rr}} + \sum_{q \neq r}^{M} \sum_{l=1}^{n_q} \frac{\gamma_{il}^{rq}}{\rho_{il}^{rq}} \right)^{-1}, \quad i \in I(1, n_r). \tag{5.46}$$

Now substituting (5.46) into (5.42) and (5.43), N_{ij}^{*rr} and N_{il}^{*rq} are represented by

$$N_{ij}^{*rr} = N_{i0}^{*rr} \frac{\frac{\sigma_{ij}^{rr}}{\rho_{ij}^{rr}}}{(1 + \sum_{k \neq i}^{n_r} \frac{\sigma_{ik}^{rr}}{\rho_{ik}^{rr}} + \sum_{q \neq r}^{M} \sum_{l=1}^{n_q} \frac{\gamma_{il}^{rq}}{\rho_{il}^{rq}})}, \quad j \in I_i^r(1, n_r), \quad \text{and} \quad N_{ij}^{*rr} = 0 \quad \text{otherwise}, \tag{5.47}$$

and

$$N_{il}^{*rq} = N_{i0}^{*rr} \frac{\frac{\gamma_{il}^{rq}}{\rho_{il}^{rq}}}{(1 + \sum_{k \neq i}^{n_r} \frac{\sigma_{ik}^{rr}}{\rho_{ik}^{rr}} + \sum_{q \neq r}^{M} \sum_{l=1}^{n_q} \frac{\gamma_{il}^{rq}}{\rho_{il}^{rq}})}, \quad l \in I_i^r(1, n_q), \quad \text{and} \quad N_{il}^{*rq} = 0 \quad \text{otherwise}. \tag{5.48}$$

Denote

$$S_{ii}^{*rr} = \left(1 + \sum_{k\neq i}^{n_r} \frac{\sigma_{ik}^{rr}}{\rho_{ik}^{rr}} + \sum_{q\neq r}^{M} \sum_{l=1}^{n_q} \frac{\gamma_{il}^{rq}}{\rho_{il}^{rq}}\right)^{-1}, \quad i \in I(1, n_r), \tag{5.49}$$

$$U_{ij}^{*rr} = \frac{\frac{\sigma_{ij}^{rr}}{\rho_{ij}^{rr}}}{(1 + \sum_{k\neq i}^{n_r} \frac{\sigma_{ik}^{rr}}{\rho_{ik}^{rr}} + \sum_{q\neq r}^{M} \sum_{l=1}^{n_q} \frac{\gamma_{il}^{rq}}{\rho_{il}^{rq}})}, \quad j \in I_i^r(1, n_r), \quad \text{and} \quad U_{ij}^{*rr} = 0 \quad \text{otherwise}, \tag{5.50}$$

$$V_{il}^{*rq} = \frac{\frac{\gamma_{ij}^{rq}}{\rho_{il}^{rq}}}{(1 + \sum_{k\neq i}^{n_r} \frac{\sigma_{ik}^{rr}}{\rho_{ik}^{rr}} + \sum_{q\neq r}^{M} \sum_{l=1}^{n_q} \frac{\gamma_{il}^{rq}}{\rho_{il}^{rq}})}, \quad l \in I_i^r(1, n_q), \quad \text{and} \quad V_{il}^{*rq} = 0 \quad \text{otherwise}. \tag{5.51}$$

The quantities in (5.46), (5.47), and (5.48) represent the equilibrium states for residents of site s_i^r present at their home site and visiting the jth site $s_j^r, j \neq i$, in their intraregional accessible domain $C_{rr}(s_i^r)$ and also visiting the lth site s_l^q in their interregional accessible domain $C_{rq}(s_i^r)$, respectively. Thus it follows that the quantities in (5.49), (5.50), and (5.51) represent the fraction of the equilibrium state for residents in the different categories present at the corresponding locations.

Observe further that for all $r, q \in I(1, M)$, $r \neq q$, persons traveling from site s_i^r to s_j^r have the per capital return rate ρ_{ij}^{rr}, and persons traveling from site s_i^r in the region C_r to site s_l^q in the region C_q, $q \neq r$, have the per capita return rate ρ_{il}^{rq}. The average length of time spent visiting sites s_j^r and s_l^q under the *exponential distribution* in (5.26) is denoted by τ_{ij}^{rr} and τ_{il}^{rq}, respectively, where

$$\tau_{ij}^{rr} = \frac{1}{\rho_{ij}^{rr}}, \quad j \in I_i^r(1, n_r),\ r \in I(1, M), \quad \text{and} \quad \tau_{ij}^{rr} = 0 \quad \text{otherwise}, \tag{5.52}$$

$$\tau_{il}^{rq} = \frac{1}{\rho_{il}^{rq}}, \quad l \in I_i^r(1, n_q),\ q \in I^r(1, M), \quad \text{and} \quad \tau_{il}^{rq} = 0 \quad \text{otherwise}. \tag{5.53}$$

In addition, for $r \in I(1, M)$, $i \in I(1, n_r)$, and $s, t \in I^r(1, M)$, let

$$K_{imn}^{rst} \equiv K_{imn}^{rst}(\tau_{im}^{rs}, \tau_{in}^{rt}) = \frac{\tau_{im}^{rs}}{\tau_{in}^{rt}}, \quad m \in I_i^r(1, n_s),\ n \in I_i^r(1, n_t), \tag{5.54}$$

be the ratio of visiting times of residents of the ith site s_i^r in the region C_r visiting the mth and nth sites in the regions C_s and C_t, respectively, where $m \neq n$ and $m, n \neq i$ whenever $s = t = r$. Now substituting (5.52), (5.53), and (5.54) into (5.49), (5.50), and (5.51) and further simplifying lead to

$$S_{ii}^{*rr} = \frac{1}{(1 + \sum_{k\neq i}^{n_r} \sigma_{ik}^{rr}\tau_{ik}^{rr} + \sum_{q\neq r}^{M} \sum_{a=1}^{n_q} \gamma_{ia}^{rq}\tau_{ia}^{rq})}, \quad i \in I(1, n_r), \tag{5.55}$$

$$U_{ij}^{*rr} = \frac{\sigma_{ij}^{rr}}{(\frac{1}{\tau_{ij}^{rr}} + \sum_{k\neq i}^{n_r} \sigma_{ik}^{rr} K_{ikj}^{rrr} + \sum_{q\neq r}^{M} \sum_{a=1}^{n_q} \gamma_{ia}^{rq} K_{iaj}^{rqr})}, \quad j \in I_i^r(1, n_r), \tag{5.56}$$

and

$$V_{il}^{*rq} = \frac{\gamma_{il}^{rq}}{(\frac{1}{\tau_{il}^{rq}} + \sum_{k\neq i}^{n_r} \sigma_{ik}^{rr} K_{ikl}^{rrq} + \sum_{q\neq r}^{M} \sum_{a=1}^{n_q} \gamma_{ia}^{rq} K_{ial}^{rqq})}, \quad l \in I_i^r(1, n_q). \tag{5.57}$$

Notation 5.5.1. For $i \in I(1, n_r)$ and $r \in I(1, M)$, denote

$$\tau_{i,\min}^{rr} = \min_{1\le j\le n_r} \tau_{ij}^{rr}, \tag{5.58}$$

$$\tau_{i,\min}^{rq} = \min_{1\le l\le n_q} \tau_{il}^{rq}, \quad q \in I^r(1, M), \tag{5.59}$$

$$\tau_{i,\min}^{r,\min} = \min_{q\neq r} \tau_{i,\min}^{rq}, \tag{5.60}$$

$$\tau_{i,\max}^{rr} = \max_{1\le j\le n_r} \tau_{ij}^{rr}, \tag{5.61}$$

$$\tau_{i,\max}^{rq} = \max_{1\le l\le n_q} \tau_{il}^{rq}, \quad q \in I^r(1, M), \tag{5.62}$$

$$\tau_{i,\max}^{r,\max} = \max_{q\neq r} \tau_{i,\max}^{rq}. \tag{5.63}$$

5.6 Special human mobility patterns in the two-scale network population

The special mobility patterns are characterized by the qualitative behavior of the mobility rates of the large-scale hierarchic regional mobility dynamics process. To understand the mobility patterns, first the qualitative behaviors of the mobility rates are classified.

Definition 5.6.1. Let f and g be real-valued functions.

1. If there exist $k > 0$ and n_0 such that $|f(n)| \le k|g(n)|$ for all $n > n_0$, then we say that f is big-o of g, which is denoted by $f(n) = O(g(n))$ or $f = O(g)$. If $f(n) \to 0$ as $n \to \infty$, that is, f tends in the limit to a zero function for sufficiently large n, then we write $f = O(\epsilon)$ or $f(n) = O(\frac{1}{n})$ for $\epsilon > 0$. If $f(n)$ is a constant function as $n \to \infty$, then we write $f(n) = O(1)$.
2. If there exist $k_1, k_2 > 0$ and n_0 such that $k_1|g(n)| \le |f(n)| \le k_2|g(n)|$ for all $n > n_0$, then we say that f is big-theta of g, which is denoted by $f(n) = \theta(g(n))$. If $f(n) \to \infty$ as $n \to \infty$, then we write $f(n) = \theta(n)$ or $f = \theta(\frac{1}{\epsilon})$ for $\epsilon > 0$.

To classify the qualitative behavior of the mobility rates, we make the following assumptions about the mobility rate functions. Assume that for fixed $r \in I(1, M)$ and $i \in I(1, n_r)$ and for all $q \in I^r(1, M)$, $j \in I_i^r(1, n_r)$, and $l \in I_i^r(1, n_q)$, the inter- and intraregional visiting times of residents of the ith site visiting the jth site within the rth region and of the ith site visiting the lth site in the qth region satisfy the following:

Hypothesis 5.6.1. Using Definition 5.6.1, we assume that
H_1: $\tau_{ij}^{rr} \to 0$ and $\tau_{il}^{rq} < \infty \iff \tau_{ij}^{rr} = 0(\epsilon)$ and $\tau_{il}^{rq} = 0(1)$.
H_2: $\tau_{il}^{rq} \to 0$ and $\tau_{ij}^{rr} < \infty \iff \tau_{il}^{rq} = 0(\epsilon)$ and $\tau_{ij}^{rr} = 0(1)$.
H_3: $\tau_{ij}^{rr} < \infty$ and $\tau_{il}^{rq} < \infty \iff \tau_{il}^{rq} = 0(1)$ and $\tau_{ij}^{rr} = 0(1)$.
H_4: $\tau_{ij}^{rr} \to 0$ and $\tau_{il}^{rq} \to 0 \iff \tau_{il}^{rq} = 0(\epsilon)$ and $\tau_{ij}^{rr} = 0(\epsilon)$.
H_5: $\tau_{ij}^{rr} \to \infty$ and $\tau_{il}^{rq} \to \infty \iff \tau_{il}^{rq} = \theta(\frac{1}{\epsilon})$ and $\tau_{ij}^{rr} = \theta(\frac{1}{\epsilon})$ for $\epsilon > 0$.

The interpretation of H_1 is that residents of site s_i^r that visit sites in their intraregional accessible domain $C_{rr}(s_i^r)$ tend to spend infinitesimally small time, whereas residents of the same site that visit sites in their interregional accessible domain $C_{rq}(s_i^r)$ tend to spend a finite amount of time. The interpretation of H_2 and H_3 is similar to that of H_1. On the other hand, H_4 means that residents of site s_i^r that visit sites in their inter- and intraregional accessible domain $C(s_i^r)$ tend to spend infinitesimally small time. Finally, H_5 means that residents of site s_i^r that visit sites in their inter- and intraregional accessible domain $C(s_i^r)$ tend to stay permanently. The following result is a characterization of the steady states as intraregional visiting time approaches zero and the interregional visiting time is finite.

Theorem 5.6.1. *If H_1 holds, then the steady state of the mobility system satisfies*

$$\begin{aligned} N_{ii}^{*rr} &\to N_i^{rr} \frac{1}{(1 + \sum_{q \neq r}^{M} \sum_{a=1}^{n_q} \gamma_{ia}^{rq} \tau_{ia}^{rq})}, \qquad N_{ij}^{*rr} \to 0, \\ N_{il}^{*rq} &\to N_i^{rr} \frac{\gamma_{il}^{rq}}{(\frac{1}{\tau_{il}^{rq}} + \sum_{q \neq r}^{M} \sum_{a=1}^{n_q} \gamma_{ia}^{rq} K_{ial}^{rqq})} \end{aligned} \tag{5.64}$$

for $j \in I_i^r(1, n_r)$, $q \in I^r(1, M)$, $l \in I_i^r(1, n_q)$.

Proof. Letting $\tau_{ij}^{rr} = 0(\epsilon)$ and $\tau_{il}^{rq} = 0(1)$ in (5.55), (5.56), and (5.57), we have

$$\begin{aligned} S_{ii}^{*rr} &\to \frac{1}{(1 + \sum_{q \neq r}^{M} \sum_{a=1}^{n_q} \gamma_{ia}^{rq} \tau_{ia}^{rq})}, \qquad U_{ij}^{*rr} \to 0, \\ V_{il}^{*rq} &\to \frac{\gamma_{il}^{rq}}{(\frac{1}{\tau_{il}^{rq}} + \sum_{q \neq r}^{M} \sum_{a=1}^{n_q} \gamma_{ia}^{rq} K_{ial}^{rqq})} \end{aligned} \tag{5.65}$$

for $l \in I_i^r(1, n_q)$. Then substituting (5.65) into (5.46), (5.47), and (5.48) gives us (5.64). □

Remark 5.6.1. Theorem 5.6.1 suggests that a fraction of residents tend to be lifelong residents of their home site and the remaining fraction of residents tend to reside for a finite time in sites of the interregional accessible domain $C_{rq}(s_i^r)$. This situation is illustrated in Figure 5.3.

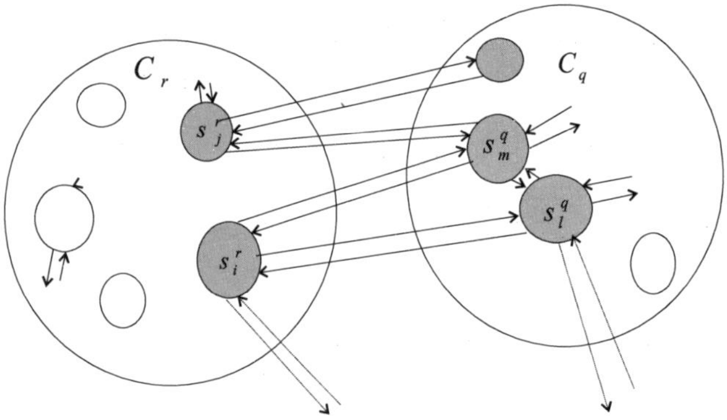

Figure 5.3: Shows that residents of site s_i^r are only present at their home site and also at those sites s_l^q that are in their interregional accessible domain $C_{rq}(s_i^r)$. The arrows represent a transport network between any two sites and regions. Furthermore, the dotted lines and arrows indicate a connection with other accessible sites in other regions.

In the next theorem, we characterize the steady states when there are finite intraregional and short interregional visiting times.

Theorem 5.6.2. *If H_2 holds, then the steady state of the mobility system satisfies*

$$N_{ii}^{*rr} \to N_i^{rr}\frac{1}{(1+\sum_{k\neq i}^{n_r}\sigma_{ik}^{rr}\tau_{ik}^{rr})}, \quad N_{il}^{*rq} \to 0, \quad \text{and} \quad N_{ij}^{*rr} \to N_i^{rr}\frac{\sigma_{ij}^{rr}}{(\frac{1}{\tau_{ij}^{rr}}+\sum_{k\neq i}^{n_r}\sigma_{ik}^{rr}K_{ikj}^{rrr})} \tag{5.66}$$

for $j \in I_i^r(1, n_r(s_i^r))$, $l \in I_i^r(1, n_q(s_i^r))$.

Proof. Similar to that of Theorem 5.6.2. □

Remark 5.6.2. Theorem 5.6.2 suggests that a fraction of residents tend to be lifelong residents of the home sites, and the remaining fraction of the residents tend to stay for a finite time in sites of their intraregional accessible domain $C_{rr}(s_i^r)$. This special pattern is illustrated in Figure 5.4.

We now characterize the steady states of the system when the inter- and intraregional visiting times are constant and the same.

Theorem 5.6.3. *If H_3 holds and $\tau_{ij}^{rr} = \tau_{il}^{rq} = \tau_i^r$ for some $\tau > 0$, then the steady states of the mobility system satisfy*

$$\begin{aligned} N_{ii}^{*rr} &\to N_i^{rr}\frac{1}{(1+\tau\sigma_i^r+\tau\gamma_i^r\gamma_{i0}^{rq})}, \quad N_{ij}^{*rr} \to N_i^{rr}\frac{\sigma_{ij}^{rr}\tau}{(1+\tau\sigma_i^r+\tau\gamma_i^r\gamma_{i0}^{rq})}, \\ N_{il}^{*rq} &\to N_i^{rr}\frac{\gamma_{il}^{rq}}{(1+\tau\sigma_i^r+\tau\gamma_i^r\gamma_{i0}^{rq})} \end{aligned} \tag{5.67}$$

for $j \in I_i^r(1, n_r)$, $q \in I^r(1, M)$, $l \in I_i^r(1, n_q)$.

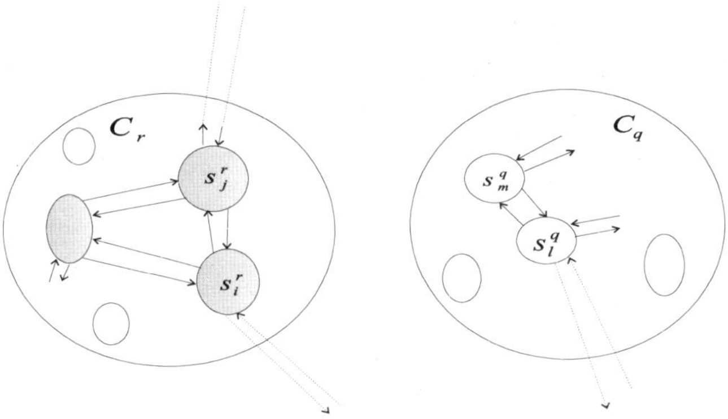

Figure 5.4: Shows that residents of site s_i^r are present at their home site, at sites in their intra- and interregional accessible domain $C(s_i^r)$ whenever the average interregional visiting times are sufficiently small. The arrows represent a transport network between any two sites and regions. Furthermore, as average interregional visiting times tends to zero, (a) the solid lines represent the intraregional mobility of the equilibrium states of the residents, which approaches finite fractions, and (b) the dotted lines indicate interregional mobility of the equilibrium states of residents, which approaches arbitrarily small fractions.

Proof. Let $\tau_{ij}^{rr} = \tau_{il}^{rq} = \tau_i^r$ in (5.55), (5.56), and (5.57) and recall that $\sigma_{ij}^{rr} = \sigma_i^r \nu_{ij}^{rr}$, $\gamma_{il}^{rq} = \gamma_i^r \gamma_{i0}^{rq} \gamma^{rq} \gamma_{0l}^{rq}$, $j \in I_i^r(1, n_r)$, $l \in I_i^r(1, n_q)$, are probabilities with sum equal to one, that is, $\sum_{k \neq i}^{n_r} \nu_{ik}^{rr} = 1$ and $\sum_{q \neq r}^{M} \sum_{l=1}^{n_q} \gamma^{rq} \gamma_{0l}^{rq} = 1$. Thus we have the following reduced formula

$$S_{ii}^{*rr} = \frac{1}{(1 + \tau_i^r \sigma_i^r + \tau_i^r \gamma_i^r \gamma_{i0}^{rq})}, \quad U_{ij}^{*rr} = \frac{\sigma_{ij}^{rr} \tau_i^r}{(1 + \tau_i^r \sigma_i^r + \tau_i^r \gamma_i^r \gamma_{i0}^{rq})},$$
$$V_{il}^{*rq} = \frac{\gamma_{il}^{rq}}{(1 + \tau_i^r \sigma_i^r + \tau_i^r \gamma_i^r \gamma_{i0}^{rq})} \tag{5.68}$$

for $j \in I_i^r(1, n_r)$, $l \in I_i^r(1, n_q)$, $q \in I_i^r(1, M)$. Hence substituting (5.68) into (5.46), (5.47), and (5.48) gives (5.67). □

Remark 5.6.3. Under the assumption of Theorem 5.6.3, almost all the residents of any site s_i^r tend to be permanent residents at their home site s_i^r, and at all sites in their intra- and interregional accessible domain $C(s_i^r)$ for finite time. Also, the fraction of the residents that reside a finite time at a given site in $C(s_i^r)$ is primarily determined by the probabilistic rate at which the residents leave their original home site s_i^r, to visit other sites in this domain $C(s_i^r)$. This special mobility pattern is illustrated in Figure 5.5.

We now consider the case where intra- and interregional visiting times decrease at the same rate.

Theorem 5.6.4. *If H_4 holds, then the steady states of the mobility system satisfy*

$$N_{ii}^{*rr} \to N_i^{rr}, \quad N_{ij}^{*rr} \to 0, \quad \textit{and} \quad N_{il}^{*rq} \to 0 \tag{5.69}$$

for $j \in I_i^r(1, n_r)$, $q \in I^r(1, M)$, $l \in I_i^r(1, n_q)$.

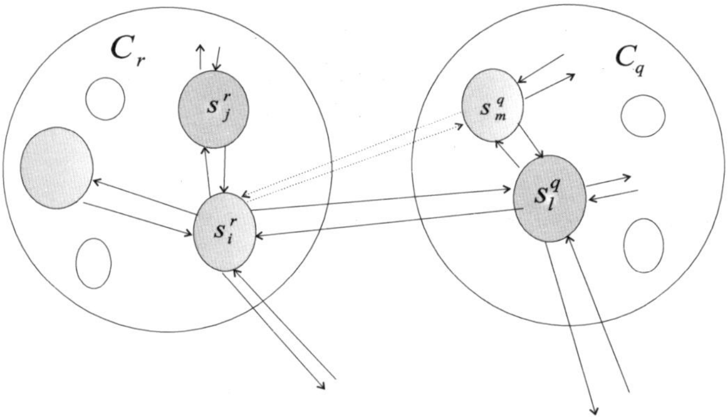

Figure 5.5: Shows that residents of site s_i^r are present at all sites in their inter- and intraregional accessible domain $C(s_i^r)$. The arrows represent a transport network between any two sites and regions. The dotted lines and arrows indicate connection with other sites in other regions.

Proof. Let $\tau_{ij}^{rr} \to 0$ and $\tau_{il}^{rq} \to 0$ in (5.55), (5.56), and (5.57). This leads us to

$$S_{ii}^{*rr} \to 1, \quad U_{ij}^{*rr} \to 0, \quad \text{and} \quad V_{il}^{*rq} \to 0, \quad l \in I_i^r(1, n_q), \, j \in I_i^r(1, n_r). \tag{5.70}$$

Substituting (5.70) into (5.46), (5.47), and (5.48) gives us (5.69). □

Remark 5.6.4. Theorem 5.6.4 suggests that the residents of site s_i^r are isolated from all other sites in their intra- and interregional accessible domain $C(s_i^r)$. That is, all the residents of site s_i^r tend to be lifelong stationary residents at their home site. However, this does not mean that site s_i^r is isolated from visitors from other sites in $C(s_i^r)$. This particular case is illustrated in Figure 5.6.

We now characterize the steady states when the intra- and interregional visiting times grow unboundedly at the same rate.

Theorem 5.6.5. *Let H_5 hold. Let $\tau_{il}^{rq} = \theta(f_0)$ and $\tau_{ij}^{rr} = \theta(g_0)$, where f_0, and g_0 are positive real-valued functions satisfying $f_0, g_0 = \theta(\frac{1}{\varepsilon})$, for $j \in I_i^r(1, n_r)$, $q \in I^r(1, M)$, $l \in I_i^r(1, n_q)$.*

1. *If $\theta(f_0) = \theta(g_0)$, then the steady state of the mobility system satisfies*

$$\begin{aligned} N_{ii}^{*rr} \to 0, \quad N_{ij}^{*rr} &\to N_i^{*rr} \frac{\sigma_{ij}^{rr}}{(\sum_{k\neq i}^{n_r} \sigma_{ik}^{rr} K_{ikj}^{rrr} + \sum_{q\neq r}^{M} \sum_{a=1}^{n_q} \nu_{ia}^{rq} K_{iaj}^{rqr})}, \\ \text{and} \quad N_{il}^{*rq} &\to N_i^{*rr} \frac{\nu_{il}^{rq}}{(\sum_{k\neq i}^{n_r} \sigma_{ik}^{rr} K_{ikl}^{rrq} + \sum_{q\neq r}^{M} \sum_{a=1}^{n_q} \nu_{ia}^{rq} K_{ial}^{rqq})}. \end{aligned} \tag{5.71}$$

2. *If $\theta(f_0) > \theta(g_0)$, then*

$$N_{ii}^{*rr} \to 0, \quad N_{ij}^{*rr} \to 0, \quad N_{il}^{*rq} \to N_i^{*rr} \frac{\nu_{il}^{rq}}{(\sum_{q\neq r}^{M} \sum_{a=1}^{n_q} \nu_{ia}^{rq} K_{ial}^{rqq})}. \tag{5.72}$$

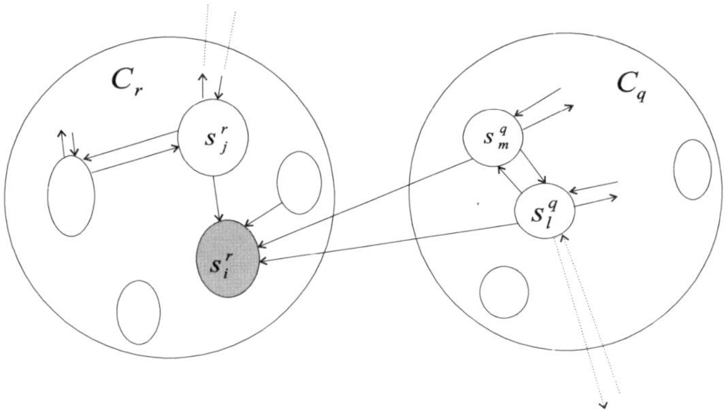

Figure 5.6: Shows that residents of site s_i^r are present only at their home site s_i^r. Hence they isolate every site from their inter- and intraregional accessible domain $C(s_i^r)$. Site s_i^r is a "sink" in the context of the compartmental system. The arrows represent a transport network between any two sites and regions. Furthermore, the dotted lines and arrows indicate connection with other sites in regions.

3. *If* $\theta(f_0) < \theta(g_0)$, *then*

$$N_{ii}^{*rr} \to 0, \quad N_{ij}^{*rr} \to N_i^{*rr} \frac{\sigma_{ij}^{rr}}{(\sum_{k\neq i}^{n_r} \sigma_{ik}^{rr} K_{ikj}^{rrr})}, \quad N_{il}^{*rq} \to 0. \tag{5.73}$$

Proof. For $j \in I_i^r(1, n_r)$, $q \in I^r(1, M)$, $l \in I_i^r(1, n_q)$, we have $\theta(K_{ilj}^{rqr}) = \frac{\theta(f_0)}{\theta(g_0)}$. If $\theta(f_0) = \theta(g_0)$, then $K_{ilj}^{rqr} = 0(1)$ and $K_{ijl}^{rrq} = 0(1)$ as $\tau_{ij}^{rr} \to \infty$ and $\tau_{il}^{rq} \to \infty$. If $\theta(f_0) > \theta(g_0)$, then $K_{iaj}^{rqr} \to \infty$ and $K_{ijl}^{rrq} \to 0$ as $\tau_{ij}^{rr} \to \infty$ and $\tau_{il}^{rq} \to \infty$. Finally, if $\theta(f_0) < \theta(g_0)$, then $K_{ilj}^{rqr} \to 0$ and $K_{ijl}^{rrq} \to \infty$ as $\tau_{ij}^{rr} \to \infty$ and $\tau_{il}^{rq} \to \infty$. By substituting each of these conditions into (5.55), (5.56), and (5.57) and then substituting the results into (5.46), (5.47), and (5.48), we obtain (5.71), (5.72), and (5.73) respectively. □

Remark 5.6.5. Under the assumption of Theorem 5.6.5, the condition $\theta(f_0) = \theta(g_0)$ states that τ_{il}^{rq} and τ_{ij}^{rr} are increasing at the same order. The significance of this relationship is that for sufficiently large values of the intra- and interregional visiting times, all the residents of site s_i^r leave their homes and emigrate to sites in their intra- and interregional accessible domain $C(s_i^r)$. Therefore at the steady state, the original home site of the residents is occupied by visitors from other sites in C_r and C_q. This particular pattern is illustrated in Figure 5.7.

Also, the results under the condition $\theta(f_0) > \theta(g_0)$ signify that for sufficiently large values of intra- and interregional visiting times, the residents of s_i^r emigrate from their home region C_r and become permanent residents at sites in their interregional accessible domain $C_{rq}(s_i^r)$. Hence their original home site s_i^r is occupied by residents from other sites in C_r and C_q. This particular pattern is illustrated in Figure 5.8.

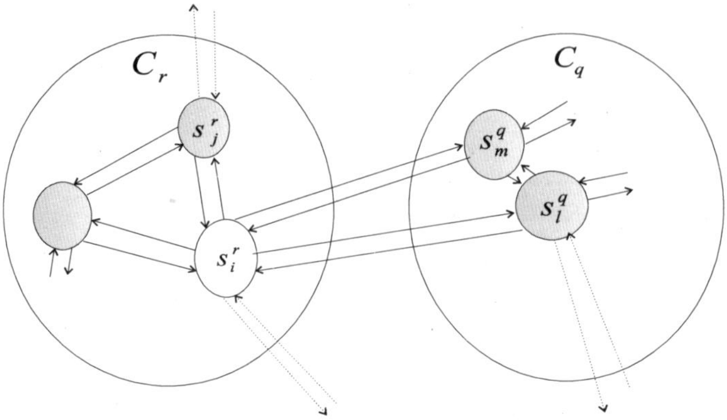

Figure 5.7: Shows that residents of site s_i^r are present only at sites s_j^r in their intraregional accessible domain and at sites s_l^r in their interregional accessible domain $C_{rq}(s_i^r)$. The arrows represent a transport network between any two sites and regions. Moreover, dotted arrows indicate connections with other sites in other regions.

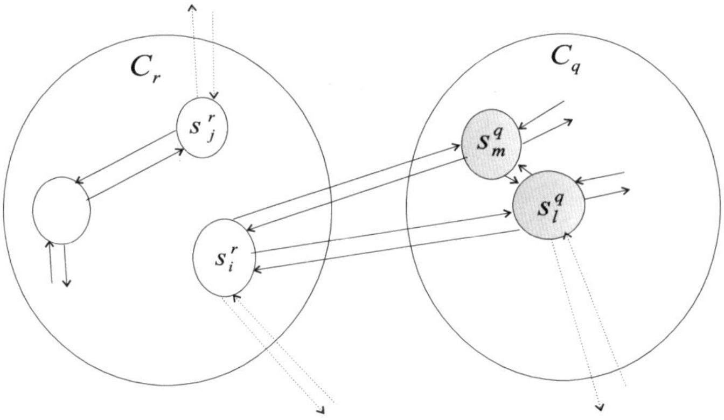

Figure 5.8: Shows that residents of site s_i^r are present only at sites s_l^r in their interregional accessible domain $C_{rq}(s_i^r)$. The arrows represent a transport network between any two sites and regions. In addition, the dotted lines and arrows indicate a connection with other sites in other regions.

The results under the condition $\theta(f_0) < \theta(g_0)$ now signify that for sufficiently large values of intra- and interregional visiting times, the residents of site s_i^r now leave their home site s_i^r and become permanent residents only at sites in their intraregional accessible domain $C_{rr}(s_i^r)$. This particular pattern is illustrated in Figure 5.9.

Assuming that the mobility rate functions change at different rates, we now characterize the steady states of the system. Consider the following assumption.

Assumption 5.6.1. *Given $\epsilon > 0$, for all $j \in I_i^r(1, n_r)$, $l \in I_i^r(1, n_q)$, let τ_{ij}^{rr} and τ_{il}^{rq} be related as follows:*

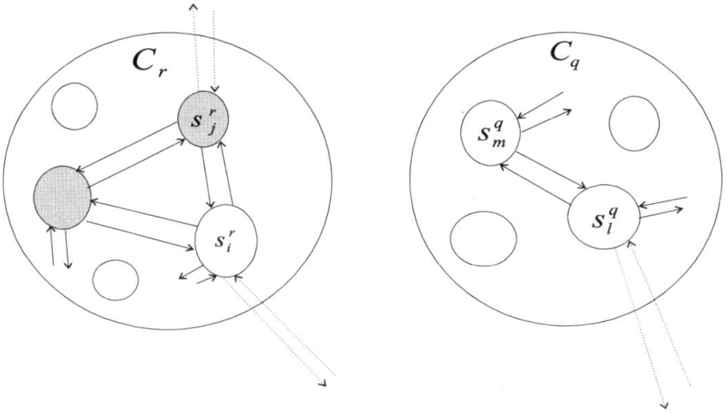

Figure 5.9: Shows that residents of site s_i^r are present only at sites s_j^r in their intraregional accessible domain $C_{rq}(s_i^r)$. The arrows represent a transport network between any two sites and regions. Furthermore, the dotted lines and arrows indicate a connection with other sites in other regions.

$$\begin{aligned}
&\tau_{ij}^{rr} = 0(f(\tau_{i,\min}^{rr})), \quad \tau_{i,\min}^{rq} = 0(h(\tau_{i,\max}^{rr})), \quad \tau_{il}^{rq} = 0(g(\tau_{i,\min}^{rq})) \\
&\quad \textit{whenever } \tau_{i,\max}^{rr} = 0(1) \textit{ and } \tau_{i,\max}^{rq} = 0(1), \textit{ and} \\
&\tau_{ij}^{rr} = \theta(f(\tau_{i,\min}^{rr})), \quad \tau_{i,\min}^{rq} = \theta(h(\tau_{i,\max}^{rr})), \quad \tau_{il}^{rq} = \theta(g(\tau_{i,\min}^{rq})) \\
&\quad \textit{whenever } \tau_{i,\min}^{rr} = \theta\left(\frac{1}{\epsilon}\right) \textit{ and } \tau_{i,\min}^{rq} = \theta\left(\frac{1}{\epsilon}\right),
\end{aligned} \tag{5.74}$$

where h has the explicit form

$$h: \,]0,\infty[\, \mapsto [0,\infty[, \; x \mapsto h(x) = x^c, \quad c > 0, \quad \textit{and} \quad f,g : \,]0,\infty[\, \mapsto [0,\infty[\tag{5.75}$$

are arbitrary positive real-valued functions.

For easy reference and simplicity, we state the following hypotheses.

Hypothesis 5.6.2. τ_{ij}^{rr} and τ_{il}^{rq} satisfy Assumption 5.6.1. Further,
H_6: $0 < \tau_{i,\max}^{rr} < 1$ and $0 < c < 1$.
H_7: $0 < \tau_{i,\max}^{rr} < 1$, and $c \geq 1$.
H_8: $\tau_{i,\max}^{rr} \geq 1$, and $0 < c < 1$.
H_9: $\tau_{i,\max}^{rr} \geq 1$, and $c \geq 1$.

In the following theorems, we describe the steady states of the mobility system under these hypotheses.

Theorem 5.6.6. *Suppose H_6 holds and for any given $\epsilon > 0$,*
1. *If $\tau_{i,\max}^{rr} \to 0^+$ and $c \to 0^+$, then for all $j \in I_i^r(1, n_r)$, $l \in I_i^r(1, n_q)$, we have $\tau_{ij}^{rr} = 0(\epsilon)$, $\tau_{il}^{rq} = 0(1)$, and $\tau_{ij}^{rr} \leq \tau_{il}^{rq}$.*

Moreover, the steady state of the mobility system satisfy

$$N_{ii}^{*rr} \to N_i^{*rr} \frac{1}{(1+\sum_{q\neq r}^{M}\sum_{a=1}^{n_q} y_{ia}^{rq}\tau_{ia}^{rq})}, \quad N_{ij}^{*rr} \to 0, \quad \text{and}$$
$$N_{il}^{*rq} \to N_i^{*rr} \frac{y_{il}^{rq}}{(\frac{1}{\tau_{il}^{rq}} + \sum_{q\neq r}^{M}\sum_{a=1}^{n_q} y_{ia}^{rq} K_{ial}^{rqq})}. \tag{5.76}$$

2. *If $\tau_{i,\max}^{rr} \to 0^+$ and $c \to 1^-$, then $\tau_{ij}^{rr} = 0(\epsilon)$ for all $j \in I_i^r(1, n_r)$, and there exist $j_0 \in I_i^r(1, n_r)$ and $l_0 \in I_i^r(1, n_q)$ such that $\tau_{il_0}^{rq} \to 0^+$ slower than $\tau_{ij_0}^{rr} \to 0^+$. Moreover,*
   i. *The steady state of the mobility system satisfies*

$$N_{ii}^{*rr} \to N_i^{*rr} \frac{1}{(1+\sum_{q\neq r}^{M}\sum_{a=1}^{n_q} y_{ia}^{rq}\tau_{ia}^{rq})}, \quad N_{ij}^{*rr} \to 0, \quad \text{and}$$
$$N_{il_0}^{*rq} \to 0, \quad N_{il}^{*rq} \to N_i^{*rr} \frac{y_{il}^{rq}}{(\frac{1}{\tau_{il}^{rq}} + \sum_{q\neq r}^{M}\sum_{a=1}^{n_q} y_{ia}^{rq} K_{ial}^{rqq})}, \quad l \neq l_0, \tag{5.77}$$

   whenever $\tau_{il}^{rq} = 0(1)$.

   ii. *The steady state is given by Theorem 5.6.4 whenever $\tau_{il}^{rq} = 0(\epsilon)$ for all $l \in I_i^r(1, n_q)$.*
3. *If $\tau_{i,\max}^{rr} \to 1^-$ and $c \to 0^+$, then for all $j \in I_i^r(1, n_r)$, $l \in I_i^r(1, n_q)$, we have $\tau_{il}^{rq} = 0(1)$. Also, there exists $j_1 \in I_i^r(1, n_r)$ such that $\tau_{ij_1}^{rr} = 0(1)$. Furthermore,*
   i. *If $\tau_{ij}^{rr} = 0(1)$ for all $j \in I_i^r(1, n_r)$ then the steady state of the mobility system is given by (5.46), (5.47), and (5.48).*
   ii. *If $\tau_{ij}^{rr} = 0(\epsilon)$, for all $j, j_1 \in I_i^r(1, n_r)$, $j \neq j_1$, then the steady state of the mobility system satisfies*

$$N_{ii}^{*rr} \to N_i^{rr} \frac{1}{(1+\sigma_{ij_1}^{rr}\tau_{ij_1}^{rr} + \sum_{q\neq r}^{M}\sum_{a=1}^{n_q} y_{ia}^{rq}\tau_{ia}^{rq})}, \quad N_{ij}^{*rr} \to 0, \quad j \neq j_1,$$
$$N_{ij_1}^{*rr} \to N_i^{*rr} \frac{\sigma_{ij_1}^{rr}}{(\frac{1}{\tau_{ij_1}^{rr}} + \sigma_{ij_1}^{rr} + \sum_{q\neq r}^{M}\sum_{a=1}^{n_q} y_{ia}^{rq} K_{iaj_1}^{rqr})}, \quad \text{and} \tag{5.78}$$
$$N_{il}^{*rq} \to N_i^{*rr} \frac{y_{il}^{rq}}{(\frac{1}{\tau_{il}^{rq}} + \sigma_{ij_1}^{rr} K_{ij_1 l}^{rrq} + \sum_{q\neq r}^{M}\sum_{a=1}^{n_q} y_{ia}^{rq} K_{ial}^{rqq})}.$$

4. *If $\tau_{i,\max}^{rr} \to 1^-$ and $c \to 1^-$, then for all $j \in I_i^r(1, n_r)$, $l \in I_i^r(1, n_q)$, we have $\tau_{il}^{rq} = 0(1)$. Also, there exists $j_2 \in I_i^r(1, n_r)$ such that $\tau_{ij_2}^{rr} = 0(1)$. In addition,*
   i. *If $\tau_{ij}^{rr} = 0(1)$ for all $j \in I_i^r(1, n_r)$ then the steady states of the mobility system is given by (5.46), (5.47), and (5.48).*
   ii. *If $\tau_{ij}^{rr} = 0(\epsilon)$ for all $j, j_2 \in I_i^r(1, n_r)$, $j \neq j_2$, then the steady state of the mobility system is given by (5.78).*

Proof. Under H_6, we have

$$\tau^{rq}_{i,\min} = 0((\tau^{rr}_{i,\max})^{c}), \quad r \neq q. \tag{5.79}$$

1. If $\tau^{rr}_{i,\max} \to 0^+$ and $c \to 0^+$, then it follows that $(\tau^{rr}_{i,\max})^c \to 1^-$. Thus from (5.74), (5.79), and $0 \le \tau^{rr}_{ij} \le \tau^{rr}_{i,\max} \to 0^+$ for all $j \in I^r_i(1, n_r)$, $l \in I^r_i(1, n_q)$ we obtain
$$\tau^{rq}_{i,\min} = 0(1), \quad \tau^{rq}_{il} = 0(g(\tau^{rq}_{i,\min})) = 0(1), \quad \text{and} \quad \tau^{rr}_{ij} = 0(\epsilon). \tag{5.80}$$
Also, $\tau^{rq}_{i,\min} = 0(1) \Rightarrow \tau^{rq}_{il} > 0$ for all $l \in I^r_i(1, n_q)$. Hence $\tau^{rq}_{il} \ge \tau^{rr}_{ij}$.
Finally, from (5.80), (5.55), and (5.56) and from (5.57), (5.46), (5.47), and (5.48) we get (5.76).
2. If $\tau^{rr}_{i,\max} \to 0^+$ and $c \to 1^-$, then from (5.79), $(\tau^{rr}_{i,\max})^c \to 0^+$ at a slower rate. Hence by taking $\tau^{rq}_{i,\min} = \tau^{rq}_{il_0}$ and $\tau^{rr}_{i,\max} = \tau^{rr}_{ij_0}$ the existence of l_0 and j_0 is verified. Now from $\tau^{rr}_{ij} = 0(\epsilon)$ and $\tau^{rq}_{il} = 0(1)$, $l \neq l_0$, the proof of the steady state follows by Theorem 5.6.6(1). From $\tau^{rr}_{ij} = 0(\epsilon)$ and $\tau^{rq}_{il} = 0(\epsilon)$ for all $l \in I^r_i(1, n_q)$ the proof of the steady state follows by Theorem 5.6.4.
3. If $\tau^{rr}_{i,\max} \to 1^-$ and $c \to 1^+$, then it suffices to note that $\tau^{rr}_{ij_1} = \tau^{rr}_{i,\max}$ for some $j_1 \in I^r_i(1, n_r)$, which proves the existence of j_1. The rest of the proof follows from parts 1 and 2.
4. The proof follows from parts 1, 2, and 3. □

Remark 5.6.6. The interpretation of Theorem 5.6.6(1) is similar to that of Theorem 5.6.1. Theorem 5.6.6(2) signifies that whenever all the residents of a given site s^r_i that travel to sites within their intraregional accessible domain $C_{rr}(s^r_i)$ spend infinitesimally small amount of time visiting the sites, and there is also one site $s^q_{l_0}$ in the interregional accessible domain $C_{rq}(s^r_i)$ where all the residents of site s^r_i that travel to this site also spend infinitesimally small amount of time visiting that site, that is, a fraction of the residents of s^r_i would remain permanent residents of their home site s^r_i, and a fraction would relocate to all sites $s^q_l \neq s^q_{l_0}$ in their interregional accessible domain $C_{rq}(s^r_i)$ and spend a finite amount of time visiting these sites.

Theorem 5.6.6(3) also states that all the residents of site s^r_i that travel to sites in their interregional accessible domain $C_{rq}(s^r_i)$, spend a finite amount of time visiting the sites, and there is a site $s^r_{j_1}$ in $C_{rr}(s^r_i)$ where the residents of site s^r_i can spend a finite amount of time visiting. From the assumption in Theorem 5.6.6(3)(ii) the residents of site s^r_i that travel in the intraregional accessible domain $C_{rr}(s^r_i)$ spend infinitesimally small amount of time visiting. This implies that the distribution of the residents of site s^r_i is such that (i) a fraction would remain permanent residents at their home site s^r_i, (ii) a fraction would migrate to site $s^r_{j_1}$ in $C_{rr}(s^r_i)$, and (iii) the remaining fraction would migrate to all sites in $C_{rq}(s^r_i)$ and become residents of those sites for a finite amount of time. Under Theorem 5.6.6(3), (i) a fraction the residents of site s^r_i would remain permanent residents at home site s^r_i, and (ii) the remaining fraction would distribute among the visiting sites for a finite visiting time. Theorem 5.6.6(4) has similar interpretations as Theorem 5.6.6(2).

The conditions on c in Theorem 5.6.6(1) and Theorem 5.6.6(2) indicate the existence of a critical value c_0 for $c \in]0,1[$. For $0 < c < c_0$, the steady state is given by (5.76), and for $c_0 < c < 1$, the steady state is given by (5.77). Similar conclusion with regards to Theorem 5.6.6(3), (4) can be drawn about a critical value c_1 for $c \in]0,1[$. Moreover, in the case of Theorem 5.6.6(3), we have $0 < c < c_1$, and the steady state is given by (5.78). In the case of Theorem 5.6.6(4), we have $c_1 < c < 1$, and the steady state is given by Theorem 5.6.6(4).

The conditions on $\tau^{rr}_{i,\max}$ in Theorem 5.6.6 also indicate the existence of a critical value for $\tau^{rr}_{i,\max} \in]0,1[$, denoted by $\tau^{rr}_{i,\max_0}$, such that for $0 < \tau^{rr}_{i,\max} < \tau^{rr}_{i,\max_0}$, whenever $0 < c < c_0$, the steady state is given by (5.76), and whenever $c_0 < c < 1$, the steady state is given by (5.77). Also, for $\tau^{rr}_{i,\max_0} < \tau^{rr}_{i,\max} < 1$, when $0 < c < c_1$, the steady state is given by (5.78), and when $c_1 < c < 1$, the steady state is given by Theorem 5.6.6(4).

Remark 5.6.7. The condition on the intraregional visiting times in (5.74) signifies that these times are of the same order. A similar conclusion can be drawn with respect to the interregional visiting times. We further note that the functions f, g, and h can take arbitrary forms such as quadratic, cubic, exponential, or logarithmic functions depending on the kind of the modeled mobility process. Under this consideration, we incorporate more detail about the mobility process. For instance, if one site $s^r_{j_0}$ in the intraregional accessible domain receives more visitors than all other sites in the domain, then we could have a hub in the domain. A possible representation of this detail about the mobility process can be $\tau^{rr}_{ij} = 0(g(\tau^{rr}_{i,\min})), j \in I^r_i(1,n_r), j \neq i, j \neq j_0, \tau^{rr}_{ij_0} = 0(g_1(\tau^{rr}_{i,\min})), j_0 \in I^r_i(1,n_r)$, with $g \neq g_1$, where g_1 is an arbitrary positive real-valued function describing the return rate of visitors to site $s^r_{j_0}$.

We now describe the steady states of the system under H_7. Observe that H_7 is composed of two conditions ($\tau^{rr}_{i,\max} \to 0^+$ and $c \geq 1$) and ($\tau^{rr}_{i,\max} \to 1^-$ and $c \geq 1$). Under $\tau^{rr}_{i,\max} \to 0^+$ and $c \geq 1$, the construction of the steady state is similar to that in Theorem 5.6.6(2). Similarly, the steady states are similar to those in Theorem 5.6.6(4) whenever $\tau^{rr}_{i,\max} \to 1^-$ and $c \geq 1$. We further remark that $\tau^{rr}_{i,\max}$ has a critical value $\tau^{rr}_{i,\max_1} \in]0,1[$ such that for $c \geq 1$, the steady states are given by Theorem 5.6.6(2) whenever $0 < \tau^{rr}_{i,\max} < \tau^{rr}_{i,\max_1}$, and the steady states are given by Theorem 5.6.6(4) whenever $\tau^{rr}_{i,\max_1} < \tau^{rr}_{i,\max} < 1$.

Finally, we describe the steady states of the system under H_8 and H_9.

Theorem 5.6.7. *Suppose H_8 and H_9 hold. Given r any $\epsilon > 0$,*

1. *If $\tau^{rr}_{i,\max} = 0(1)$ and $0 < c < 1$ for all $j \in I^r_i(1,n_r)$, and $l \in I^r_i(1,n_q)$, then $\tau^{rq}_{il} = 0(1)$, there exist $j_3 \in I^r_i(1,n_q)$ and $l_3 \in I^r_i(1,n_r)$ such that $\tau^{rr}_{ij_3} = 0(1)$ and $\tau^{rq}_{il_3} \leq \tau^{rr}_{ij_3}$, and, furthermore,*
   i. *If $\tau^{rr}_{ij} = 0(1)$ for all $j \in I^r_i(1,n_r)$, then the steady state of the mobility system satisfies (5.46), (5.47) and (5.48).*
   ii. *If $\tau^{rr}_{ij} = 0(\epsilon)$ for all $j \neq j_3, j, j_3 \in I^r_i(1,n_r)$, then the steady state of the mobility system satisfies*

$$
\begin{aligned}
N_{ii}^{*rr} &\to N_i^{rr}\frac{1}{(1+\sigma_{ij_3}^{rr}\tau_{ij_3}^{rr}+\sum_{q\neq r}^{M}\sum_{a=1}^{n_q}\gamma_{ia}^{rq}\tau_{ia}^{rq})}, \qquad N_{ij}^{*rr}\to 0,\ j\neq j_3,\\
N_{ij_3}^{*rr} &\to N_i^{rr}\frac{\sigma_{ij_3}^{rr}}{(\frac{1}{\tau_{ij_3}^{rr}}+\sigma_{ij_3}^{rr}+\sum_{q\neq r}^{M}\sum_{a=1}^{n_q}\gamma_{ia}^{rq}K_{iaj_3}^{rqr})},\\
N_{il}^{*rq} &\to N_i^{rr}\frac{\gamma_{il}^{rq}}{(\frac{1}{\tau_{il}^{rq}}+\sigma_{ij_3}^{rr}K_{ij_3l}^{rrq}+\sum_{q\neq r}^{M}\sum_{a=1}^{n_q}\gamma_{ia}^{rq}K_{ial}^{rqq})}.
\end{aligned}
\tag{5.81}
$$

2. *If $\tau_{i,\max}^{rr}=0(1)$ and $c\geq 1$, then (a) for all $l\in I_i^r(1,n_r)$, $\tau_{il}^{rq}=0(1)$, (b) there exists $j_4\in I_i^r(1,n_r)$ such that $\tau_{ij_4}^{rr}=0(1)$ and $\tau_{ij}^{rr}\leq\tau_{il}^{rq}$ for all $j\in I_i^r(1,n_r)$, $l\in I_i^r(1,n_q)$, and, furthermore,*
   i. *If $\tau_{ij}^{rr}=0(1)$ for all $j\in I_i^r(1,n_r)$ then the steady state of the mobility system satisfies (5.46), (5.47), and (5.48).*
   ii. *If $\tau_{ij}^{rr}=0(\epsilon)$ for all $j,j_4\in I_i^r(1,n_r)$, $j\neq j_4$, then the steady state of the mobility system satisfies (5.81) (where we replace j_3 with j_4).*
3. *If $\tau_{i,\max}^{rr}=\theta(\frac{1}{\epsilon})$ and $0<c<1$, then $\tau_{il}^{rq}=\theta(\frac{1}{\epsilon})$ for all $j\in I_i^r(1,n_r)$, $l\in I_i^r(1,n_q)$, there exists $j_5\in I_i^r(1,n_r)$ such that $\tau_{ij_5}^{rr}=\theta(\frac{1}{\epsilon})$, and, furthermore,*
   i. *For $\tau_{ij}^{rr}=0(1)$ or $\tau_{ij}^{rr}=0(\epsilon)$ for all $j\neq j_5$, if $\theta(f)>\theta(g)$, then*

$$
N_{ii}^{*rr}\to 0,\quad N_{ij_5}^{*rr}\to N_i^{*rr},\quad N_{ij}^{*rr}\to 0,\ j\neq j_5,\quad N_{il}^{*rq}\to 0,
\tag{5.82}
$$

   and if $\theta(f)<\theta(g)$, then

$$
N_{ii}^{*rr}\to 0,\quad N_{ij}^{*rr}\to 0,\quad N_{il}^{*rq}\to N_i^{rr}\frac{\gamma_{il}^{rq}}{(\sum_{q\neq r}^{M}\sum_{a=1}^{n_q}\gamma_{ia}^{rq}K_{ial}^{rqq})}.
\tag{5.83}
$$

   Finally, if $\theta(f)=\theta(g)$, then

$$
\begin{aligned}
&N_{ii}^{*rr}\to 0,\quad N_{ij_5}^{*rr}\to N_i^{rr}\frac{\sigma_{ij_5}^{rr}}{(\sigma_{ij_5}^{rr}+\sum_{q\neq r}^{M}\sum_{a=1}^{n_q}\gamma_{ia}^{rq}K_{iaj_5}^{rqr})},\\
&N_{ij}^{*rr}\to 0,\ j\neq j_5,\quad N_{il}^{*rq}\to N_i^{rr}\frac{\gamma_{il}^{rq}}{(\sigma_{ij_5}^{rr}K_{ij_5l}^{rrq}+\sum_{q\neq r}^{M}\sum_{a=1}^{n_q}\gamma_{ia}^{rq}K_{ial}^{rqq})}.
\end{aligned}
\tag{5.84}
$$

   ii. *For $\tau_{ij}^{rr}=0(\frac{1}{\epsilon})$ for all $j\in I_i^r(1,n_r)$, the steady states are similar to those in Theorem 5.6.5.*
4. *If $\tau_{i,\max}^{rr}=\theta(\frac{1}{\epsilon})$ and $c\geq 1$, then $\tau_{il}^{rq}=\theta(\frac{1}{\epsilon})$ and $\tau_{ij}^{rr}\leq\tau_{il}^{rq}$ for all $j\in I_i^r(1,n_r)$, $l\in I_i^r(1,n_q)$. Also, there exists $j_6\in I_i^r(1,n_r)$ such that $\tau_{ij_6}^{rr}=\theta(\frac{1}{\epsilon})$. Furthermore, for $\tau_{ij}^{rr}=0(1)$ or $\tau_{ij}^{rr}=0(\epsilon)$, for all $j\neq j_6$, if $\theta(f)>\theta(g)$, then the steady state of the mobility system satisfies (5.82); if $\theta(f)<\theta(g)$, then the steady state satisfies (5.83); and if $\theta(f)=\theta(g)$, then the steady state satisfies (5.84).*

Proof. Parts 1 and 2 follow from Theorem 5.6.6. Parts 3 and 4 follow from Theorem 5.6.5. □

Remark 5.6.8. A remark similar to Remark 5.6.5 and Remark 5.6.6 can be formulated with regards to Theorem 5.6.7.

We also examine the situation where the growth rate of the minimum intraregional visiting time is compared with a power function of the maximum interregional visiting time. We will consider the cases where the power function is a fractional power and a polynomial function.

Assumption 5.6.2. *Given $\epsilon > 0$, let τ_{ij}^{rr} and τ_{il}^{rq} for all $j \in I_i^r(1,n_r)$, $l \in I_i^r(1,n_q)$ be related as follows:*

$$\begin{aligned}
&\tau_{ij}^{rr} = 0(g(\tau_{i,\min}^{rr})), \quad \tau_{i,\min}^{rr} = 0(h(\tau_{i,\max}^{r,\max})), \quad \tau_{il}^{rq} = 0(f(\tau_{i,\min}^{rq})),\\
&\quad \text{whenever } \tau_{i,\max}^{r,\max} = 0(1) \text{ and } \tau_{i,\max}^{rr} = 0(1), \text{ and}\\
&\tau_{ij}^{rr} = \theta(g(\tau_{i,\min}^{rr})), \quad \tau_{i,\min}^{rr} = \theta(h(\tau_{i,\max}^{r,\max})), \quad \tau_{il}^{rq} = \theta(f(\tau_{i,\min}^{rq}))\\
&\quad \text{whenever } \tau_{i,\min}^{rr} = \theta\left(\frac{1}{\epsilon}\right) \text{ and } \tau_{i,\min}^{rq} = \theta\left(\frac{1}{\epsilon}\right),
\end{aligned} \tag{5.85}$$

where f, g, and h are defined in Assumption 5.6.1.

We state the following hypotheses.

Hypothesis 5.6.3. The resident times τ_{ij}^{rr} and τ_{il}^{rq} satisfy Assumption 5.6.2, and, moreover,
H_{10}: $0 < \tau_{i,\max}^{r,\max} < 1$, and $0 < c < 1$.
H_{11}: $0 < \tau_{i,\max}^{r,\max} < 1$, and $c \geq 1$.
H_{12}: $\tau_{i,\max}^{r,\max} \geq 1$, and $0 < c < 1$.
H_{13}: $\tau_{i,\max}^{r,\max} \geq 1$, and $c \geq 1$.

We will characterize the steady states of the mobility process under H_{13}. The static behavior of the system under the other hypotheses can be derived in a similar manner.

Theorem 5.6.8. *Suppose H_{13} holds. Given $\epsilon > 0$,*
1. *If $\tau_{i,\max}^{r,\max} = 0(1)$ and $c \geq 1$, then $\tau_{ij}^{rr} = 0(1)$ and $\tau_{il}^{rq} \leq \tau_{ij}^{rr}$ for all $q \in I^r(1,M)$, $j \in I_i^r(1,n_r)$, and $l \in I_i^r(1,n_q)$. Furthermore,*
   i. *If $\tau_{il}^{rq} = 0(\epsilon)$ for all $q \in I^r(1,M)$ and $l \in I_i^r(1,n_r)$, then the steady states of the mobility system satisfy*

$$N_{ii}^{*rr} \to N_i^{*rr} \frac{1}{(1+\sum_{j\neq i}^{n_r} \sigma_{ij}^{rr}\tau_{ij}^{rr})}, \quad N_{ij}^{*rr} \to N_i^{*rr} \frac{\sigma_{ij}^{rr}}{(\frac{1}{\tau_{ij}^{rr}} + \sum_{k\neq i}^{n_r} \sigma_{ik}^{rr} K_{ikj}^{rrr})}, \quad j \in I_i^r(1,n_r), \tag{5.86}$$
$$N_{il}^{*rq} \to 0.$$

   ii. *If $\tau_{il}^{rq_1} = 0(\epsilon)$ for some $q_1 \in I^r(1,M)$ and $l \in I_i^r(1,n_{q_1})$ and $\tau_{il}^{rq} = 0(1)$ for all $q \in I^r(1,M)$ and $l \in I_i^r(1,n_q)$, $q \neq q_1$, then the steady state of the mobility system satisfies*

$$
\begin{aligned}
N_{ii}^{*rr} &\to N_i^{*rr}\frac{1}{(1+\sum_{j\neq i}^{n_r}\sigma_{ij}^{rr}\tau_{ij}^{rr}+\sum_{q\neq r,q_1}^{M}\sum_{l=1}^{n_q}\gamma_{il}^{rq}\tau_{il}^{rq})},\\
N_{ij}^{*rr} &\to N_i^{*rr}\frac{\sigma_{ij}^{rr}}{(\frac{1}{\tau_{ij}^{rr}}+\sum_{k\neq i}^{n_r}\sigma_{ik}^{rr}K_{ikj}^{rrr}+\sum_{q\neq r,q_1}^{M}\sum_{a=1}^{n_q}\gamma_{ia}^{rq}K_{ilj}^{rqr})},\\
N_{il}^{*rq_1} &\to 0,\quad l\in I_i^r(1,n_{q_1}),\quad \textit{and for } q\neq q_1,\ l\in I_i^r(1,n_q),\\
N_{il}^{*rq} &\to N_i^{*rr}\frac{\gamma_{il}^{rq}}{(\frac{1}{\tau_{il}^{rq}}+\sum_{k\neq i}^{n_r}\sigma_{ik}^{rr}K_{ikj}^{rrr}+\sum_{q\neq r,q_1}^{M}\sum_{a=1}^{n_q}\gamma_{ia}^{rq}K_{ilj}^{rqr})}.
\end{aligned}
\tag{5.87}
$$

iii. *If* $\tau_{il}^{rq}=0(1)$ *for all* $q\in I^r(1,M)$ *and* $l\in I_i^r(1,n_r)$*, then the steady state of the mobility system satisfies (5.46), (5.47), and (5.48).*

2. *If* $\tau_{i,\max}^{r,\max}=\theta(\frac{1}{\epsilon})$ *and* $c\geq 1$*, then (a)* $\tau_{ij}^{rr}=\theta(\frac{1}{\epsilon})$ *for all* $j\in I_i^r(1,n_r)$ *and (b) for all* $l\in I_i^r(1,n_q)$*,*

   i. *If* $\tau_{il}^{rq}=0(\epsilon)$ *or* $\tau_{il}^{rq}=0(1)$ *for all* $q\in I(1,M)$ *and* $l\in I_i^r(1,n_q)$*, then for* $j\in I_i^r(1,n_r)$*, the steady state of the mobility system satisfies*

$$
N_{ii}^{*rr}\to 0,\quad N_{ij}^{*rr}\to N_i^{rr}\frac{\sigma_{ij}^{rr}}{(\sum_{k\neq i}^{n_r}\sigma_{ik}^{rr}K_{ikj}^{rrr})},\quad N_{il}^{*rq}\to 0.
\tag{5.88}
$$

   ii. *If* $\tau_{il}^{rq_2}=0(\epsilon)$ *for some* $q_2\in I^r(1,M)$ *and* $l\in I_i^r(1,n_{q_2})$ *and* $\tau_{il}^{rq}=0(1)$ *for all* $q\in I^r(1,M)$*,* $q\neq q_2$*, and* $l\in I_i^r(1,n_q)$*, then the steady state of the mobility system satisfies*

$$
\begin{aligned}
&N_{ii}^{*rr}\to 0,\quad N_{ij}^{*rr}\to N_i^{rr}\frac{\sigma_{ij}^{rr}}{(\sum_{k\neq i}^{n_r}\sigma_{ik}^{rr}K_{ikj}^{rrr})},\\
&N_{il}^{*rq}\to 0\quad \textit{for all } q\in I^r(1,M),\ l\in I_i^r(1,n_q(s_i^r)).
\end{aligned}
\tag{5.89}
$$

   iii. *If* $\tau_{il}^{rq}=\theta(\frac{1}{\epsilon})$ *for all* $q\in I^r(1,M)$ *and* $l\in I_i^r(1,n_q)$*, then for* $j\in I_i^r(1,n_r)$*, if* $\theta(g)<\theta(f)$*, then the steady state of the mobility system satisfies*

$$
N_{ii}^{*rr}\to 0,\quad N_{ij}^{*rr}\to N_i^{rr}\frac{\sigma_{ij}^{rr}}{(\sum_{k\neq i}^{n_r}\sigma_{ik}^{rr}K_{ikj}^{rrr})},\quad N_{il}^{*rq}\to 0;
\tag{5.90}
$$

   if $\theta(g)>\theta(f)$*, then the steady state of the mobility system satisfies*

$$
\begin{aligned}
&N_{ii}^{*rr}\to 0,\quad N_{ij}^{*rr}\to 0,\\
&N_{il}^{*rq}\to N_i^{rr}\frac{\gamma_{il}^{rq}}{(\sum_{q\neq r}^{M}\sum_{a=1}^{n_q}\gamma_{ia}^{rq}K_{ilj}^{rqr})};
\end{aligned}
\tag{5.91}
$$

   and if $\theta(g)=\theta(f)$*, then the steady state of the mobility system satisfies*

$$\begin{aligned}
&N_{ii}^{*rr} \to 0, \quad N_{ij}^{*rr} \to N_i^{rr} \frac{\sigma_{ij}^{rr}}{(\sum_{k \neq i}^{n_r} \sigma_{ik}^{rr} K_{ikj}^{rrr} + \sum_{q \neq r}^{M} \sum_{a=1}^{n_q} \gamma_{ia}^{rq} K_{ilj}^{rqr})}, \\
&N_{il}^{*rq} \to N_i^{rr} \frac{\gamma_{il}^{rq}}{(\sum_{k \neq i}^{n_r} \sigma_{ik}^{rr} K_{ikj}^{rrr} + \sum_{q \neq r}^{M} \sum_{a=1}^{n_q} \gamma_{ia}^{rq} K_{ilj}^{rqr})}.
\end{aligned} \tag{5.92}$$

iv. *Assume that there exists* $q_3 \in I^r(1, M)$ *such that for all* $q \neq q_3$, $\tau_{il}^{rq} = \theta(\frac{1}{\epsilon})$ *and* $\tau_{il}^{rq_3} = 0(\epsilon)$ *or* $\tau_{il}^{rq_3} = 0(1)$, $l \in I_i^r(1, n_{q_3})$*).*
(a) If $\theta(g) < \theta(f)$*, then the steady state of the mobility system satisfies (5.90). (b) If* $\theta(g) > \theta(f)$*, then the steady state of the mobility system satisfies*

$$\begin{aligned}
&N_{ii}^{*rr} \to 0, \quad N_{ij}^{*rr} \to 0, \\
&N_{il}^{*rq_3} \to 0, \quad N_{il}^{*rq} \to N_i^{rr} \frac{\gamma_{il}^{rq}}{(\sum_{q \neq r}^{M} \sum_{a=1}^{n_q} \gamma_{ia}^{rq} K_{ilj}^{rqr})}, \quad q \neq q_3;
\end{aligned} \tag{5.93}$$

and (c) if $\theta(g) = \theta(f)$*, then the steady state of the mobility system satisfies*

$$\begin{aligned}
&N_{ii}^{*rr} \to 0, \quad N_{ij}^{*rr} \to N_i^{rr} \frac{\sigma_{ij}^{rr}}{(\sum_{k \neq i}^{n_r} \sigma_{ik}^{rr} K_{ikj}^{rrr} + \sum_{q \neq r}^{M} \sum_{a=1}^{n_q} \gamma_{ia}^{rq} K_{ilj}^{rqr})}, \\
&N_{il}^{*rq_3} \to 0, \quad N_{il}^{*rq} \to N_i^{rr} \frac{\gamma_{il}^{rq}}{(\sum_{k \neq i}^{n_r} \sigma_{ik}^{rr} K_{ikj}^{rrr} + \sum_{q \neq r}^{M} \sum_{a=1}^{n_q} \gamma_{ia}^{rq} K_{ilj}^{rqr})}
\end{aligned} \tag{5.94}$$

for $q \neq q_3$.

v. *Finally, assume that there exist* $q_4, q_5 \in I(1, M)$ *such that, for all* $q \neq q_4, q_5$, $\tau_{il}^{rq} = \theta(\frac{1}{\epsilon})$ *for* $l \in I_i^r(1, n_q)$, $\tau_{il}^{rq_4} = 0(\epsilon)$ *for* $l \in I_i^r(1, n_{q_4})$, *and* $\tau_{il}^{rq_5} = 0(1)$ *for* $l \in I_i^r(1, n_{q_5})$. *Then (a) if* $\theta(g) < \theta(f)$*, then the steady states of the mobility system are given by (5.91), (b) if* $\theta(g) > \theta(f)$*, then the steady states of the mobility system are given by (5.93), and (c) if* $\theta(g) = \theta(f)$*, then the steady states of the mobility system are given by (5.94). Here* N^{*rq_4} *and* N^{*rq_5} *take the same value as* N^{*rq_3} *in each case.*

Proof. The proof follows immediately from the definition of τ_{ij}^{rr} and τ_{il}^{rq} in Assumption 5.6.2 and from the proofs of Theorems 5.6.5 and 5.6.7. □

Remark 5.6.9. The interpretations of the results of Theorem 5.6.8 are those of Theorem 5.6.7(3).

5.7 Simulation results for a prototype two-scale network human mobility dynamic process

In this section the behavior of the steady-state population in the two-scale network human dynamic population model for some special real-life human mobility scenarios are characterized. Employing (i) a three-community single-scale model, (ii) the mobility dynamic structure determined by the simulated data set for the three communities in the district of Central Manitoba, Canada, and (iii) the data set for the interdistrict movement of the people in the West Indian Island of Dominica [31], we derive a two-scale human mobility model. Note that the estimates of the underlying parameters under both simulated and real data sets are recorded in [31].

The development of this example is based on the following assumptions:

(a) In the absence of data and without loss of generality, we assume that the human mobility pattern represented by the simulated data of the three communities in the district of Central Manitoba, Canada, is the structure of a two-scale human mobility model at the interregional level.

(b) The three districts in the West Indian Island of Dominica represent "sites" for each of the three regions. Conceptually, this assumption is not representative in the sense of geography/size but in the sense of sample drawn from small community or vice versa, that is, representative of the big or small region. In short, we identify the three communities in the district of Central Manitoba, Canada as regions, and the districts of the West Indian Island of Dominica as sites. Therefore every region has three sites, and we denote the regions by C_r, $r = 1, 2, 3$, and the sites by s_i^r, $i = 1, 2, 3$.

(c) The intraregional visiting/travel rates are displayed in Table 5.1. This table was rerecorded from the data set under the column of "Dominica mobility" in [31, p. 14].

Table 5.1: Intraregional visiting/travel rates for sites in the regions C_r, $r = 1, 2, 3$. The intraregional travel rates within each of the three regions C_r, $r = 1, 2, 3$, are assumed to be the same. The estimates of the parameters are derived from data [31] collected from the Island of Dominica in 1991. The parameter estimates reflect the rates of travel that can be obtained in the regions that have a low technological development.

Parameter	Intraregional mobility
σ_1^{rr}	0.00147
σ_2^{rr}	0.03695
σ_3^{rr}	0.03754
ν_{12}^{rr}	0.7432
ν_{13}^{rr}	0.2568
ν_{21}^{rr}	0.9860
ν_{23}^{rr}	0.014
ν_{31}^{rr}	0.8852
ν_{32}^{rr}	0.1147

(d) We used the following information about the structure of the one-scale (interregional) mobility simulated data set (NH mobility [31, p. 14]) for the above-defined three regions and generate the interregional visiting/travel rates as follows: (1) communication patterns between two of the three regions is completely symmetric, and

(2) the third region is partially symmetric with one of the two regions. This is because of the fact that there is a zero flow rate into itself from one of the two regions; that is, one of the two complete symmetric regions is a "sink" for the third region. From this description we conclude that the travel pattern in [31] includes the human mobility structure of our presented model as a particular case. Using the structural and probabilistic understanding, we constructed the interregional visiting/travel rates. We display the interregional travel rates in Table 5.2. Furthermore, the large-scale two-level population structure and the underlying human mobility pattern are exhibited in Figure 5.10.

Table 5.2: The interregional visiting/travel rates between the three regions. The derivation of these rates is based on the structural understanding of the data set under the column "NH mobility" in [31, p. 14] and probabilistic understanding of our presented model.

Parameter	Interregional mobility
$(\gamma_1^1, \gamma_2^1, \gamma_3^1)$	(0.5, 0.3.24, 0.176)
$(\gamma_{10}^{12}, \gamma_{20}^{12}, \gamma_{30}^{12})$	(1, 1, 1)
$(\gamma_{10}^{13}, \gamma_{20}^{13}, \gamma_{30}^{13})$	(0, 0, 0)
$(\gamma_1^2, \gamma_2^2, \gamma_3^2)$	(0.16, 0.06, 0.03)
$(\gamma_{10}^{21}, \gamma_{20}^{21}, \gamma_{30}^{21})$	(0.222, 0.12, 0.658)
$(\gamma_{10}^{23}, \gamma_{20}^{23}, \gamma_{30}^{23})$	(0.778, 0.88, 0.342)
$(\gamma_1^3, \gamma_2^3, \gamma_3^3)$	(0.12, 0.011, 0.001)
$(\gamma_{10}^{31}, \gamma_{20}^{31}, \gamma_{30}^{31})$	(0.200, 0.080, 0.006)
$(\gamma_{10}^{32}, \gamma_{20}^{32}, \gamma_{30}^{32})$	(0.800, 0.920, 0.994)
$(\gamma^{12}, \gamma^{13})$	(3, 0)
$(\gamma^{21}, \gamma^{23})$	(0.6, 2)
$(\gamma^{31}, \gamma^{32})$	(0.286, 2.714)
$(\gamma_{01}^{13}, \gamma_{02}^{13}, \gamma_{03}^{13})$	(0, 0, 0)
$(\gamma_{01}^{rq}, \gamma_{02}^{rq}, \gamma_{03}^{rq})_{r,q=2,3,r\neq q}$	(0.333, 0.27, 0.397)

(e) We further assume that for every site s_i^r, $i = 1, 2, 3$, in any region C_r, $r = 1, 2, 3$, for all $j \neq i$, intraregional visiting times τ_{ij}^{rr} of residents are the same. Moreover, we assume that the corresponding interregional visiting times τ_{il}^{rq} of residents of site s_i^r in the region C_r are also the same for all $l = 1, 2, 3$ and $q \neq r$.

5.7.1 Mathematical algorithm and simulation results

In this section, simulation results for three general scenarios labeled "Case 1, Case 2, and Case 3" are presented. Each scenario is based on restrictions on the intra- and interregional visiting times between zero and hundred days. We recall the fraction of the steady state population of residents of site s_i^r in the region C_r at the home site, intra-

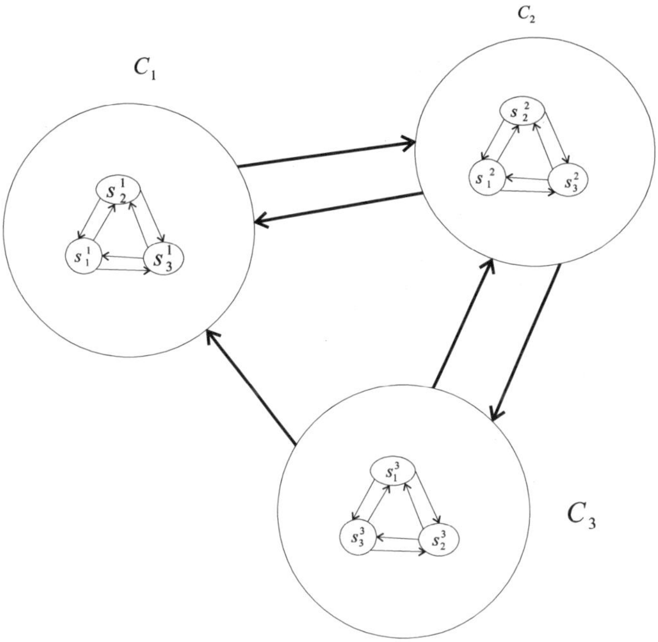

Figure 5.10: A two-scale network of three spatial regions C_r, $r = 1, 2, 3$, of human habitation and three interconnected sites s_i^r, $i = 1, 2, 3$, in each region. The arrows represent direction of human mobility and summarize the heterogeneities in the epidemic process at each site and region. C_1 and C_2 as well as C_2 and C_3 are symmetric in the human mobility process. C_1 is a sink for C_3 in human mobility. All sites in each region are completely symmetric in the human mobility process.

and interregional accessible domains are given in (5.55), (5.56), and (5.57), respectively. Therefore all categories of the steady-state population of the residents of site s_i^r in the region C_r subject to continuous changes in the intra- and interregional visiting times can be characterized by the following functions:

$$S_{ii}^{*rr}, U_{ij}^{*rr}, V_{il}^{*rq} : [0,\infty] \times [0,\infty] \to [0,\infty], \quad j \in I_i^r(1,n_r),\ l \in I_i^r(1,n_q),$$

where

$$S_{ii}^{*rr}(\tau_{ij}^{rr}, \tau_{il}^{rq}) = \frac{1}{(1 + \sum_{k\neq i}^{n_r} \sigma_{ik}^{rr}\tau_{ik}^{rr} + \sum_{q\neq r}^{M}\sum_{a=1}^{n_q} \nu_{ia}^{rq}\tau_{ia}^{rq})}, \qquad i \in I(1,n_r), \tag{5.95}$$

$$U_{ij}^{*rr}(\tau_{ij}^{rr}, \tau_{il}^{rq}) = \frac{\sigma_{ij}^{rr}}{(\frac{1}{\tau_{ij}^{rr}} + \sum_{k\neq i}^{n_r} \sigma_{ik}^{rr}K_{ikj}^{rrr} + \sum_{q\neq r}^{M}\sum_{a=1}^{n_q} \nu_{ia}^{rq}K_{iaj}^{rqr})}, \qquad j \in I_i^r(1,n_r), \tag{5.96}$$

and

$$V_{il}^{*rq}(\tau_{ij}^{rr}, \tau_{il}^{rq}) = \frac{\nu_{il}^{rq}}{(\frac{1}{\tau_{il}^{rq}} + \sum_{k\neq i}^{n_r} \sigma_{ik}^{rr}K_{ikl}^{rrq} + \sum_{q\neq r}^{M}\sum_{a=1}^{n_q} \nu_{ia}^{rq}K_{ial}^{rqq})}, \qquad l \in I_i^r(1,n_q). \tag{5.97}$$

Furthermore, by $W_{ia}^{*ru}(\tau_{ij}^{rr}, \tau_{il}^{rq})$, $W \in \{S, U, V\}$, $u \in I(1, M)$, $a \in I(1, n_r)$, we denote the two hierarchic population interaction levels, which should not be understood as representing a function of two variables. In the following, we present the three scenarios. We fix $r = 1$ and $i = 1$.

Case 1: Constant interregional visiting time and varying intraregional visiting time. Suppose that the assumptions of the example presented in the previous section are satisfied. We further assume that interregional visiting time τ_{1l}^{rq} of residents of a given site s_1^1 in community C_1 is 10 days and that the intraregional visiting time of the residents τ_{1j}^{11}, $j = 2, 3$, equally vary between zero and 100 days. Utilizing (5.95), (5.96), (5.97), and the methods of graph sketching [30], we obtain values and graphs for the steady-state population for the different categories of the residents of site s_1^1 over all $0 \leq \tau_{1j}^{11} \leq 100$, $j = 2, 3$. The results are displayed in Figure 5.11.

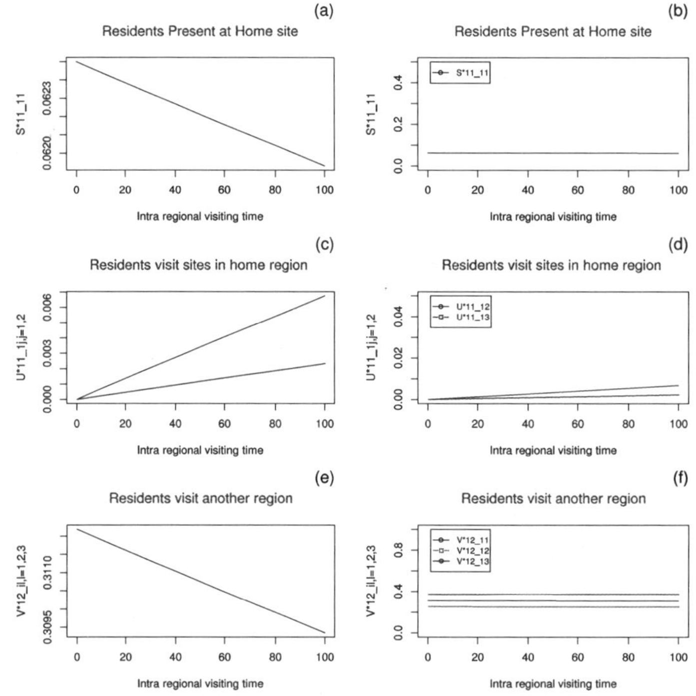

Figure 5.11: Diagrams (b), (d), and (f) corresponding to (a), (c), and (e), respectively, under magnified scales. Note that the axis labels for (a) and (b), (c) and (d), and (e) and (f), respectively, are the same, and the axes represent quantities on different scale units. The figure displays the behavior of the proportions S_{11}^{*11}, U_{1j}^{*11}, $j = 2, 3$, and V_{1l}^{*12}, $l = 1, 2, 3$, of the different classes of residents of site s_1^1 subject to continuous changes in intraregional visiting time over the interval from zero to 100 days, given that interregional visiting time is fixed at 10 days. More comments about this figure are given in Remark 5.7.1.

Remark 5.7.1. We observe from Figure 5.11 that as the intraregional visiting time continuously changes value from zero to 100, diagrams (a) and (b) for S_{11}^{*11} and (e) and (f) for V_{1l}^{*12}, $l = 1, 2, 3$, show a smooth decrease, indicating that for larger values of intraregional visiting time, S_{11}^{*11} and V_{1l}^{*12}, $l = 1, 2, 3$, turn to be smaller values. This qualitative behavior of S_{11}^{*11} and V_{1l}^{*12}, $l = 1, 2, 3$, is exhibited in diagrams (a) and (e), respectively. Diagram (d) for U_{1j}^{*11}, $j = 2, 3$, on the other hand has a smooth rise with intraregional visiting time, also illustrating the growth of U_{1j}^{*11}, $j = 2, 3$, with larger values of intraregional visiting time. Furthermore, for low intraregional visiting time, (i) S_{11}^{*11} and V_{1l}^{*12} are maximum for $l = 1, 2, 3$, and (ii) U_{1j}^{*11}, $j = 2, 3$, is minimum; that is, residents of site s_1^1 distribute themselves between home sites and sites in other regions.

Case 2: Varying interregional visiting time and constant intraregional visiting time. Suppose the intraregional visiting time of residents of a given site s_1^1 in the region C_1 is 20 days and the corresponding interregional visiting time of the residents τ_{1l}^{12}, $l = 1, 2, 3$, equally vary between zero and 100 days. We utilize (5.95), (5.96), (5.97), and the basic methods of graph sketching [30] to obtain values and graphs for the steady-state population for the different categories of the residents of site s_1^1 over all $0 \leq \tau_{1l}^{12} \leq 100$, $l = 1, 2, 3$. The results are illustrated in Figure 5.12.

Remark 5.7.2. We observe from Figure 5.12 that as the interregional visiting time continuously changes value from zero to 100, diagrams (a) and (b) for S_{11}^{*11} and (b) and (c) for U_{1j}^{*11}, $j = 2, 3$, show a continuous decrease, thus indicating that for larger values of interregional visiting time, S_{11}^{*11} and U_{1j}^{*11}, $j = 2, 3$, tend to be very small values. This qualitative behavior of S_{11}^{*11} and U_{1j}^{*11}, $j = 2, 3$, is exhibited in diagrams (a) and (c), respectively. Diagram (e) for V_{1l}^{12}, $l = 1, 2, 3$, has a continuous rise with an intraregional visiting time. Thus there is a tendency for V_{1l}^{*12} to increase with larger values of interregional visiting time. Also, for this example, the rising of V_{1l}^{*12}, $l = 1, 2, 3$, approaches one, for large values of interregional visiting time. These observations signify that for sufficiently larger values of interregional visiting time, the residents of site s_1^1 tend to distribute themselves only among sites in other regions. For this specific scenario, it is clear that for sufficiently large values of interregional visiting time, the residents of site s_1^1 totally isolate their home region. Therefore the fixed intraregional visiting time does not change the residents' decision to emigrate to another region, that is, after spending more that 100 days in another region, all residents of site s_1^1 would become permanent residents of other sites in those regions C_r.

Case 3: Varying inter- and intraregional visiting times. Suppose the intraregional τ_{1j}^{11}, $j = 2, 3$, and interregional τ_{1l}^{12}, $l = 1, 2, 3$, visiting times of residents of the site s_1^1 in the region C_1 equally vary between zero and 100 days. From (5.95), (5.96), (5.97), and the basic methods of graph sketching [30] we obtain values and graphs for the

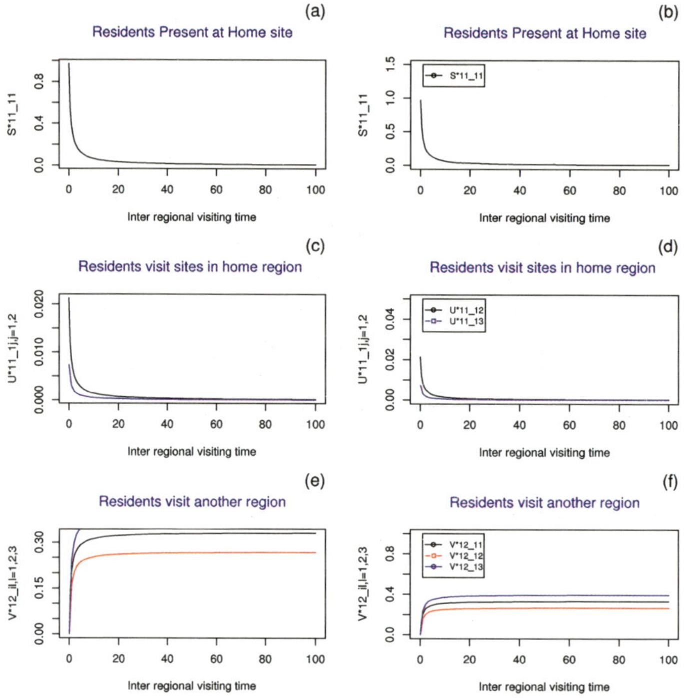

Figure 5.12: Diagrams (b), (d), and (f) corresponding to (a), (c), and (e) respectively, under magnified scales. Note that the axis labels for (a) and (b), (c) and (d), and (e) and (f), respectively, are the same and that the axes represent quantities on different scale units. The figure shows the behavior of the proportions S_{11}^{*11}, U_{1j}^{*11}, $j = 2, 3$, and V_{1l}^{*12}, $l = 1, 2, 3$, of the different classes of residents of site s_1^1 when the proportions are subject to continuous changes in interregional visiting time over the interval from zero to 100 days. The intraregional visiting time is 20 days. More comments about this figure are given in Remark 5.7.2.

steady-state population for the different categories of the residents of site s_1^1 over all $0 \le \tau_{1j}^{11}, \tau_{1l}^{12} \le 100, j = 2, 3, l = 1, 2, 3$. The results are shown in Figure 5.13. Note that the results under these scenarios exhibit the behavior of the steady-state population as visitors are allowed to spend up to 100 days at their destinations.

Remark 5.7.3. Figure 5.13 exhibits that as both visiting times continuously change values from 0 to 100, for S_{11}^{11}, graph (b) shows a smooth decrease. Diagrams (d) for $U_{1j}^{*11}, j = 2, 3$, and (f) for $V_{1l}^{*12}, l = 1, 2, 3$, show a continuous rise with intra- and interregional visiting times and vice versa. This qualitative behavior of $U_{1j}^{*11}, j = 2, 3$, and $V_{1l}^{*12}, l = 1, 2, 3$, is exhibited in diagrams (d) and (f), respectively. This suggests that more residents are found visiting other sites that are within their regions or in other regions when intra- and interregional visiting times are large. This also signifies that for smaller values of intra- and interregional visiting times, more residents are found in their home site s_1^1, which can lead to an isolation of home site.

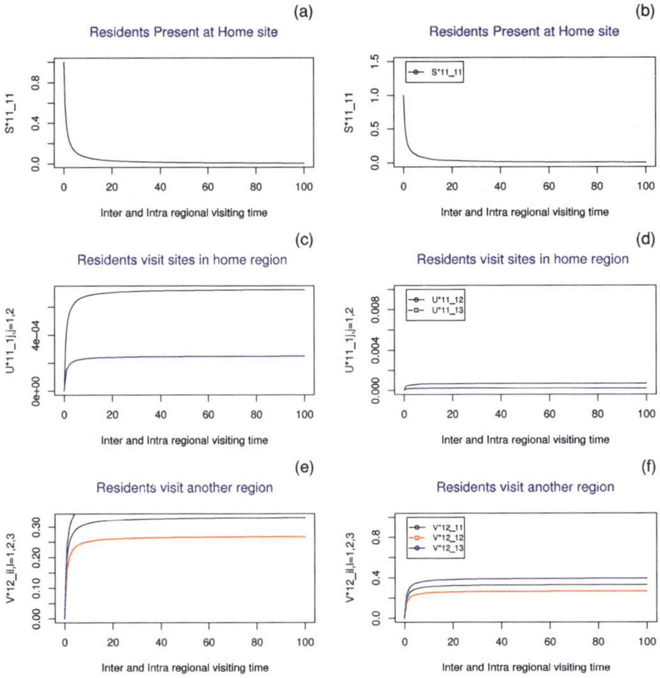

Figure 5.13: Diagrams (b), (d), and (f) corresponding to (a), (c), and (e), respectively, under magnified scales. Note that the axes labels for (a) and (b), (c) and (d), and (e) and (f), respectively, are the same, and the axes represent quantities on different scale units. The figure further exhibits the behavior of the proportions S_{11}^{*11}, U_{1j}^{*11}, $j = 2, 3$, and V_{1l}^{*12}, $l = 1, 2, 3$, of the different classes of residents of site s_1^1 when the proportions are subject to continuous change in inter- and intraregional visiting times simultaneously over the interval from zero to 100 days. The inter- and intraregional visiting times are assumed to be equal. More comments about this figure are given in Remark 5.7.3.

For this particular example, with the increase of visiting times, the rise of both U_{1j}^{*11}, $j = 2, 3$, and V_{1l}^{*12}, $l = 1, 2, 3$, saturates at different values less than one. This signifies that there is always a fraction of residents of site s_1^1 at other sites within their region and in other regions.

5.8 Expanding applications: investigating communicable disease epidemics through mobility models

To conclude this chapter, it is important to note that the two-scale human mobility process studied here serves as an underlying driving force for new infectious disease outbreaks, epidemics, and pandemics, such as those caused by COVID-19 and the H1N1 virus. The interconnectedness inherent in the human mobility network not only facilitates the rapid spread of infectious agents across different regions but also provides a crucial framework for understanding how such dynamics can be mitigated or controlled.

In the subsequent chapters the dynamic modeling of infectious disease epidemics influenced by the two-scale human mobility network will be explored. These models will enable detailed investigations into how spatial patterns of mobility impact the prevalence, persistence, and extinction of communicable diseases in finite-scale network populations. Additionally, the hereditary features of diseases, immunity dynamics, and other key aspects of disease behavior will be examined within the framework of the two-scale network population.

The insights gained from these studies will contribute to a broader understanding of the interaction between mobility and disease dynamics, which is essential for designing effective public health interventions. For instance, mobility-based models can inform strategies for vaccination deployment, quarantine measures, and resource allocation during outbreaks. They can also shed light on the role of hierarchical population structures in shaping the trajectory of epidemics, thus highlighting critical nodes or regions that act as hubs for transmission. This approach ultimately paves the way for developing comprehensive models that account for both local and global scales of human interaction in combating infectious diseases.

6 Stochastic modeling of communicable disease epidemics dynamics in finite-scale network populations

6.1 Introduction

In this modern era of efficient communication and transportation services worldwide, local endemic communicable diseases have found easy access to new endemic zones across various sites and regions globally. This has been evident in recent pandemics, such as COVID-19 [26, 6] and the 2009 H1N1 flu pandemic [15], with outbreaks originating in *Wuhan, China*, and *La Gloria, Mexico*, respectively. The dynamic human mobility framework introduced in Chapter 5 motivates two fundamental questions:

(a) To what extent does an infectious disease outbreak in a given site s_i^r of a region C_r propagate into a pandemic across the two-scale network population $C = C_{00}^{00}$ under the influence of human mobility?

(b) What are the epidemiological drivers and forces influencing infectious disease dynamics, and how do these factors impact disease eradication or persistence in the two-scale network population?

These questions are broad and require extensive investigation over multiple chapters. In the spirit of Chapter 5, the modeling of infectious disease dynamics in the two-scale network population employs compartmental epidemic models such as SIS (susceptible–infected–susceptible), SEIRS (susceptible–exposed–infected–removed–susceptible), and SIR (susceptible–infected–removed) models (see [49, 46, 48, 32, 39, 37, 36]). For instance, the SEIRS model characterizes transitions among the compartments: susceptible (S), exposed (E), infectious (I), and removed or recovered (R), with individuals potentially returning to the susceptible state.

Within this framework, epidemic dynamics can be modeled using various approaches, including deterministic and stochastic models (see [33, 39, 47, 36, 40, 41]). Among stochastic models, there are:

- **Discrete-time and discrete-state models**, such as *discrete-time Markov chain (DTMC)* models [40, 41, 51].
- **Continuous-time and discrete-state models**, such as *continuous-time Markov chains (CTMC)* and *renewal* models [40, 44].
- **Continuous-time and continuous-state models**, such as diffusion and jump processes, including *Brownian motion*, *Ornstein–Uhlenbeck*, and *Lévy jump* processes (see [23, 2]).

This chapter focuses on stochastic modeling using *Itô–Doob stochastic differential equations*, which represent random fluctuations in disease dynamics through diffusion processes, particularly Brownian motion. General compartmental models, including SIR and SIRS, are considered without focusing on specific diseases. The modeling techniques

https://doi.org/10.1515/9783111613215-006

are detailed in the literature [23] and have been applied in ecological and epidemiological studies by Wanduku (see [49, 46, 48, 32, 39, 37, 36, 23]).

Stochastic models are crucial for analyzing epidemic dynamics as they capture the inherent randomness and uncertainty in disease transmission, recovery, and environmental factors. Deterministic models often rely on average behavior and fail to account for the variability that may significantly alter disease trajectories, especially in small populations or during the initial stages of an outbreak. In contrast, stochastic models can describe events such as the probability of disease extinction or outbreak, fluctuations in infection levels, and the impacts of rare or extreme events.

The importance of stochastic approaches lies in their ability to incorporate real-world complexities, such as:

- **Noise and variability:** Random perturbations in transmission rates due to environmental or social factors.
- **Threshold effects:** Capturing conditions under which diseases die out or persist.
- **Interaction dynamics:** Modeling interactions at different spatial and temporal scales to reflect the heterogeneity in human populations.

Stochastic methods also provide tools for simulating scenarios and testing interventions, helping policymakers design robust strategies to control outbreaks. For instance, diffusion processes like Brownian motion can simulate the effects of continuous random variations in transmission, whereas hybrid models incorporating jumps or switches can handle sudden changes, such as policy shifts or environmental disasters.

The derivation of dynamic models for infectious disease outbreaks and their progression in the two-scale network population $C = C_{00}^{00}$, under the influence of the dynamic human mobility process in Chapter 5, involves the following steps:

Step 1: Establish the scale-structured population framework in Chapter 2, defining hierarchical interactions within the metapopulation. These include first- and second-level partitions given by $C = \bigcup_{r=1}^{M} \bigcup_{i=1}^{n_r} \{s_i^r\}$ and $C = \bigcup_{r=1}^{M} C_r$, respectively.

Step 2: Represent interconnections within the network population through the human mobility network and the dynamic human mobility process described in Chapter 5.

Step 3: Derive interpersonal contact mechanisms leading to disease transmission. This includes:
- Resident–resident interactions within site s_i^r.
- Resident–visitor interactions, either intraregional ($s_j^r, j \neq i$) or interregional ($s_l^q, q \neq r$).

Step 4: Incorporate random environmental perturbations in the disease dynamics. These include stochastic effects from transmission, recovery, and mortality processes, modeled using diffusion processes like *Brownian motion*. This chapter focuses on white noise perturbations, whereas hybrid stochastic differential equations with colored noise (e. g., Lévy jump or Markovian switching processes) are deferred to future work.

Basic concepts in modeling dynamic processes with Markov jump processes (i. e., CTMC) and ordinary differential equations (ODEs) representing mean-field dynamics are assumed in this chapter. Readers are referred to Section 5.4, classic *Mathematical Biology* texts (e. g., [29]) for a review of ODE-based epidemic models, and Wanduku et al. [38, 51, 50, 54] for CTMC modeling. Familiarity with the *law of mass action* and its application in average incidence rate derivations is also assumed (see [38, Theorem 3.1]). For stochastic differential equations in biological and physical systems, we refer to Ladde [23].

6.2 Derivation of an epidemic model with Itô–Doob stochastic differential equations

Recall the following facts in Section 5.4. Suppose in a complete probability $(\Omega, \{\mathfrak{F}_t\}_{t\geq 0}, \mathbb{P})$ a stochastic process $\{x(t), t \geq 0\}$ represents a population $x(t) \in \mathbb{R}^m$ at time $t \geq 0$, where the mean-field dynamic model for the underlying *birth-and-death process* in $x(t)$ is defined by

$$\frac{dx(t)}{dt} = f(x(t), t), \tag{6.1}$$

where $x(t) \equiv \mathbb{E}[x(t)]$, the function $f(x(t), t)$ also denotes the expected value $f(x(t), t) \equiv \mathbb{E}[f(x(t), t)]$ of a stochastic process $\{f(x(t), t), 0 \leq t \leq T\}$ for some $T > 0$, and $f(x(t), \cdot) : [0, T) \to \mathbb{R}^m$ is an $\mathfrak{F}_t$*-adapted process*, i. e., the random variable $f(x(t), t)$ is $\mathfrak{F}_t$-measurable, almost surely continuously differentiable with respect to $x(t)$, and continuous with respect to t. In addition, the expectation

$$\mathbb{E}\left[\int_0^T \|f(x(s), s)\| ds\right] < \infty. \tag{6.2}$$

From (6.1) an Itô–Doob stochastic differential equation (see [23]) with respect to a *Wiener or Brownian motion process* $\{w(t), t \geq 0\}$ is defined by the expression

$$dx(t) = f(x(t), t)dt + g(x(t), t)dw(t), \tag{6.3}$$

where the function $g(x(t), t)$ is also the average of a stochastic process $\{g(x(t), t), 0 \leq t \leq T\}$ for some $T > 0$, where $g(x(t), \cdot) : [0, T) \to \mathbb{R}^m$ is $\mathfrak{F}_t$-measurable, almost surely twice continuously differentiable with respect to $x(t)$, and continuous with respect to t. Furthermore, the expectation

$$\mathbb{E}\left[\int_0^T \|g(x(s), s)\|^2 ds\right] < \infty. \tag{6.4}$$

Based on this simple characterizations of $f(x(t), t)$ and $g(x(t), t)$ in (6.3) as averages in the underlying birth-death processes in the system (see Ladde [23, Observation 3.1.1]), this implies that they are smooth functions with respect to the average $x(t)$ and time t. Thus, from (6.3), a new stochastic process with respect to the Wiener process $\{w(t), t \geq 0\}$ emerges and over small time interval $[t, t + \Delta t]$ is given by

$$x(t + \Delta t) = x(t) + \int_{t}^{t+\Delta t} f(x(s), s)ds + \int_{t}^{t+\Delta t} g(x(s), s)dw(s), \quad 0 \leq t \leq t + \Delta t \leq T, \tag{6.5}$$

and at any time $t \in [t_0, T)$, $t_0 \geq 0$, the stochastic process $\{x(t), t \in [t_0, T)\}$ given by

$$x(t) = x(t_0) + \int_{t}^{t} f(x(s), s)ds + \int_{t_0}^{t} g(x(s), s)dw(s), \quad t_0 \leq t \leq T, \tag{6.6}$$

is called an *Itô–Doob stochastic process*, where the stochastic integral

$$\int_{t_0}^{t} g(x(s), s)dw(s) = \lim_{\mu(\mathcal{P})\to 0} \sum_{k=0}^{n-1} g(x(t_k), t_k)[w(t_{k+1}) - w(t_k)], \tag{6.7}$$

and $\mu(\mathcal{P}) = \max_{0\leq k\leq n} (t_{k+1} - t_k)$ is the mesh of the partition

$$\mathcal{P} = \{t_0 < t_1 < \ldots, t_k < t_{k+1} < \ldots, t_{n-1} < t_n = t\}. \tag{6.8}$$

In this monograph, the approach to derive the Itô–Doob stochastic differential equations (SDEs) is similar to that of Ladde [23, Section 3.1]. So it is assumed that there exists a mean-field dynamic model for the epidemic dynamics given by (6.1), where the rate of change of the mean $x(t)$ of the population with time t is given by a smooth function $f(x(t), t)$. It is assumed that besides the underlying birth-and-death process of the model (6.1), there are other sources of variability due to *environmental random fluctuations or impacts that are independent and identical to each other and occur at discrete times* in the partition $\mathcal{P}$ and, consequently, suitably represented by a *discrete-time Gaussian white noise process*

$$\{Z_k, k \geq 0\} = \{Z_1, Z_2, \ldots\}. \tag{6.9}$$

The *central limit theorem* will be applied to the Gaussian process to obtain the Itô–Doob stochastic integral for the process. Example 6.2.1 illustrates the process of deriving an SDE for a simple SIS epidemic model.

Example 6.2.1 (Deriving an Itô–Doob SDE). Consider a simple SIS epidemic model with vital dynamics[1] in a fixed population $N(t) = S(t)+I(t) = N$, where $N > 0$ is constant, and

1 Compartmental epidemic models with vital dynamics assume that there is birth and natural death and that the epidemic persist over a time frame comparable with a lifespan of an individual.

susceptible (S) individuals are infected by infectious (I) individuals at the exponential rate β_{IS} and recover at the exponential rate α_{SI}. There is birthrate and natural death rate μ. Assume that there is no disease-related death. The mean-field dynamic equations for the underlying continuous-time Markov chain (CTMC) process is given by the system of ODEs

$$\begin{aligned} \frac{dS(t)}{dt} &= \mu N - \beta_{IS}\frac{I(t)}{N}S(t) + \alpha_{IS}I(t) - \mu S(t), \\ \frac{dI(t)}{dt} &= \beta_{IS}\frac{I(t)}{N}S(t) - \alpha_{IS}I(t) - \mu I(t), \end{aligned} \tag{6.10}$$

where $x(t) \equiv \mathbb{E}[x(t)]$ for all $x \in \{S, I\}$. Also, since $S(t) = N - I(t)$, it follows that expressing system (6.10) in terms of dimensionless quantities by letting $i(t) = \frac{I(t)}{N}$ and $s(t) = \frac{s(t)}{N}$ leads to a reduced form of system (6.10):

$$\begin{aligned} \frac{di(t)}{dt} &= \beta_{IS}i(t)[1 - i(t)] - (\alpha_{IS} + \mu)i(t), \\ s(t) &= 1 - i(t), \quad s(t) + i(t) = 1. \end{aligned} \tag{6.11}$$

Denote (6.11) by

$$di(t) = f(i(t), t)dt, \tag{6.12}$$

where $f(i(t), t)$ is the right-hand-side of (6.11).

Consider a general source of variability in system (6.10), which does not originate in the jumps of the CTMC process that lead to the mean-field dynamic equations (6.10), but rather from other subprocesses where the exponential rates β, α_{IS}, and μ are no longer constants over time but random variables that change values over time due to random environmental influences, e. g., seasonal changes that affect the disease transmission rate β_{IS} or changes in government health policies that are unpredictable over time thereby turning the recovery rate α_{IS} to a time series with a Gaussian white-noise perturbation, etc.

Indeed, let $(\Omega, \{\mathfrak{F}_t\}_{t\geq 0}, \mathbb{P})$ be a complete probability space, and consider a partition of a small time interval $[t, t + \Delta t]$ given by

$$\mathcal{P} = \{t = t_0 < t_1 < \cdots < t_k < t_{k+1} < \cdots < t_{n-1} < t_n = t + \Delta t\}. \tag{6.13}$$

Assume that (1) the aggregate of all random impacts in (6.11) (aggregate impacts imply there is particular identification of whether the source of variability is from β_{IS}, α_{SI}, or μ) are felts in the system at the time points t_k, $k \in I(1, n-1)$; (2) The impacts are independent and identical to each other; and (3) The magnitudes of the impacts are proportional to the mean of the state of the process, that is, the random impacts have magnitudes proportional to the mean $i(t)$.

Without loss of generality, assume that the random impacts over $\mathcal{P}$ are represented by the *simple random walk process* $\{Z_{t_k}, k \geq 0\}$, where

$$Z_{t_k} = \pm i(t)\Delta x, \quad \mathbb{P}(Z_{t_k} = i(t)\Delta x) = p, \quad \mathbb{P}(Z_{t_k} = -i(t)\Delta x) = 1 - p. \tag{6.14}$$

It follows from (6.13) that over any small time interval $[t_k, t_{k+1}] = [t_k, t_k + \Delta t]$,

$$i(t_{k+1}) = i(t_k) + [f(i(t_k), t_k)]\Delta t + Z_{t_k}\Delta t, \quad k = 0, 1, \ldots, n-1, \tag{6.15}$$

further expressed as

$$i(t_{k+1}) = i(t_k) + [f(i(t_k), t_k)]\Delta t + i(t)\Delta x \Delta t \bar{Z}_{t_k}, \quad k = 0, 1, \ldots, n-1, \tag{6.16}$$

where

$$\bar{Z}_{t_k} = \begin{cases} 1 & \text{when } Z_{t_k} = i(t)\Delta x, \\ -1 & \text{when } Z_{t_k} = -i(t)\Delta x. \end{cases} \tag{6.17}$$

Also,

$$\mathbb{E}[\bar{Z}_{t_k}] = 2p - 1, \quad \mathbb{V}ar[\bar{Z}_{t_k}] = 1 - (2p-1)^2. \tag{6.18}$$

For simplicity, take $p = 1/2$. It follows from (6.18) that $\mathbb{E}[\bar{Z}_{t_k}] = 0$ and $\mathbb{V}ar[\bar{Z}_{t_k}] = 1$. This further implies from (6.14) that

$$\mathbb{E}[Z_{t_k}] = (2p-1)i(t)\Delta x = 0, \quad \mathbb{V}ar[Z_{t_k}] = [1 - (2p-1)^2](i(t)\Delta x)^2 = (i(t)\Delta x)^2. \tag{6.19}$$

By iterative computations over $\mathcal{P}$ it follows from (6.16) that

$$i(t + \Delta t) - i(t) = \sum_{k=0}^{n-1} [f(i(t_k), t_k)]\Delta t + i(t)\Delta x \Delta t \sum_{k=0}^{n-1} \bar{Z}_{t_k}. \tag{6.20}$$

Taking the limit of (6.20) as the mesh $\mu(\mathcal{P}) \to 0$, observe that the first term on the right-hand side of (6.20) is the *Riemann sum* leading to the Riemann integral, which is further approximated by the **fundamental theorem of integral calculus** as follows:

$$\lim_{\mu(\mathcal{P}) \to 0} \sum_{k=0}^{n-1} [f(i(t_k), t_k)]\Delta t = \int_t^{t+\Delta t} [f(i(t), t)]dt \approx [f(i(t), t)]\Delta t. \tag{6.21}$$

The second term on the right-hand-side of (6.20) is a **random walk process** that ***converges in distribution*** to the **Wiener or Brownian motion process** by the ***central limit theorem***, that is, for large n,

$$i(t)\Delta x \Delta t \sum_{k=0}^{n-1} \bar{Z}_{t_k} \xrightarrow{\text{distribution}} n(\Delta x) i(t)(\Delta t \mathbb{N}), \tag{6.22}$$

where $\mathbb{N}$ has the standard Gaussian or normal distribution with mean 0 and variance 1. Observe that the random variable $\Delta t\mathbb{N}$ is equivalently distributed as the increment of the Wiener process $\{w(t), t \geq 0\}$ over the interval $[t, t+\Delta t]$, that is,

$$\Delta t\mathbb{N} \sim w_I(t+\Delta t) - w_I(t). \tag{6.23}$$

Thus, taking $\Delta x = \frac{\sigma_I}{n}$ with constant $\sigma_I > 0$, it follows from (6.22) that for large n,

$$n(\Delta x)i(t)(\Delta t\mathbb{N}) \sim \sigma_I i(t)\big[w_I(t+\Delta t) - w_I(t)\big]. \tag{6.24}$$

It is easy to see from (6.24), (6.21), and (6.20) that

$$i(t+\Delta t) - i(t) = \big[f(i(t),t)\big]\Delta t + \sigma_I i(t)\big[w_I(t+\Delta t) - w_I(t)\big]. \tag{6.25}$$

Now taking the limit of (6.25) as $\mu(\mathcal{P}) \to 0$ gives the Itô–Doob stochastic differential equation

$$di(t) = \big[f(i(t),t)\big]dt + \sigma_I i(t)dw_I(t). \tag{6.26}$$

From (6.19) note that $\sigma_I i(t)$ is the standard deviation of the random environmental fluctuations in the infectious state I, and σ_I is called the intensity of the noise.

Similarly, considering the random fluctuations in the state S proportional to the state of the system, it follows from (6.11) that the SDE for $s(t)$ is given by

$$ds(t) = -\big[f(i(t),t)\big]dt + \sigma_S s(t)dw_S(t), \tag{6.27}$$

where $f(i(t),t)$ is defined earlier, and $\sigma_S s(t)$ is the standard deviation of the random environmental fluctuations in the susceptible state s; σ_S is called the intensity of the noise in S. In addition, the random variables $w_I(t)$ and $w_S(t)$ are independent for all $t \geq 0$.

Therefore from (6.11) and (6.12) the Itô–Doob SDE for the SIS epidemic model is given by

$$\begin{aligned} ds(t) &= -\big[f(i(t),t)\big]dt + \sigma_S s(t)dw_S(t), \\ di(t) &= \big[f(i(t),t)\big]dt + \sigma_I i(t)dw_I(t), \\ f(i(t),t) &= \big[\beta_{IS} i(t)[1-i(t)] - (\alpha_{IS}+\mu)i(t)\big], \\ s(t) &= 1 - i(t), \quad s(t) + i(t) = 1. \end{aligned} \tag{6.28}$$

Note that (6.28) can be rewritten in terms of a Gaussian white noise as follows:

$$\begin{aligned} ds(t) &= -\big[f(i(t),t)\big]dt + \sigma_S s(t)\xi_S(t)dt, \\ di(t) &= \big[f(i(t),t)\big]dt + \sigma_I i(t)\xi_I(t)dt, \\ f(i(t),t) &= \big[\beta_{IS} i(t)[1-i(t)] - (\alpha_{IS}+\mu)i(t)\big], \\ s(t) &= 1 - i(t), \quad s(t) + i(t) = 1, \end{aligned} \tag{6.29}$$

where $\{\xi_I(t), t \geq 0\}$ and $\{\xi_S(t), t \geq 0\}$ are independent Gaussian white noise processes interrelated with the Brownian motion $\{w_x(t), t \geq 0\}$ by

$$dw_x(t) = \xi_x(t)dt, \quad x \in \{S, I\}. \tag{6.30}$$

Based on the derivation of the Itô–Doob SDE in Example 6.2.1, the following steps will be adopted to derive the SDEs for any disease epidemic dynamics in this monograph.

Algorithm 6.2.1 (Algorithm for deriving epidemic models with Itô–Doob SDEs).

Step 1: Derive the mean-field dynamic equations for the disease model in the absence of any environmental noise in the system using the methods in Section 5.4,

$$\frac{dx(t)}{dt} = f(\vec{\alpha}x(t), t), \tag{6.31}$$

where $\vec{\alpha}$ is a parameter vector in the model, e. g., $\vec{\alpha} = (\beta_{IS}, \alpha_{IS}, \mu)$ in the mean-field model (6.10).

Step 2: Identify the source of additional random environmental fluctuations in the model (6.31) obtained in **Step 1** and derive a Gaussian white noise model for the random fluctuations, e. g., assume that the random environmental fluctuations lead to randomness in the parameter vector $\vec{\alpha}$ over time t. Denote the change in the random vector over a small time interval $[t, t + dt]$ by $d\tilde{\vec{\alpha}}(t) = \tilde{\vec{\alpha}}dt$. It follows that over the small time interval $[t, t + dt]$, the random perturbations are given by the model

$$d\tilde{\vec{\alpha}}(t) \equiv \tilde{\vec{\alpha}}dt = \vec{\alpha}dt + \vec{\sigma}d\vec{w}(t), \tag{6.32}$$

where $\{\vec{w}(t), t \geq 0\}$ is the multivariate Wiener or Brownian motion process with coordinates that are independent random variables, and the expected value vector and covariance matrix are given by

$$\mathbb{E}[\tilde{\vec{\alpha}}] = \vec{\alpha}, \quad \mathbb{V}ar[d\tilde{\vec{\alpha}}(t)] = \mathbb{V}ar[\vec{\sigma}d\vec{w}(t)] = \vec{\sigma}\mathbb{I}\vec{\sigma}^T dt; \tag{6.33}$$

$\mathbb{I}$ is the identity matrix, and $\vec{\sigma}$ is a constant vector.

Step 3: Substitute (6.32) into (6.31) to obtain an Itô–Doob SDE.

6.3 Description of the SIRS disease epidemic in the two-scale network population

In this chapter, we consider a SIRS disease epidemic model. The model describes the dynamics of a general disease epidemic within the two-scale network population, denoted by $C = C_{00}^{00}$, under the influence of the two-scale network dynamic human mobility process explored in Chapter 5. Furthermore, the model incorporates random environmental fluctuations, which introduce variability into the disease dynamics and are represented by the Gaussian white noise process. The detailed description of the disease epidemic in the two-scale network population follows subsequently.

The total population in (5.5) and Definition 5.4.1 will be further decomposed according to the compartments of the SIRS epidemic model as shown below.

Definition 6.3.1 (Disease endemic population decomposition and aggregation). For each $r \in I(1, M)$, let $i \in I_i^r(1, n_r)$. From (5.5) and Definition 5.4.1, the total population N_{i0}^{rr} of residents of site s_i^r at time t, distributed among the sites in their intra- and interregional domain $C(s_i^r)$, is further partitioned into three general disease compartments, namely, susceptible (S), infectious (I), and removals (R), those previously sick and recovered with acquired immunity from the disease. That is, A_{il}^{rq} is the number of residents of site s_i^r with disease status $A \in \{S, I, R\}$ visiting site s_l^q, $l \in I_i^q(1, n_q)$, in the region C_q, $q \in I^r(1, M)$. Furthermore, when $r = q$, A_{ik}^{rr} is the number of residents of site s_i^r with disease status $A \in \{S, I, R\}$ visiting site s_k^r, $k \in I_i^r(1, n_r)$, in their home region C_r. Moreover, when $k = i$, A_{ii}^{rr} is the number of residents of site s_i^r with disease status $A \in \{S, I, R\}$ remaining as permanent residents at their home site. Hence N_i^{rr} is given by

$$N_{i0}^{rr} = S_{i0}^{rr} + I_{i0}^{rr} + R_{i0}^{rr}, \tag{6.34}$$

where

$$S_{i0}^{rr} = \sum_{q=1}^{M} \sum_{k=1}^{n_q} S_{ik}^{rq}, \quad I_{i0}^{rr} = \sum_{q=1}^{M} \sum_{k=1}^{n_q} I_{ik}^{rq}, \quad \text{and} \quad R_{i0}^{rr} = \sum_{q=1}^{M} \sum_{k=1}^{n_q} R_{ik}^{rq}. \tag{6.35}$$

Remark 6.3.1. Note that the effective population $eff(N_{i0}^{rr})$ present at the site s_i^r at anytime is different from the census population or the total number of residents N_{i0}^{rr} (6.34) with permanent residence site s_i^r. At anytime t, the effective community size of site s_i^r is made up of the permanent residents of site s_i^r and all visitors of to site s_i^r. This is given as follows:

$$eff(N_i^{rr}) = \sum_{q=1}^{M} \sum_{k=1}^{n_q} S_{ki}^{qr} + \sum_{q=1}^{M} \sum_{k=1}^{n_q} I_{ki}^{qr} + \sum_{q=1}^{M} \sum_{k=1}^{n_q} R_{ki}^{qr}. \tag{6.36}$$

So $eff(N_i^{rr})$ represents the population that is at risk for infection at site s_i^r and is the population size resulted by the mobility process in the two-scale network structure.

Definition 6.3.2 (Disease transmission process). The disease transmission process in any site s_i^r in the region C_r in a mobile population necessitates (1) a susceptible person to travel from site s_k^u in the region C_u to site s_i^r ($u = r$ and $k = i$ if there is no traveling), (2) an infectious person traveling from site s_l^q in the region $C_q, q \neq r$, to site s_i^r, (3) the susceptible and infectious persons meeting at a contact zone z (which may be the home, market place, recreational facility, etc.) in site s_i^r with probability p of a person being at a zone z at anytime t, and (4) the probability β of the infectious agent being transmitted from the infectious person to the susceptible person knowing that the contact between the susceptible and infectious individuals took place.

Let n_{ri} be the number of contact zones denoted by $z^r_{i_b}$, $b \in \{1, 2, \dots, n_{r_i}\} \equiv I(1, n_{r_i})$, at each site s^r_i. Furthermore, let $p^r_{i_b}$ be the probability that a member of the effective population would be in a zone $z^r_{i_b}$ at time t; in addition, we assume that the events of visiting contact zones are independent and that the probability $p^r_{i_b}$ of being in a given zone $z^r_{i_b}$ is independent of the permanent residence of the individual. In each zone $z^r_{i_b}$, there is random mixing and transmission of the infectious agent from an infectious person to a susceptible person via a direct contact between the two individuals. Moreover, let β^{ruv*}_{ikj} be the probability that the transmission takes place given that the contact occurs in any zone $z^r_{i_b}$, $b \in I(1, n_{r_i})$, in site s^r_i between a susceptible S^{ur}_{ki} from site s^u_k in the region C_u and an infectious individual I^{vr}_{mi} from site s^v_m in the region C_v. Then the infectious rate (average number of contacts per individual per unit time required to transmit the disease) $\beta^{ruv*}_{i_bkm}$ in a zone $z^r_{i_b}$ between S^{ur}_{ki} and I^{vr}_{mi} is given by

$$\beta^{ruv*}_{i_bkm} = (p^r_{i_b})^2 \beta^{ruv*}_{ikm} \tag{6.37}$$

for $v, u \in I(1, M)$, $v \neq u$. The infection process in a zone $z^r_{i_b}$ is illustrated by the transition

$$S^{ur}_{ki} + I^{vr}_{mi} \xrightarrow{\beta^{ruv*}_{i_bkm}} I^{ur}_{ki} + I^{vr}_{mi}. \tag{6.38}$$

Hence the net conversion rate to the infectious class from the susceptible class during the disease transmission process at the site s^r_i in the region C_r of the metapopulation with M regions is given by

$$\sum_{v=1}^{M} \sum_{u=1}^{M} \sum_{m=1}^{n_v} \sum_{k=1}^{n_u} \sum_{b=1}^{n_{r_i}} \beta^{ruv*}_{i_bkm} I^{vr}_{mi} S^{ur}_{ki}. \tag{6.39}$$

We set

$$\beta^{ruv}_{ikm} = \sum_{b=1}^{n_{r_i}} \beta^{ruv*}_{i_bkm}. \tag{6.40}$$

We further assume that the disease status of an individual in the population does not affect the travel rates and mobility pattern.

A diagram illustrating the disease transmission and mobility processes in the two-scale dynamic structure described in Definition 6.3.2 is exhibited in Figure 6.1.

Definition 6.3.3 (Acquisition and loss of immunity Process). The changes in environmental conditions influence the immunity systems of individuals in the large-scale two-level population dynamic structure. This leads to dependence of the acquisition and loss of immunity rates of residents of all sites in all regions in the two-scale structured population on the current locations of the residents in the population dynamic structure. In each site s^r_i, let $\frac{1}{\varrho^r_i}$ be the average active infectious period of infected individual (I) who recovered from the disease and acquired immunity (R) immediately after the infectious period. Also, let $\frac{1}{\eta^r_i}$ be the average infectious period of infected person in site s^r_i

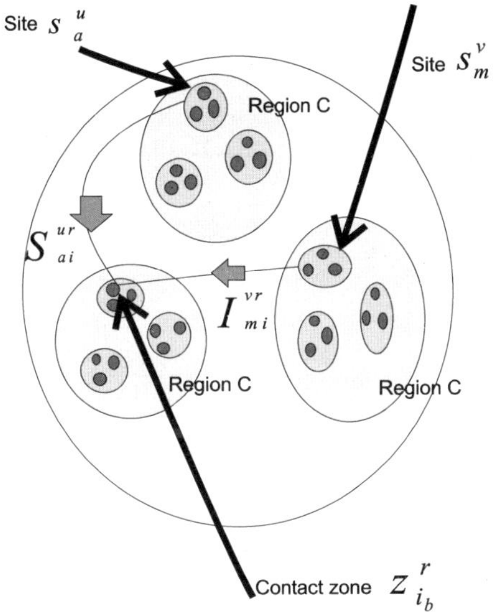

Figure 6.1: The movement of susceptible (S_{ai}^{ur}) and infective (I_{mi}^{vr}) from an arbitrary home site s_a^u in the region C_u and from site s_m^v in the region C_v to visit an arbitrary contact zone $z_{i_b}^r$ in site s_i^r in the region C_r. Disease transmission takes place in the zone $z_{i_b}^r$.

who recovered from the disease and became susceptible (S) immediately after the infectious period. Furthermore, let $\frac{1}{\alpha_i^r}$ be the average immunity period of removal person (R) in site s_i^r who has lost his/her immunity and became susceptible (S) again immediately after the immunity period. The recovery process of an infected person in site s_i^r as well as the loss of immunity of a removal person is illustrated by the following disease transition processes:

$$I_{ki}^{ur} \xrightarrow{\varrho_i^r} R_{ki}^{ur}, \quad I_{ki}^{ur} \xrightarrow{\eta_i^r} S_{ki}^{ur}, \quad R_{ki}^{ur} \xrightarrow{\alpha_i^r} S_{ki}^{ur} \tag{6.41}$$

for $u \in I(1, M)$ and $k \in I(1, n_u)$.

Definition 6.3.4 (Population demography). The current SIRS infectious disease involves time scales comparable with the life time of individuals in the population. Furthermore, all births occur at home site, and deaths occur at current locations of residents in the two-scale population structure. Let B_i^r be a constant birthrate of the human population at site s_i^r and at time t. We assume that every new born is a susceptible and becomes a resident of the site of birth. Let δ_i^r be the per capita natural mortality rate, and let d_i^r be the per capita disease-related mortality rate of all members of the effective population at site s_i^r.

6.4 Derivation of the mean-field two-scale network epidemic dynamic model

In this section, the Step 1 of Algorithm 6.2.1 is addressed, that is, the dynamic equations for the mean-field two-scale network SIRS epidemic model are obtained. Applying a method similar to that in Section 5.4, the mean-field epidemic dynamic model is derived as a system of ordinary differential equations representing the rate of change of the mean of different disease states S, I, R in the compartmental system comprising of the exponential transition rates between the states S, I, R exhibited in the framework given in Figure 6.2, where it is assumed that the transition times between compartments are exponentially distributed and that the underlying process for the transitions between the states is a CTMC.

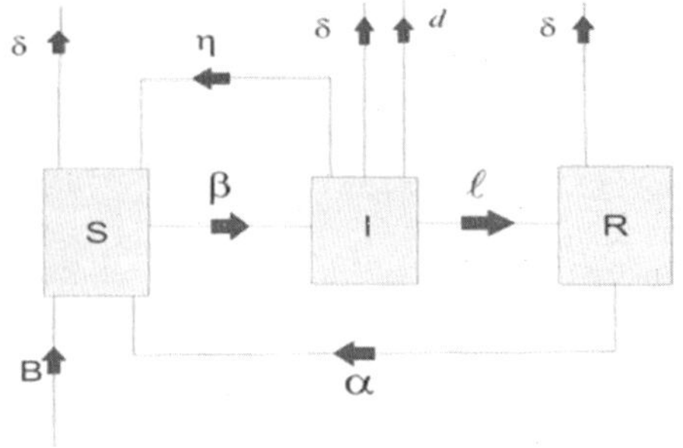

Figure 6.2: Compartmental framework summarizing the transition stages in the SIRS epidemic process. All the parameters presented in this figure are define in Section 6.3 for particular sites and regions.

From Definitions 6.3.1–6.3.4 the complete SIRS epidemic model under the influence of the two-scale dynamic human mobility process in Section 5.4 is given by

$$\frac{dS_{il}^{rq}}{dt} = \begin{cases} [B_i^r + \sum_{k=1}^{n_r} \rho_{ik}^{rr} S_{ik}^{rr} + \sum_{q\neq r}^{M} \sum_{a=1}^{n_q} \rho_{ia}^{rq} S_{ia}^{rq} + \eta_i^r I_{ii}^{rr} + \alpha_i^r R_{ii}^{rr} \\ \quad - (\gamma_i^r + \sigma_i^r + \delta_i^r) S_{ii}^{rr} - \sum_{u=1}^{M} \sum_{a=1}^{n_u} \beta_{iia}^{rru} S_{ii}^{rr} I_{ai}^{ur}], \quad q = r, l = i, \\ [\sigma_{ij}^{rr} S_{ii}^{rr} + \eta_j^r I_{ij}^{rr} + \alpha_j^r R_{ij}^{rr} - (\rho_{ij}^{rr} + \delta_j^r) S_{ij}^{rr} - \sum_{u=1}^{M} \sum_{a=1}^{n_u} \beta_{jia}^{rru} S_{ij}^{rr} I_{aj}^{ur}], \\ \quad q = r, l = j, j \neq i, \\ [\gamma_{il}^{rq} S_{ii}^{rr} + \eta_l^q I_{il}^{rq} + \alpha_l^q R_{il}^{rq} - (\rho_{il}^{rq} + \delta_l^q) S_{il}^{rq} - \sum_{u=1}^{M} \sum_{a=1}^{n_u} \beta_{lia}^{qru} S_{il}^{rq} I_{al}^{uq}], \\ \quad q \neq r, \end{cases} \tag{6.42}$$

$$\frac{dI_{il}^{rq}}{dt} = \begin{cases} [\sum_{k=1}^{n_r} \rho_{ik}^{rr} I_{ik}^{rr} + \sum_{q\neq r}^{M} \sum_{a=1}^{n_q} \rho_{ia}^{rq} I_{ia}^{rq} - \eta_i^r I_{ii}^{rr} - \varrho_i^r I_{ii}^{rr} \\ \quad - (\gamma_i^r + \sigma_i^r + \delta_i^r + d_i^r) I_{ii}^{rr} + \sum_{u=1}^{M} \sum_{a=1}^{n_u} \beta_{iia}^{rru} S_{ii}^{rr} I_{ai}^{ur}], \quad q = r, l = i, \\ [\sigma_{ij}^{rr} I_{ii}^{rr} - \eta_j^r I_{ij}^{rr} - \varrho_j^r I_{ij}^{rr} - (\rho_{ij}^{rr} + \delta_j^r + d_j^r) I_{ij}^{rr} + \sum_{u=1}^{M} \sum_{a=1}^{n_u} \beta_{jia}^{rru} S_{ij}^{rr} I_{aj}^{ur}], \\ \quad q = r, l = j, i \neq j, \\ [\gamma_{il}^{rq} I_{ii}^{rr} - \eta_l^q I_{il}^{rq} - \varrho_l^q I_{il}^{rq} - (\rho_{il}^{rq} + \delta_l^q + d_l^q) I_{il}^{rq} + \sum_{u=1}^{M} \sum_{a=1}^{n_u} \beta_{lia}^{qru} S_{il}^{rq} I_{al}^{uq}], \\ \quad q \neq r, \end{cases} \tag{6.43}$$

$$\frac{dR_{il}^{rq}}{dt} = \begin{cases} [\sum_{k=1}^{n_r} \rho_{ik}^{rr} R_{ik}^{rr} + \sum_{q\neq r}^{M} \sum_{l=1}^{n_q} \rho_{il}^{rq} R_{il}^{rq} + \varrho_i^r I_{ii}^{rr} - (\gamma_i^r + \sigma_i^r + \alpha_i^r + \delta_i^r) R_{ii}^{rr}], \\ \quad q = r, l = i \\ [\sigma_{ij}^{rr} R_{ii}^{rr} + \varrho_j^r I_{ij}^{rr} - (\rho_{ij}^{rr} + \alpha_j^r + \delta_j^r) R_{ij}^{rr}], \quad q = r, l = j, i \neq j, \\ [\gamma_{il}^{rq} R_{ii}^{rr} + \varrho_l^q I_{il}^{rq} - (\rho_{il}^{rq} + \alpha_l^q + \delta_l^q) R_{il}^{rq}], \quad q \neq r, \end{cases} \tag{6.44}$$

where $i \in I(1,n_r)$, $l \in I_i^r(1,n_q)$, $r \in I(1,M)$, $q \in I^r(1,M)$. Furthermore, the parameters B_i^r, η_a^u, a_a^u, δ_a^u, and d_a^u are nonnegative, and ϱ_a^u is positive for $r,u \in I(1,M)$, $i \in I(1,n_r)$, and $a \in I(1,n_u)$. Also, at time $t = t_0$, for all $r \in I(1,M)$ and $i \in I(1,n_r)$, $(S_{ii}^{rr}(t_0), S_{ij}^{rr}(t_0), S_{il}^{rq}(t_0)) = (S_{ii0}^{rr}, S_{ij0}^{rr}, S_{il0}^{rq})$, $(I_{ii}^{rr}(t_0), I_{ij}^{rr}(t_0), I_{il}^{rq}(t_0)) = (I_{ii0}^{rr}, I_{ij0}^{rr}, I_{il0}^{rq})$, and $(R_{ii}^{rr}(t_0), R_{ij}^{rr}(t_0), R_{il}^{rq}(t_0)) = (R_{ii0}^{rr}, R_{ij0}^{rr}, R_{il0}^{rq})$ whenever $j \in I_i^r(1,n_r)$ and $l \in I_i^r(1,n_q)$. Furthermore, we denote $n = \sum_{u=1}^M n_u$.

The state of system (6.42)–(6.44) in vector form is given below. We denote

$$\begin{aligned} x_{ia}^{ru} &= (S_{ia}^{ru}, I_{ia}^{ru}, R_{ia}^{ru})^T \in \mathbb{R}^3, \\ x_{i0}^{ru} &= (x_{i1}^{ruT}, x_{i2}^{ruT}, \ldots, x_{i,nu}^{ruT})^T \in \mathbb{R}^{3n_u}, \\ x_{00}^{ru} &= (x_{10}^{ruT}, x_{20}^{ruT}, \ldots, x_{n_r0}^{ruT})^T \in \mathbb{R}^{3n_r n_u}, \\ x_{00}^{r0} &= (x_{00}^{r1T}, x_{00}^{r2T}, \ldots, x_{00}^{rMT})^T \in \mathbb{R}^{3n_r \sum_{u=1}^M n_u}, \\ x_{00}^{00} &= (x_{00}^{10}, x_{00}^{20}, \ldots, x_{00}^{M0})^T \in \mathbb{R}^{3(\sum_{r=1}^M n_r)(\sum_{u=1}^M n_u)}, \end{aligned} \tag{6.45}$$

where $r,u \in I(1,M)$, $i \in I(1,n_r)$, $a \in I_i^r(1,n_u)$. We set $n = \sum_{u=1}^M n_u$.

Definition 6.4.1.

1. ***p*-norm in $\mathbb{R}^{3n^2}$:**
   Let $z_{00}^{00} \in \mathbb{R}^{3n^2}$ be an arbitrary vector defined in (6.45), where $z_{ia}^{ru} = (z_{ia1}^{ru0}, z_{ia2}^{ru0}, z_{ia3}^{ru0})^T$ for $r,u \in I(1,M)$, $i \in I(1,n_r)$, $a \in I_i^r(1,n_u)$. The p-norm on $\mathbb{R}^{3n^2}$ is defined as

$$\|z_{00}^{00}\|_p = \left(\sum_{r=1}^M \sum_{u=1}^M \sum_{i=1}^{n_r} \sum_{a=1}^{n_u} \sum_{j=1}^3 |z_{iaj}^{ru0}|^p \right)^{\frac{1}{p}} \tag{6.46}$$

   for $1 \le p < \infty$ and

$$\bar{z} \equiv \|z_{00}^{00}\|_p = \max_{1 \le r,u \le M, 1 \le i \le n_r, 1 \le a \le n_u, 1 \le j \le 3} |z_{iaj}^{ru0}| \tag{6.47}$$

   for $p = \infty$. Let

$$\underline{k} \equiv k_{00\min}^{00} = \min_{1 \le r,u \le M, 1 \le i \le n_r, 1 \le a \le n_u} |k_{ia}^{ru}|. \tag{6.48}$$

2. **Closed Ball in $\mathbb{R}^{3n^2}$:**
   Let $z_{00}^{*00} \in \mathbb{R}^{3n^2}$. The closed ball in $\mathbb{R}^{3n^2}$ with center at z_{00}^{*00} and radius $r > 0$ is the set

$$\bar{\mathfrak{B}}_{\mathbb{R}^{3n^2}}(z_{00}^{*00}; r) = \{z_{00}^{00} \in \mathbb{R}^{3n^2} : \|z_{00}^{00} - z_{00}^{*00}\|_p \le r\}. \tag{6.49}$$

6.5 Derivation of the Itô–Doob SDE for the SIRS epidemic dynamic model

In this section, we address Steps 2 and 3 of Algorithm 6.2.1. Specifically, we detail the derivation of white noise perturbations in the SIRS epidemic dynamics and their incorporation into the dynamic mean-field equations for the epidemic model (6.42)–(6.44).

6.5.1 Step 2: identification of the sources of variability and derivation of the white noise processes

The disease subprocesses governed by exponential rates, namely, the birth rate or influx rate B, the death rate δ, the disease transmission rate β, the recovery rates η and α, and the departure, visiting, and return rates σ and ρ, are subject to random environmental noise. These environmental perturbations introduce variability into the exponential rates, which are central in the epidemic model (6.42)–(6.44). Consequently, this variability is incorporated into the epidemic dynamic process described by the mean-field dynamic model.

To simplify the analysis and minimize model complexity, we assume, without loss of generality, that random environmental fluctuations primarily manifest as variations in the effective infection rate or disease transmission rate β. Denoting the random infection rate by $\tilde{\beta}$, the variability in $\tilde{\beta}$ over a small interval $[t, t+dt]$ is represented by a white noise process as follows:

$$\tilde{\beta}dt = \beta dt + \nu\xi(t)dt, \quad dw(t) = \xi(t)dt, \quad \mathbb{V}ar(\tilde{\beta}dt) = \nu^2 dt, \quad \mathbb{E}(\tilde{\beta}dt) = \beta dt, \tag{6.50}$$

where $\xi(t)$ is a standard white noise process, and $w(t)$ is the corresponding normalized Wiener process, also known as a homogeneous Brownian motion, with the following properties: $w(0) = 0$, $\mathbb{E}(w(t)) = 0$, and $\mathbb{V}ar(w(t)) = t$.

For all $r \in I(1, M)$ and $i \in I(1, n_r)$, the variability in the disease transmission process at sites s_i^r, s_j^r, and s_l^q—arising from interactions between a *susceptible* individual from site s_k^u and an *infective* individual from an arbitrary site s_m^v—results in the infection of the susceptible individual over the interval $[t, t+dt]$ at rates expressed in terms of white noise processes as follows:

$$\begin{aligned}
\tilde{\beta}_{ikm}^{ruv}dt &= \beta_{ikm}^{ruv}dt + \nu_{ikm}^{ruv}dw_{ikm}^{ruv}(t), \quad dw_{ikm}^{ruv}(t) = \xi_{ikm}^{ruv}(t)dt,\\
\tilde{\beta}_{jkm}^{ruv}dt &= \beta_{jkm}^{ruv} + \nu_{jkm}^{ruv}dw_{jkm}^{ruv}(t), \quad dw_{jkm}^{ruv}(t) = \xi_{jkm}^{ruv}(t)dt,\\
\tilde{\beta}_{lkm}^{quv}dt &= \beta_{lkm}^{quv} + \nu_{lkm}^{quv}dw_{lkm}^{quv}(t), \quad dw_{lkm}^{quv}(t) = \xi_{lkm}^{quv}(t)dt,
\end{aligned} \tag{6.51}$$

and

$$\mathbb{V}ar(\tilde{\beta}_{ikm}^{ruv}dt) = (\nu_{ikm}^{ruv})^2 dt, \quad \mathbb{V}ar(\tilde{\beta}_{jkm}^{ruv}dt) = (\nu_{jkm}^{ruv})^2 dt, \quad \mathbb{V}ar(\tilde{\beta}_{lkm}^{quv}dt) = (\nu_{lkm}^{quv})^2 dt, \tag{6.52}$$

where $q, u, v \in I^r(1, M)$, $k \in I_i^u(1, n_u)$, $m \in I_i^v(1, n_v)$, and $l \in I_i^r(1, n_q)$.

6.5.2 Step 3: derivation of the Itô–Doob SDE SIRS model

We substitute (6.51) into the mean-field dynamic model (6.42)–(6.44) by replacing the constant disease transmission rate βdt with the random variable that depends on time expressed as $\tilde{\beta} dt = \beta dt + \nu dw(t)$ and obtain the following system of Itô–Doob SDEs for the SIRS epidemic dynamics in the two-scale network population under the influence of the dynamic human mobility process in Chapter 5:

$$dS_{il}^{rq} = \begin{cases} [B_i^r + \sum_{k=1}^{n_r} \rho_{ik}^{rr} S_{ik}^{rr} + \sum_{q \neq r}^{M} \sum_{a=1}^{n_q} \rho_{ia}^{rq} S_{ia}^{rq} + \eta_i^r I_{ii}^{rr} + \alpha_i^r R_{ii}^{rr} \\ \quad - (\gamma_i^r + \sigma_i^r + \delta_i^r) S_{ii}^{rr} - \sum_{u=1}^{M} \sum_{a=1}^{n_u} \beta_{iia}^{rru} S_{ii}^{rr} I_{ai}^{ur}]dt \\ -[\sum_{u=1}^{M} \sum_{a=1}^{n_u} \nu_{iia}^{rru} S_{ii}^{rr} I_{ai}^{ur} dw_{iia}^{rru}(t)], \quad q = r, l = i, \\ [\sigma_{ij}^{rr} S_{ii}^{rr} + \eta_j^r I_{ij}^{rr} + \alpha_j^r R_{ij}^{rr} - (\rho_{ij}^{rr} + \delta_j^r) S_{ij}^{rr} - \sum_{u=1}^{M} \sum_{a=1}^{n_u} \beta_{jia}^{rru} S_{ij}^{rr} I_{aj}^{ur}]dt \\ \quad - [\sum_{u=1}^{M} \sum_{a=1}^{n_u} \nu_{jia}^{rru} S_{ij}^{rr} I_{aj}^{ur} dw_{jia}^{rru}(t)], \quad q = r, l = j, j \neq i, \\ [\gamma_{il}^{rq} S_{ii}^{rr} + \eta_l^q I_{il}^{rq} + \alpha_l^q R_{il}^{rq} - (\rho_{il}^{rq} + \delta_l^q) S_{il}^{rq} \\ \quad - \sum_{u=1}^{M} \sum_{a=1}^{n_u} \beta_{lia}^{qru} S_{il}^{rq} I_{al}^{uq}]dt - [\sum_{u=1}^{M} \sum_{a=1}^{n_u} \nu_{lia}^{qru} S_{il}^{rq} I_{al}^{uq} dw_{lia}^{qru}(t)], \quad q \neq r, \end{cases} \tag{6.53}$$

$$dI_{il}^{rq} = \begin{cases} [\sum_{k=1}^{n_r} \rho_{ik}^{rr} I_{ik}^{rr} + \sum_{q \neq r}^{M} \sum_{a=1}^{n_q} \rho_{ia}^{rq} I_{ia}^{rq} - \eta_i^r I_{ii}^{rr} - \varrho_i^r I_{ii}^{rr} \\ \quad - (\gamma_i^r + \sigma_i^r + \delta_i^r + d_i^r) I_{ii}^{rr} + \sum_{u=1}^{M} \sum_{a=1}^{n_u} \beta_{iia}^{rru} S_{ii}^{rr} I_{ai}^{ur}]dt \\ \quad + [\sum_{u=1}^{M} \sum_{a=1}^{n_u} \nu_{iia}^{rru} S_{ii}^{rr} I_{ai}^{ur} dw_{iia}^{rru}(t)], \quad q = r, l = i, \\ [\sigma_{ij}^{rr} I_{ii}^{rr} - \eta_j^r I_{ij}^{rr} - \varrho_j^r I_{ij}^{rr} - (\rho_{ij}^{rr} + \delta_j^r + d_j^r) I_{ij}^{rr} + \sum_{u=1}^{M} \sum_{a=1}^{n_u} \beta_{jia}^{rru} S_{ij}^{rr} I_{aj}^{ur}]dt \\ \quad + [\sum_{u=1}^{M} \sum_{a=1}^{n_u} \nu_{jia}^{rru} S_{ij}^{rr} I_{aj}^{ur} dw_{jia}^{rru}(t)], \quad q = r, l = j, j \neq i, \\ [\gamma_{il}^{rq} I_{ii}^{rr} - \eta_l^q I_{il}^{rq} - \varrho_l^q I_{il}^{rq} - (\rho_{il}^{rq} + \delta_l^q + d_l^q) I_{il}^{rq} \\ \quad + \sum_{u=1}^{M} \sum_{a=1}^{n_u} \beta_{lia}^{qru} S_{il}^{rq} I_{al}^{uq}]dt + [\sum_{u=1}^{M} \sum_{a=1}^{n_u} \nu_{lia}^{qru} S_{il}^{rq} I_{al}^{uq} dw_{lia}^{qru}(t)], \quad q \neq r, \end{cases} \tag{6.54}$$

$$dR_{il}^{rq} = \begin{cases} [\sum_{k=1}^{n_r} \rho_{ik}^{rr} R_{ik}^{rr} + \sum_{q \neq r}^{M} \sum_{l=1}^{n_q} \rho_{il}^{rq} R_{il}^{rq} + \varrho_i^r I_{ii}^{rr} - (\gamma_i^r + \sigma_i^r + \alpha_i^r + \delta_i^r) R_{ii}^{rr}]dt, \\ \quad \text{for } q = r, l = i \\ [\sigma_{ij}^{rr} R_{ii}^{rr} + \varrho_j^r I_{ij}^{rr} - (\rho_{ij}^{rr} + \alpha_j^r + \delta_j^r) R_{ij}^{rr}]dt, \quad q = r, l = j, j \neq i, \\ [\gamma_{il}^{rq} R_{ii}^{rr} + \varrho_l^q I_{il}^{rq} - (\rho_{il}^{rq} + \alpha_l^q + \delta_l^q) R_{il}^{rq}]dt, \quad q \neq r, \end{cases} \tag{6.55}$$

where $i \in I(1, n_r)$, $l \in I_i^r(1, n_q)$,; $r \in I(1, M)$, $q \in I^r(1, M)$, and all parameters are as defined before.

At time $t = t_0$, for all $r \in I(1, M)$ and $i \in I(1, n_r)$,

$$\begin{aligned} (S_{ii}^{rr}(t_0), S_{ij}^{rr}(t_0), S_{il}^{rq}(t_0)) &= (S_{ii0}^{rr}, S_{ij0}^{rr}, S_{il0}^{rq}), \\ (I_{ii}^{rr}(t_0), I_{ij}^{rr}(t_0), I_{il}^{rq}(t_0)) &= (I_{ii0}^{rr}, I_{ij0}^{rr}, I_{il0}^{rq}), \\ (R_{ii}^{rr}(t_0), R_{ij}^{rr}(t_0), R_{il}^{rq}(t_0)) &= (R_{ii0}^{rr}, R_{ij0}^{rr}, R_{il0}^{rq}) \end{aligned} \tag{6.56}$$

for $j \in I_i^r(1,n_r)$ and $l \in I_i^r(1,n_q)$, where the random variables $(S_{ii}^{rr}(t_0), S_{ij}^{rr}(t_0), S_{il}^{rq}(t_0))$, $(I_{ii}^{rr}(t_0), I_{ij}^{rr}(t_0), I_{il}^{rq}(t_0))$, and $(R_{ii}^{rr}(t_0), R_{ij}^{rr}(t_0), R_{il}^{rq}(t_0))$ are $\mathcal{F}_0$-measurable and are independent of $w(t)$ for $t \geq t_0$.

6.6 Model validation results for the two-scale network stochastic epidemic dynamic model

In this section, we validate the epidemic dynamic model (6.53)–(6.55) by showing that the system leads to a unique stochastic process with almost surely continuous positive paths. Furthermore, it is essential that the paths of the stochastic process do not have unbounded paths to lead to a meaningful characterization of a real disease epidemic dynamics. That is, we show that the initial value problem associated with system (6.53)–(6.55) has a unique solution. Observe that the rate functions of the system are nonlinear and locally Lipschitz continuous with respect to x_{00}^{00} but do not satisfy the linear growth condition. As a result, the classical existence and uniqueness results [24] are not applicable. Therefore we use the Lyapunov energy function method (see [10, 11, 24, 17]) to prove the existence and uniqueness of a solution process of the system. We first state and prove two lemmas that are useful for the proof of the existence and uniqueness. From (6.53)–(6.55), define the vector $y_{00}^{00} \in \mathbb{R}^{n^2}$ as follows: For $i \in I(1,n_r)$, $l \in I_i^r(1,n_q)$, $r \in I(1,M)$, and $q \in I^r(1,M)$,

$$\begin{aligned}
y_{ia}^{ru} &= S_{ia}^{ru} + I_{ia}^{ru} + R_{ia}^{ru} \in \mathbb{R}_+ = [0,\infty),\\
y_{i0}^{ru} &= (y_{i1}^{ru}, y_{i2}^{ru}, \ldots, y_{i,nu}^{ru})^T \in \mathbb{R}_+^{n_u},\\
y_{00}^{ru} &= (y_{10}^{ruT}, y_{20}^{ruT}, \ldots, y_{n_r0}^{ruT})^T \in \mathbb{R}_+^{n_r n_u},\\
y_{00}^{r0} &= (y_{00}^{r1T}, y_{00}^{r2T}, \ldots, y_{00}^{rMT})^T \in \mathbb{R}_+^{n_r \sum_{u=1}^{M} n_u},\\
y_{00}^{00} &= (y_{00}^{10T}, y_{00}^{20T}, \ldots, y_{00}^{M0T})^T \in \mathbb{R}_+^{(\sum_{r=1}^{M} n_r)(\sum_{u=1}^{M} n_u)}.
\end{aligned} \tag{6.57}$$

Then we obtain

$$dy_{il}^{rq} = \begin{cases} [B_i^r + \sum_{k\neq i}^{n_r} \rho_{ik}^{rr} y_{ik}^{rr} + \sum_{q\neq r}^{M} \sum_{a=1}^{n_q} \rho_{ia}^{rq} y_{ia}^{rq} - (\gamma_i^r + \sigma_i^r + \delta_i^r) y_{ii}^{rr} - d_i^r I_{ii}^{rr}]dt, \\ \quad q = r, l = i \\ [\sigma_{ij}^{rr} y_{ii}^{rr} - (\rho_{ij}^{rr} + \delta_j^r) y_{ij}^{rr} - d_j^r I_{ij}^{rr}]dt, \quad q = r, a = j, i \neq j, \\ [\gamma_{il}^{rq} y_{ii}^{rr} - (\rho_{il}^{rq} + \delta_l^q) y_{il}^{rq} - d_l^q I_{il}^{rq}]dt, \quad q \neq r, y_{il}^{rq}(t_0) \geq 0. \end{cases} \tag{6.58}$$

In the following, we show that the solution process of the initial value problem (6.58) is nonnegative, that is, $y_{ia}^{ru}(t) \geq 0$ for all $t \geq 0$ whenever $y_{ia}^{ru}(t_0) \geq 0$.

Lemma 6.6.1. *Let $r,u \in I(1,M)$, $i \in I^r(1,n_r)$, and $a \in I_i^r(1,n_u)$. If $y_{ia}^{ru}(t_0) \geq 0$, then $y_{ia}^{ru}(t) \geq 0$ for all $t \geq t_0$.*

Proof. It follows from (6.57) and (6.53)–(6.55) that system (6.58) is of the form $u' = A(t,u)w(t,u)$, $u(t_0) \geq 0$, in [22, equation (8)] and satisfies the quasimonotonicity condition. Furthermore, from Remark 4 in [22] we state that this system (6.58) has nonnegative solutions whenever $y_{il}^{rq}(0) \geq 0$, $i \in I(1,n_r)$, $l \in I_i^r(1,n_q)$, $r \in I(1,M)$, $q \in I^r(1,M)$. □

Remark 6.6.1. From the decomposition described in (6.34) we observe that $y_{ia}^{ru}(t) = N_{ia}^{ru} = S_{ia}^{ru}(t) + I_{ia}^{ru}(t) + R_{ia}^{ru}(t)$. Furthermore, $N_{i0}^{rr} = \sum_{u=1}^{M}\sum_{a=1}^{n_u} y_{ia}^{ru}$. Therefore Lemma 6.6.1 establishes that for any nonnegative initial endemic population, the number of residents of site s_i^r present at home, y_{ii}^{rr}, or visiting any given site s_{ia}^{ru} in any other region C_u, y_{ia}^{ru}, is nonnegative. This implies that the total population of residents of site s_i^r present at home and also visiting sites in regions in their intra- and interregional accessible domains, $N_{i0}^{rr}(t)$, is nonnegative. Moreover, Lemma 6.6.1 exhibits that the effective population at any site in any region given by (6.36) is nonnegative at all times $t \geq t_0$. Furthermore, $R_+^{n^2} = \{y \in R^{n^2} : y \geq 0\}$ is a self-invariant set with respect to (6.58).

In the following lemma, we use Lemma 6.6.1 to find an upper bound for the solution process of (6.53)–(6.55).

Lemma 6.6.2. *Let* $\mu = \min_{1\leq u\leq M, 1\leq a\leq n_u}(\delta_a^u)$. *If*

$$\sum_{r=1}^{M}\sum_{u=1}^{M}\sum_{i=1}^{n_r}\sum_{a=1}^{n_u} y_{ia}^{ru}(t_0) \leq \frac{1}{\mu}\sum_{r=1}^{M}\sum_{i=1}^{n_r} B_i^r, \tag{6.59}$$

then

$$\sum_{r=1}^{M}\sum_{u=1}^{M}\sum_{i=1}^{n_r}\sum_{a=1}^{n_u} y_{ia}^{ru}(t) \leq \frac{1}{\mu}\sum_{r=1}^{M}\sum_{i=1}^{n_r} B_i^r \quad \text{for } t \geq 0 \quad \text{a. s.} \tag{6.60}$$

Proof. From (6.57), define

$$\sum_{r=1}^{M}\sum_{u=1}^{M}\sum_{i=1}^{n_r}\sum_{a=1}^{n_u} dy_{ia}^{ru} = \sum_{r=1}^{M}\sum_{i=1}^{n_r}\left[dy_{ii}^{rr} + \sum_{a\neq i}^{n_r} dy_{ia}^{rr} + \sum_{u\neq r}^{M}\sum_{a=1}^{n_r} dy_{ia}^{ru}\right]. \tag{6.61}$$

From (6.53)–(6.55) and (6.61) we can see that

$$\sum_{r=1}^{M}\sum_{u=1}^{M}\sum_{i=1}^{n_r}\sum_{a=1}^{n_u} dy_{ia}^{ru} = \left[\sum_{r=1}^{M}\sum_{i=1}^{n_r} B_i^r - \sum_{r=1}^{M}\sum_{i=1}^{n_r}\sum_{u=1}^{M}\sum_{a=1}^{n_u}(\delta_a^u y_{ia}^{ru} + d_a^u I_{ia}^{ru})\right]dt. \tag{6.62}$$

From Lemma 6.6.1 and (6.62) we have

$$d\left\{\sum_{r=1}^{M}\sum_{u=1}^{M}\sum_{i=1}^{n_r}\sum_{a=1}^{n_u} y_{ia}^{ru}\right\} \leq \left[\sum_{r=1}^{M}\sum_{i=1}^{n_r} B_i^r - \mu\sum_{r=1}^{M}\sum_{i=1}^{n_r}\sum_{u=1}^{M}\sum_{a=1}^{n_u} y_{ia}^{ru}\right]dt \tag{6.63}$$

for a nonnegative differential of t. We note that (6.63) is a first-order deterministic differential inequality [24], and its solution is given by

$$\sum_{r=1}^{M}\sum_{u=1}^{M}\sum_{i=1}^{n_r}\sum_{a=1}^{n_u} y_{ia}^{ru}(t) \leq \frac{1}{\mu}\sum_{r=1}^{M}\sum_{i=1}^{n_r} B_i^r + \left[\sum_{r=1}^{M}\sum_{i=1}^{n_r}\sum_{u=1}^{M}\sum_{a=1}^{n_u} y_{ia}^{ru}(t_0)\right] e^{-\mu t}. \tag{6.64}$$

Therefore (6.60) is satisfied, provided that (6.59) is valid. □

Remark 6.6.2. From Lemma 6.6.2 we conclude that a closed ball in R^{3n^2} under the sum norm with radius $r = \frac{1}{\mu}\sum_{r=1}^{M}\sum_{i=1}^{n_r} B_i^r$ is self-invariant with regard to a two-scale network dynamic of human epidemic process that is under the influence of human mobility process [45].

Before presenting the model validation result, we need to establish an auxiliary result. This result provides a fundamental tool in the context of the energy Lyapunov function approach. We will use the following definition.

Definition 6.6.1 (The Itô formula). Let $\{x(t), t \geq 0\}$, where $x \in \mathbb{R}^{3n}$, be a stochastic process that satisfies the SDE equation

$$dx = f(x,t)dt + g(x,t)dw, \tag{6.65}$$

where $\{w(t), t \geq 0\}$ is a Wiener process. Furthermore, let $V \in C^{2,1}(\mathbb{R}^{3n} \times \mathbb{R}^+, \mathbb{R}^+)$, where $(x(t), t) \mapsto V(x(t), t)$ is a twice continuously differentiable function with respect to $x(t)$ and continuously differentiable with respect to t. The Itô–Doob stochastic differential of V with respect to (6.65) is given by

$$dV = V_t dt + V_x dx + \frac{1}{2} dx^T V_{xx} dx \tag{6.66}$$

with gradient vectors

$$V_t = \frac{\partial V}{\partial t}, \quad V_x = \left(\frac{\partial V}{\partial x_1(t)}, \frac{\partial V}{\partial x_2(t)}, \ldots, \frac{\partial V}{\partial x_{3n}(t)}\right), \tag{6.67}$$

and

$$V_{xx} = \left(\frac{\partial V_x}{\partial x_1(t)}, \frac{\partial V_x}{\partial x_2(t)}, \ldots, \frac{\partial V_x}{\partial x_{3n}(t)}\right). \tag{6.68}$$

Lemma 6.6.3. *Let us assume that the hypotheses of Lemma* 6.6.2 *be satisfied. Let V be the function defined by* $V : \mathbb{R}_+^{3n^2} \times \mathbb{R}_+ \to \bar{\mathbb{R}}_+$ *as follows:*

$$V(x_{00}^{00}) = \sum_{r=1}^{M}\sum_{i=1}^{n_r}\sum_{u=1}^{M}\sum_{a=1}^{n_u} V_{ia}^{ru}(x_{ia}^{ru}), \tag{6.69}$$

where

$$V_{ia}^{ru}(x_{ia}^{ru}) = [(S_{ia}^{ru} - 1 - \log S_{ia}^{ru}) + (I_{ia}^{ru} - 1 - \log I_{ia}^{ru}) + (R_{ia}^{ru} - 1 - \log R_{ia}^{ru})]. \tag{6.70}$$

Furthermore, let us denote

$$
\begin{aligned}
M_{001}^{00} &= \max_{1\le r,q\le M, q\ne r, 1\le l\le n_q} 1 + \frac{S_{il}^{rq}}{I_{il}^{rq}},\\
M_{002}^{00} &= \max_{1\le r,q\le M, q\ne r, 1\le l\le n_q} 1 + \frac{(S_{il}^{rq})^2}{(I_{il}^{rq})^2},\\
N_{00}^{00} &= \max_{1\le r,q\le M, q\ne r, 1\le l\le n_q} 1 + S_{il}^{rq},\\
\beta_{000}^{000} &= \max_{1\le r,q,u\le M, q\ne r, 1\le l,a\le n_{q,u}} \beta_{ail}^{urq},\\
\nu_{000}^{000} &= \max_{1\le r,q,u\le M, q\ne r, 1\le l,a\le n_{q,u}} \nu_{ail}^{urq},\\
(\rho_{00}^{00}, \alpha_0^0, \delta_0^0, d_0^0, \sigma_{i0}^{00}, \varrho_0^0) &= \max_{1\le r,u\le M, 1\le a\le n_u} (\rho_{ia}^{ru}, \alpha_a^u, \delta_a^u, d_a^u, \sigma_{ia}^{ru}, \varrho_a^u).
\end{aligned}
\tag{6.71}
$$

Then there exists $\tilde{K} > 0$ such that

$$
dV(x_{00}^{00}) \le \tilde{K}dt + \sum_{r=1}^{M}\sum_{i=1}^{n_r}\sum_{u=1}^{M}\sum_{a=1}^{n_u}\sum_{v=1}^{M}\sum_{b=1}^{n_v}\left(1 - \frac{S_{ia}^{ru}}{I_{ia}^{ru}}\right)\nu_{aib}^{urv} I_{ba}^{vu}\, dw_{aib}^{urv}. \tag{6.72}
$$

Proof. For $r, u \in I(1, M)$, $i \in I^r(1, n_r)$, and $a \in I_i^r(1, n_u)$, under the assumptions of Lemma 6.6.2 and the definitions of S_{ia}^u, I_{ia}^u, and R_{ia}^u, the function V defined in (6.69) belongs to $C^{2,1}(\mathbb{R}_+^{3n^2} \times \mathbb{R}_+, \bar{\mathbb{R}}_+)$. Moreover, we rewrite (6.69) as

$$
\begin{aligned}
V(x_{00}^{00}) &= \sum_{r=1}^{M}\sum_{i=1}^{n_r}\sum_{u=1}^{M}\sum_{a=1}^{n_u} V_{ia}^{ru}(x_{00}^{00})\\
&= \sum_{r=1}^{M}\sum_{i=1}^{n_r}\left\{V_{ii}^{rr}(x_{00}^{00}) + \sum_{a\ne i}^{n_r} V_{ia}^{rr}(x_{00}^{00}) + \sum_{u\ne r}^{M}\sum_{a=1}^{n_u} V_{ia}^{ru}(x_{00}^{00})\right\},
\end{aligned}
\tag{6.73}
$$

where

$$
V_{ia}^{ru}(x_{00}^{00}) = (S_{ia}^{ru} - 1 - \log S_{ia}^{ru}) + (I_{ia}^{ru} - 1 - \log I_{ia}^{ru}) + (R_{ia}^{ru} - 1 - \log R_{ia}^{ru}). \tag{6.74}
$$

From (6.73) and (6.74) it follows that

$$
dV(x_{00}^{00}) = \sum_{r=1}^{M}\sum_{i=1}^{n_r}\left\{dV_{ii}^{rr}(x_{00}^{00}) + \sum_{a\ne i}^{n_r} dV_{ia}^{rr}(x_{00}^{00}) + \sum_{u\ne r}^{M}\sum_{a=1}^{n_u} dV_{ia}^{ru}(x_{00}^{00})\right\}, \tag{6.75}
$$

where

$$
\begin{aligned}
dV_{ia}^{ru}(x_{00}^{00}) = {}& \left[\left(1 - \frac{1}{S_{ia}^{ru}}\right)dS_{ia}^{ru} + \frac{1}{2(S_{ia}^{ru})^2}(dS_{ia}^{ru})^2\right] + \left[\left(1 - \frac{1}{I_{ia}^{ru}}\right)dI_{ia}^{ru} + \frac{1}{2(I_{ia}^{ru})^2}(dI_{ia}^{ru})^2\right]\\
& + \left[\left(1 - \frac{1}{R_{ia}^{ru}}\right)dR_{ia}^{ru} + \frac{1}{2(R_{ia}^{ru})^2}(dR_{ia}^{ru})^2\right].
\end{aligned}
\tag{6.76}
$$

In the following, by considering the positive differential of t $(0 < \Delta t \approx dt)$, using the nature of the rate coefficients of (6.53)–(6.55) and definitions (6.71), we carefully estimate the three terms in the right-hand side of (6.76). This is achieved by the usage of nested argument process.

Site level

The estimates of terms in the right-hand side of (6.76) for the case of $u = r$ and $a = i$:

$$\begin{aligned}
&\left[\left(1-\frac{1}{S_{ii}^{rr}}\right)dS_{ii}^{rr} + \frac{1}{2(S_{ii}^{rr})^2}\left(dS_{ii}^{rr}\right)^2\right] \\
&\quad \le \left\{B_i^r + \sum_{b=1}^{n_r}\rho_{ib}^{rr}S_{ib}^{rr} + \sum_{v\neq r}^{M}\sum_{b=1}^{n_v}\rho_{ib}^{rv}S_{ib}^{rv} + \eta_i^r I_{ii}^{rr} + \alpha_i^r R_{ii}^{rr} + (\gamma_i^r + \sigma_i^r + \delta_i^r + d_i^r)\right. \\
&\qquad \left. + \sum_{v=1}^{M}\sum_{b=1}^{n_v}\beta_{iib}^{rrv}I_{bi}^{vr} + \frac{1}{2}\sum_{v=1}^{M}\sum_{b=1}^{n_v}(\nu_{iib}^{rrv})^2(I_{bi}^{vr})^2\right\}dt + \left[(1-S_{ii}^{rr})\sum_{v=1}^{M}\sum_{b=1}^{n_v}\nu_{iib}^{rrv}I_{bi}^{vr}dw_{iib}^{rrv}\right], \qquad (6.77)
\end{aligned}$$

$$\begin{aligned}
&\left[\left(1-\frac{1}{I_{ii}^{rr}}\right)dI_{ii}^{rr} + \frac{1}{2(I_{ii}^{rr})^2}\left(dI_{ii}^{rr}\right)^2\right] \\
&\quad \le \left\{\sum_{b=1}^{n_r}\rho_{ib}^{rr}I_{ib}^{rr} + \sum_{v\neq r}^{M}\sum_{b=1}^{n_v}\rho_{ib}^{rv}I_{ib}^{rv} + (\varrho_i^r + \eta_i^r + \gamma_i^r + \sigma_i^r + \delta_i^r + d_i^r) + \sum_{v=1}^{M}\sum_{b=1}^{n_v}\beta_{iib}^{rrv}\frac{S_{ii}^{rr}I_{bi}^{vr}}{I_{ii}^{rr}}\right. \\
&\qquad \left. + \frac{1}{2(I_{ii}^{rr})^2}(S_{ii}^{rr})^2\sum_{v=1}^{M}\sum_{b=1}^{n_v}(\nu_{iib}^{rrv})^2(I_{bi}^{vr})^2\right\}dt + \left(S_{ii}^{rr} - \frac{S_{ii}^{rr}}{I_{ii}^{rr}}\right)\left[\sum_{v=1}^{M}\sum_{b=1}^{n_v}\nu_{iib}^{rrv}I_{bi}^{vr}dw_{iib}^{rrv}\right], \qquad (6.78)
\end{aligned}$$

and

$$\left(1-\frac{1}{R_{ii}^{rr}}\right)dR_{ii}^{rr} \le \left[\sum_{b=1}^{n_r}\rho_{ib}^{rr}R_{ib}^{rr} + \sum_{v\neq r}^{M}\sum_{b=1}^{n_v}\rho_{ib}^{rv}R_{ib}^{rv} + \varrho_i^r I_{ii}^{rr} + (\gamma_i^r + \alpha_i^r + \sigma_i^r + \delta_i^r + d_i^r)\right]dt. \qquad (6.79)$$

Regional level

The estimated terms in the right-hand side of (6.76) for the case of $u = r$ and $a \neq i$:

$$\begin{aligned}
\sum_{a\neq i}^{n_r}\left[\left(1-\frac{1}{S_{ia}^{rr}}\right)dS_{ia}^{rr} + \frac{1}{2(S_{ia}^{rr})^2}\left(dS_{ia}^{rr}\right)^2\right] \le \sum_{a\neq i}^{n_r}&\left\{\left[\sigma_{ia}^{rr}S_{ii}^{rr} + \eta_a^r I_{ia}^{rr} + \alpha_a^r R_{ia}^{rr} + \sum_{v=1}^{M}\sum_{b=1}^{n_v}\beta_{aib}^{rrv}I_{ba}^{vr}\right.\right. \\
&+ (\sigma_{ia}^{rr} + \delta_a^r + d_a^r) \\
&\left. + \frac{1}{2}\left(\sum_{v=1}^{M}\sum_{b=1}^{n_v}(\nu_{aib}^{rrv})^2(I_{bi}^{vr})^2\right)\right]dt \\
&\left. + (1-S_{ia}^{rr})\sum_{v=1}^{M}\sum_{b=1}^{n_v}\nu_{aib}^{rrv}I_{bi}^{vr}dw_{aib}^{rrv}\right\}, \qquad (6.80)
\end{aligned}$$

$$\sum_{a\neq i}^{n_r}\left[\left(1-\frac{1}{I_{ia}^{rr}}\right)dI_{ia}^{rr} + \frac{1}{2(I_{ia}^{rr})^2}\left(dI_{ia}^{rr}\right)^2\right] \le \sum_{a\neq i}^{n_r}\left\{\left[\sigma_{ia}^{rr}I_{ii}^{rr} + \sum_{v=1}^{M}\sum_{b=1}^{n_v}\beta_{aib}^{rrv}S_{ia}^{rr}I_{ba}^{vr}\right.\right.$$

$$+ (\varrho_a^r + \eta_a^r + \rho_{ia}^{rr} + \delta_a^r + d_a^r)$$
$$+\frac{(S_{ia}^{rr})^2}{2(I_{ia}^{rr})^2}\sum_{v=1}^{M}\sum_{b=1}^{n_v}(v_{aib}^{rrv})^2(I_{ba}^{vr})^2\Bigg]dt$$
$$+\left(S_{ia}^{rr} - \frac{S_{ia}^{rr}}{I_{ii}^{rr}}\right)\left[\sum_{v=1}^{M}\sum_{b=1}^{n_v} v_{aib}^{rrv} I_{ba}^{vr} dw_{aib}^{rrv}\right]\Bigg\}, \tag{6.81}$$

and

$$\sum_{a\neq i}^{n_r}\left(1 - \frac{1}{R_{ia}^{rr}}\right)dR_{ia}^{rr} \leq \sum_{a\neq i}^{n_r}[\sigma_{ia}^{rr}R_{ii}^{rr} + \varrho_a^r I_{ia}^{rr} + (\rho_{ia}^{rr} + \alpha_a^r + \delta_a^r + d_a^r)]dt. \tag{6.82}$$

Interregional level

The estimate of terms in the right-hand side of (6.76) for the case of $u \neq r$, $a \in I(1, n_u)$:

$$\sum_{u\neq r}^{M}\sum_{a=1}^{n_u}\left[\left(1 - \frac{1}{S_{ia}^{ru}}\right)dS_{ia}^{ru} + \frac{1}{2(S_{ia}^{ru})^2}(dS_{ia}^{ru})^2\right] \leq \sum_{u\neq r}^{M}\sum_{a=1}^{n_u}\{[\gamma_{ia}^{ru}S_{ii}^{rr} + \eta_a^u I_{ia}^{rr} + \alpha_a^u R_{ia}^{ru}$$
$$+ (\rho_{ia}^{ru} + \delta_a^u + d_a^u)$$
$$+ \sum_{v=1}^{M}\sum_{b=1}^{n_v}\beta_{aib}^{urv}I_{ba}^{vr} + \frac{1}{2}\left[\sum_{v=1}^{M}\sum_{b=1}^{n_v}(v_{aib}^{urv})^2(I_{ba}^{vr})^2\right]\Bigg]dt$$
$$+ (1 - S_{ia}^{ru})\sum_{v=1}^{M}\sum_{b=1}^{n_v}v_{aib}^{urv}I_{ba}^{vr}dw_{aib}^{urv}\Bigg\}, \tag{6.83}$$

$$\sum_{u\neq r}^{M}\sum_{a=1}^{n_u}\left[\left(1 - \frac{1}{I_{ia}^{ru}}\right)dI_{ia}^{ru} + \frac{1}{2(I_{ia}^{ru})^2}(dI_{ia}^{ru})^2\right] \leq \sum_{u\neq r}^{M}\sum_{a=1}^{n_u}\Bigg\{\Bigg[\gamma_{ia}^{ru}I_{ii}^{rr} + \sum_{v=1}^{M}\sum_{b=1}^{n_v}\beta_{aib}^{urv}S_{ia}^{ru}I_{ba}^{vu}$$
$$+ (\eta_a^r + \varrho_a^u + \rho_{ia}^{ru} + \delta_a^u + d_a^u)$$
$$+ \frac{(S_{ia}^{ru})^2}{2(I_{ia}^{ru})^2}\left(\sum_{v=1}^{M}\sum_{b=1}^{n_v}(v_{aib}^{urv})^2(I_{ba}^{vr})^2\right)\Bigg]dt$$
$$+\left(S_{ia}^{ru} - \frac{S_{ia}^{ru}}{I_{ia}^{ru}}\right)\sum_{v=1}^{M}\sum_{b=1}^{n_v}v_{aib}^{urv}I_{ba}^{vr}dw_{aib}^{urv}\Bigg\}, \tag{6.84}$$

and

$$\sum_{u\neq r}^{M}\sum_{a=1}^{n_u}\left(1 - \frac{1}{R_{ia}^{ru}}\right)dR_{ia}^{ru} \leq \sum_{u\neq r}^{M}\sum_{a=1}^{n_u}[\gamma_{ia}^{ru}R_{ii}^{rr} + \varrho_a^u I_{ia}^{ru} + (\rho_{ia}^{ru} + \alpha_a^u\delta_a^u + d_a^u)]dt. \tag{6.85}$$

From (6.76) and (6.77)–(6.79), the first term in the right-hand side of (6.73) can be estimated as follows:

$$\sum_{r=1}^{M}\sum_{i=1}^{n_r}dV_{ii}^{rr}(x_{00}^{00}) = \sum_{r=1}^{M}\sum_{i=1}^{n_r}\left\{\left[\left(1 - \frac{1}{S_{ii}^{rr}}\right)dS_{ii}^{rr} + \frac{1}{2(S_{ii}^{rr})^2}(dS_{ii}^{rr})^2\right]\right.$$

$$
\begin{aligned}
&+ \left[\left(1-\frac{1}{I_{ii}^{rr}}\right)dI_{ii}^{rr} + \frac{1}{2(I_{ii}^{rr})^2}(dI_{ii}^{rr})^2\right] \\
&+ \left[\left(1-\frac{1}{R_{ii}^{rr}}\right)dR_{ii}^{rr} + \frac{1}{2(R_{ii}^{rr})^2}(dR_{ii}^{rr})^2\right]\Bigg\} \\
\leq &\sum_{r=1}^{M}\sum_{i=1}^{n_r}\Bigg\{\Bigg[B_i^r + \sum_{b=1}^{n_r}\rho_{ib}^{rr}(S_{ib}^{rr}+I_{ib}^{rr}+R_{ib}^{rr}) + \sum_{b\neq r}^{M}\sum_{b=1}^{n_v}\rho_{ib}^{rv}(S_{ib}^{rv}+I_{ib}^{rv}+R_{ib}^{rv}) \\
&+ (\varrho_i^r+\eta_i^r+\alpha_i^r)(S_{ii}^{rr}+I_{ii}^{rr}+R_{ii}^{rr}) + 3(\gamma_i^r+\sigma_i^r+\alpha_i^r+\delta_i^r+d_i^r) \\
&+ \left(1+\frac{S_{ii}^{rr}}{I_{ii}^{rr}}\right)\sum_{v=1}^{M}\sum_{b=1}^{n_v}\beta_{iib}^{rrv}(S_{bi}^{vr}+I_{bi}^{vr}+R_{bi}^{vr}) \\
&+ \frac{1}{2}\left(1+\frac{(S_{ii}^{rr})^2}{(I_{ii}^{rr})^2}\right)\sum_{v=1}^{M}\sum_{b=1}^{n_v}(\nu_{iib}^{rrv})^2(S_{bi}^{vr}+I_{bi}^{vr}+R_{bi}^{vr})^2\Bigg]dt \\
&+ \left(1-\frac{S_{ii}^{rr}}{I_{ii}^{rr}}\right)\left[\sum_{v=1}^{M}\sum_{b=1}^{n_v}\nu_{iib}^{rrv}I_{bi}^{vr}dw_{iib}^{rrv}\right]\Bigg\} \\
\leq &\Bigg\{\Bigg[\sum_{r=1}^{M}\sum_{i=1}^{n_r}B_i^r + \rho_{00}^{00}\sum_{r=1}^{M}\sum_{i=1}^{n_r}\sum_{v=1}^{M}\sum_{b=1}^{n_v}(S_{ib}^{rv}+I_{ib}^{rv}+R_{ib}^{rv}) \\
&+ (\varrho_0^0+\eta_0^0+\alpha_0^0)\sum_{r=1}^{M}\sum_{i=1}^{n_r}(S_{ii}^{rr}+I_{ii}^{rr}+R_{ii}^{rr}) \\
&+ 3\sum_{r=1}^{M}\sum_{i=1}^{n_r}(\gamma_0^0+\sigma_0^0+\alpha_0^0+\delta_0^0+d_0^0) \\
&+ M_{001}^{00}\beta_{000}^{000}\sum_{r=1}^{M}\sum_{i=1}^{n_r}\sum_{v=1}^{M}\sum_{b=1}^{n_v}(S_{bi}^{vr}+I_{bi}^{vr}+R_{bi}^{vr}) \\
&+ M_{002}^{00}(\nu_{000}^{000})^2\sum_{r=1}^{M}\sum_{i=1}^{n_r}\sum_{v=1}^{M}\sum_{b=1}^{n_v}(S_{bi}^{vr}+I_{bi}^{vr}+R_{bi}^{vr})^2\Bigg]dt \\
&+ \sum_{r=1}^{M}\sum_{i=1}^{n_r}\left(1-\frac{S_{ii}^{rr}}{I_{ii}^{rr}}\right)\left[\sum_{v=1}^{M}\sum_{b=1}^{n_v}\nu_{iib}^{rrv}I_{bi}^{vr}dw_{iib}^{rrv}\right]\Bigg\}. \qquad (6.86)
\end{aligned}
$$

By Lemma 6.6.2, (6.86) becomes

$$
\sum_{r=1}^{M}\sum_{i=1}^{n_r}dV_{ii}^{rr} \leq \tilde{K}_1 dt + \sum_{r=1}^{M}\sum_{i=1}^{n_r}\left(1-\frac{S_{ii}^{rr}}{I_{ii}^{rr}}\right)\left[\sum_{v=1}^{M}\sum_{b=1}^{n_v}\nu_{iib}^{rrv}I_{bi}^{vr}dw_{iib}^{rrv}\right], \qquad (6.87)
$$

where

$$
\begin{aligned}
\tilde{K}_1 = &\Bigg\{\Bigg[\sum_{r=1}^{M}\sum_{i=1}^{n_r}B_i^r + \rho_{00}^{00}\frac{1}{\mu}\sum_{r=1}^{M}\sum_{i=1}^{n_r}B_i^r(\varrho_0^0+\eta_0^0+\alpha_0^0)\frac{1}{\mu}\sum_{r=1}^{M}\sum_{i=1}^{n_r}B_i^r \\
&+ 3\sum_{r=1}^{M}\sum_{i=1}^{n_r}(\gamma_0^0+\sigma_0^0+\alpha_0^0+\delta_0^0+d_0^0) + M_{001}^{00}\beta_{000}^{000}\frac{1}{\mu}\sum_{r=1}^{M}\sum_{i=1}^{n_r}B_i^r
\end{aligned}
$$

$$+ M_{002}^{00}(v_{000}^{000})^2 \frac{1}{\mu^2}\left(\sum_{r=1}^{M}\sum_{i=1}^{n_r} B_i^r\right)^2\Bigg]\Bigg\} > 0. \tag{6.88}$$

Similarly, by (6.76) and (6.80)–(6.82) the second term in the righthand side of (6.73) is estimated as

$$\begin{aligned}
&\sum_{r=1}^{M}\sum_{i=1}^{n_r}\sum_{a\neq i}^{n_r} dV_{ia}^{rr}\\
&= \sum_{r=1}^{M}\sum_{i=1}^{n_r}\sum_{a\neq i}^{n_r}\left[\left(1-\frac{1}{S_{ia}^{rr}}\right)dS_{ia}^{rr} + \frac{1}{2(S_{ia}^{rr})^2}(dS_{ia}^{rr})^2\right] + \sum_{a\neq i}^{n_r}\left[\left(1-\frac{1}{I_{ia}^{rr}}\right)dI_{ia}^{rr} + \frac{1}{2(I_{ia}^{rr})^2}(dI_{ia}^{rr})^2\right]\\
&\quad + \sum_{a\neq i}^{n_r}\left[\left(1-\frac{1}{R_{ia}^{rr}}\right)dR_{ia}^{rr} + \frac{1}{2(R_{ia}^{rr})^2}(dR_{ia}^{rr})^2\right]\\
&\le \sum_{r=1}^{M}\sum_{i=1}^{n_r}\sum_{a\neq i}^{n_r}\{\sigma_{00}^{00}(S_{ii}^{rr}+I_{ii}^{rr}+R_{ii}^{rr}) + (\eta_0^0+\alpha_0^0+\varrho_0^0)(S_{ia}^{rr}+I_{ia}^{rr}+R_{ia}^{rr})\\
&\quad + \beta_{000}^{000}N_{00}^{00}\sum_{v=1}^{M}\sum_{b=1}^{n_v}(S_{ba}^{vr}+I_{ba}^{vr}+R_{ba}^{vr})\\
&\quad + 3(\rho_{00}^{00}+\varrho_0^0+\eta_0^0+\alpha_0^0+\delta_0^0+d_0^0) + \frac{M_{002}^{00}(v_{000}^{000})^2}{2}\left[\sum_{v=1}^{M}\sum_{b=1}^{n_v}(S_{ba}^{vr}+I_{ba}^{vr}+R_{ba}^{vr})^2\right]\Bigg\}\,dt\\
&\quad + \sum_{r=1}^{M}\sum_{i=1}^{n_r}\sum_{a\neq i}^{n_r}\left(1-\frac{S_{ia}^{rr}}{I_{ia}^{rr}}\right)\left[\sum_{v=1}^{M}\sum_{b=1}^{n_v} v_{aib}^{rrv}I_{ba}^{vr}dw_{aib}^{rrv}\right]\\
&= \Bigg\{\sigma_{00}^{00}\sum_{r=1}^{M}\sum_{i=1}^{n_r}\sum_{a\neq i}^{n_r}(S_{ii}^{rr}+I_{ii}^{rr}+R_{ii}^{rr}) + (\eta_0^0+\alpha_0^0+\varrho_0^0)\sum_{r=1}^{M}\sum_{i=1}^{n_r}\sum_{a\neq i}^{n_r}(S_{ia}^{rr}+I_{ia}^{rr}+R_{ia}^{rr})\\
&\quad + \beta_{000}^{000}N_{00}^{00}\sum_{r=1}^{M}\sum_{i=1}^{n_r}\sum_{a\neq i}^{n_r}\sum_{v=1}^{M}\sum_{b=1}^{n_v}(S_{ba}^{vr}+I_{ba}^{vr}+R_{ba}^{vr}) + \sum_{r=1}^{M}\sum_{i=1}^{n_r}\sum_{a\neq i}^{n_r}3(\rho_{00}^{00}+\varrho_0^0+\eta_0^0+\alpha_0^0+\delta_0^0+d_0^0)\\
&\quad + \frac{M_{002}^{00}(v_{000}^{000})^2}{2}\sum_{r=1}^{M}\sum_{i=1}^{n_r}\sum_{a\neq i}^{n_r}\left[\sum_{v=1}^{M}\sum_{b=1}^{n_v}(S_{ba}^{vr}+I_{ba}^{vr}+R_{ba}^{vr})^2\right]\Bigg\}\,dt\\
&\quad + \sum_{r=1}^{M}\sum_{i=1}^{n_r}\sum_{a\neq i}^{n_r}\left(1-\frac{S_{ia}^{rr}}{I_{ia}^{rr}}\right)\left[\sum_{v=1}^{M}\sum_{b=1}^{n_v} v_{aib}^{rrv}I_{ba}^{vr}dw_{aib}^{rrv}\right].
\end{aligned} \tag{6.89}$$

Again by Lemma 6.6.2, the above random differential inequality reduces to

$$\sum_{r=1}^{M}\sum_{i=1}^{n_r}\sum_{a\neq i}^{n_r} dV_{ia}^{rr} \le \tilde{K}_2dt + \sum_{r=1}^{M}\sum_{i=1}^{n_r}\sum_{a\neq i}^{n_r}\left(1-\frac{S_{ia}^{rr}}{I_{ia}^{rr}}\right)\left[\sum_{v=1}^{M}\sum_{b=1}^{n_v} v_{aib}^{rrv}I_{ba}^{vr}dw_{aib}^{rrv}\right], \tag{6.90}$$

where

$$\tilde{K}_2 = \left\{\sigma_{00}^{00}\frac{1}{\mu}\sum_{r=1}^{M}\sum_{i=1}^{n_r} B_i^r + (\eta_0^0 + \alpha_0^0 + \varrho_0^0)\frac{1}{\mu}\sum_{r=1}^{M}\sum_{i=1}^{n_r} B_i^r + \beta_{000}^{000} N_{00}^{00}\frac{1}{\mu}\sum_{r=1}^{M}\sum_{i=1}^{n_r} B_i^r \right.$$
$$\left. + \sum_{r=1}^{M}\sum_{i=1}^{n_r}\sum_{a\neq i}^{n_r} 3(\rho_{00}^{00} + \varrho_0^0 + \eta_0^0 + \alpha_0^0 + \delta_0^0 + d_0^0) + \frac{M_{002}^{00}(\nu_{000}^{000})^2}{2}\frac{1}{\mu^2}\left(\sum_{r=1}^{M}\sum_{i=1}^{n_r} B_i^r\right)^2\right\}. \quad (6.91)$$

Finally, by (6.73), (6.76), and (6.83)–(6.85) the third term in (6.73) is estimated as follows:

$$\sum_{r=1}^{M}\sum_{i=1}^{n_r}\sum_{u\neq r}^{M}\sum_{a=1}^{n_u} dV_{ia}^{ru}$$
$$= \sum_{r=1}^{M}\sum_{i=1}^{n_r}\sum_{u\neq r}^{M}\sum_{a=1}^{n_u}\left[\left(1 - \frac{1}{S_{ia}^{ru}}\right)dS_{ia}^{ru} + \frac{1}{2(S_{ia}^{ru})^2}(dS_{ia}^{ru})^2\right]$$
$$+ \sum_{u\neq r}^{M}\sum_{a=1}^{n_u}\left[\left(1 - \frac{1}{I_{ia}^{ru}}\right)dI_{ia}^{ru} + \frac{1}{2(I_{ia}^{ru})^2}(dI_{ia}^{ru})^2\right]$$
$$+ \sum_{r=1}^{M}\sum_{i=1}^{n_r}\sum_{u\neq r}^{M}\sum_{a=1}^{n_u}\left[\left(1 - \frac{1}{R_{ia}^{ru}}\right)dR_{ia}^{ru} + \frac{1}{2(R_{ia}^{ru})^2}(dR_{ia}^{ru})^2\right]$$
$$\leq \sum_{r=1}^{M}\sum_{i=1}^{n_r}\sum_{u\neq r}^{M}\sum_{a=1}^{n_u}\{\gamma_{ia}^{ru}(S_{ii}^{rr} + I_{ii}^{rr} + R_{ii}^{rr}) + (\eta_a^u + \alpha_a^u + \varrho_a^u)(S_{ia}^{ru} + I_{ia}^{ru} + R_{ia}^{ru})$$
$$+ 3(\eta_a^u + \varrho_a^u + \alpha_a^u + \rho_{ia}^{ru} + \delta_a^u + d_a^u)$$
$$+ (1 + S_{ia}^{ru})\sum_{v=1}^{M}\sum_{b=1}^{n_v}\beta_{aib}^{urv}(S_{ba}^{vu} + I_{ba}^{vu} + R_{ba}^{vu})$$
$$+\frac{1}{2}\left(1 + \frac{(S_{ia}^{ru})^2}{(I_{ia}^{ru})^2}\right)\sum_{v=1}^{M}\sum_{b=1}^{n_v}(\nu_{aib}^{urv})^2(S_{ba}^{vu} + I_{ba}^{vu} + R_{ba}^{vu})^2\Bigg\}\,dt$$
$$+ \sum_{r=1}^{M}\sum_{i=1}^{n_r}\sum_{u\neq r}^{M}\sum_{a=1}^{n_u}\sum_{v=1}^{M}\sum_{b=1}^{n_v}\left(1 - \frac{S_{ia}^{ru}}{I_{ia}^{ru}}\right)\nu_{aib}^{urv} I_{ba}^{vu}\, dw_{aib}^{urv}$$
$$\leq \left\{\gamma_{00}^{00}\sum_{r=1}^{M}\sum_{i=1}^{n_r}\sum_{u\neq r}^{M}\sum_{a=1}^{n_u}(S_{ii}^{rr} + I_{ii}^{rr} + R_{ii}^{rr}) + (\eta_0^0 + \alpha_0^0 + \varrho_0^0)\sum_{r=1}^{M}\sum_{i=1}^{n_r}\sum_{u\neq r}^{M}\sum_{a=1}^{n_u}(S_{ia}^{ru} + I_{ia}^{ru} + R_{ia}^{ru})\right.$$
$$+ \sum_{r=1}^{M}\sum_{i=1}^{n_r}\sum_{u\neq r}^{M}\sum_{a=1}^{n_u} 3(\eta_0^0 + \varrho_0^0 + \alpha_0^0 + \rho_{00}^{00} + \delta_0^0 + d_0^0)$$
$$+ N_{00}^{00}\beta_{000}^{000}\sum_{r=1}^{M}\sum_{i=1}^{n_r}\sum_{u\neq r}^{M}\sum_{a=1}^{n_u}\sum_{v=1}^{M}\sum_{b=1}^{n_v}(S_{ba}^{vu} + I_{ba}^{vu} + R_{ba}^{vu})$$
$$\left. + \frac{M_{002}^{00}(\nu_{000}^{000})^2}{2}\sum_{r=1}^{M}\sum_{i=1}^{n_r}\sum_{u\neq r}^{M}\sum_{a=1}^{n_u}\sum_{v=1}^{M}\sum_{b=1}^{n_v}(S_{ba}^{vu} + I_{ba}^{vu} + R_{ba}^{vu})^2\right\} dt$$
$$+ \sum_{r=1}^{M}\sum_{i=1}^{n_r}\sum_{u\neq r}^{M}\sum_{a=1}^{n_u}\sum_{v=1}^{M}\sum_{b=1}^{n_v}\left(1 - \frac{S_{ia}^{ru}}{I_{ia}^{ru}}\right)\nu_{aib}^{urv} I_{ba}^{vu}\, dw_{aib}^{urv}. \quad (6.92)$$

By Lemma 6.6.2 differential inequality (6.93) becomes

$$\sum_{r=1}^{M}\sum_{i=1}^{n_r}\sum_{u\neq r}^{M}\sum_{a=1}^{n_u} dV_{ia}^{ru} \leq \tilde{K}_3 dt + \sum_{r=1}^{M}\sum_{i=1}^{n_r}\sum_{u\neq r}^{M}\sum_{a=1}^{n_u}\sum_{v=1}^{M}\sum_{b=1}^{n_v}\left(1-\frac{S_{ia}^{ru}}{I_{ia}^{ru}}\right)v_{aib}^{urv} I_{ba}^{vu} dw_{aib}^{urv}, \tag{6.93}$$

where

$$\begin{aligned}\tilde{K}_3 = \Bigg\{ & \gamma_{00}^{00}\frac{1}{\mu}\sum_{r=1}^{M}\sum_{i=1}^{n_r} B_i^r + (\eta_0^0 + \alpha_0^0 + \varrho_0^0)\frac{1}{\mu}\sum_{r=1}^{M}\sum_{i=1}^{n_r} B_i^r \\ & + \sum_{r=1}^{M}\sum_{i=1}^{n_r}\sum_{u\neq r}^{M}\sum_{a=1}^{n_u} 3(\eta_0^0 + \varrho_0^0 + \alpha_0^0 + \rho_{00}^{00} + \delta_0^0 + d_0^0) \\ & + N_{00}^{00}\beta_{000}^{000}\frac{1}{\mu}\sum_{r=1}^{M}\sum_{i=1}^{n_r} B_i^r + \frac{M_{002}^{00}(v_{000}^{000})^2}{2}\frac{1}{\mu^2}\left(\sum_{r=1}^{M}\sum_{i=1}^{n_r} B_i^r\right)^2\Bigg\}. \end{aligned} \tag{6.94}$$

Hence from (6.87), (6.90), and (6.93) we arrive at the following stochastic differential inequality:

$$\begin{aligned} dV(x_{00}^{00}(t)) \leq {} & (\tilde{K}_1 + \tilde{K}_2 + \tilde{K}_3)\, dt + \sum_{r=1}^{M}\sum_{i=1}^{n_r}\left(1-\frac{S_{ii}^{rr}}{I_{ii}^{rr}}\right)\left[\sum_{v=1}^{M}\sum_{b=1}^{n_v} v_{iib}^{rrv} I_{bi}^{vr} dw_{iib}^{rrv}\right] \\ & + \sum_{r=1}^{M}\sum_{i=1}^{n_r}\sum_{a\neq i}^{n_r}\left(1-\frac{S_{ia}^{rr}}{I_{ia}^{rr}}\right)\left[\sum_{v=1}^{M}\sum_{b=1}^{n_v} v_{aib}^{rrv} I_{ba}^{vr} dw_{aib}^{rrv}\right] \\ & + \sum_{r=1}^{M}\sum_{i=1}^{n_r}\sum_{u\neq r}^{M}\sum_{a=1}^{n_u}\sum_{v=1}^{M}\sum_{b=1}^{n_v}\left(1-\frac{S_{ia}^{ru}}{I_{ia}^{ru}}\right)v_{aib}^{urv} I_{ba}^{vu} dw_{aib}^{urv}. \end{aligned} \tag{6.95}$$

Therefore choosing $\tilde{K} = \tilde{K}_1 + \tilde{K}_2 + \tilde{K}_3 > 0$ and combining the last three summations conclude the proof of the theorem. □

We now show the existence of a unique solution of system (6.53)–(6.55).

Theorem 6.6.4. *Given any initial condition $x_{00}^{00}(t_0) \in \mathbb{R}_+^{3n^2}$, under the assumptions of* Lemma 6.7.1, *there is a unique solution process of system* (6.53)–(6.55) *in $\mathbb{R}_+^{3n^2}$ for $t \geq t_0$ almost surely.*

Proof. Given that the rate functions of the system are locally Lipschitz continuous in x_{00}^{00}, it follows that for any initial value $x_{00}^{00}(t_0) \in \mathbb{R}_+^{3n^2}$, there is a unique local solution $x_{00}^{00}(t)$ of system (6.53)–(6.55) for $t \in (t_0, t_e)$, where $t = t_e$ is the first exit time of x_{00}^{00}. Therefore to show the solution process of the system exists for all $t \geq t_0$, it suffices to show that $t_e = \infty$.

Let $k_{00}^{00} \in \mathbb{R}_+^{n^2}$. From (6.47) and (6.48) we have

$$\|k_{00}^{00}\|_\infty = \max_{1\leq r,u\leq M, 1\leq i\leq n_r, 1\leq a\leq n_u} |k_{ia}^{ru}|, \quad k_{00\,\min}^{00} = \min_{1\leq r,u\leq M, 1\leq i\leq n_r, 1\leq a\leq n_u} |k_{ia}^{ru}|. \tag{6.96}$$

We denote

$$k \equiv k_{00\,\min}^{00}. \tag{6.97}$$

We choose $k_{00}^{*00} \in \mathbb{R}_+^{n^2}$ with each component k_{ia}^{*ru} sufficiently large such that $S_{ia}^{ru}(t_0)$, $I_{ia}^{ru}(t_0), R_{ia}^{ru}(t_0) \in [\frac{1}{k_{ia}^{*ru}}, k_{ia}^{*ru}] \equiv \bar{\mathfrak{B}}_{\mathbb{R}}(\frac{\frac{1}{k_{ia}^{*ru}}+k_{ia}^{*ru}}{2}; \frac{\frac{1}{k_{ia}^{*ru}}-k_{ia}^{*ru}}{2})$ for $i \in I(1,n_r)$, $a \in I(1,n_u)$, and $r,u \in I(1,M)$. In other words, from (6.45) we have $x_{00}^{00}(t_0) \in \prod_{r=1}^{M}\prod_{u=1}^{M}\prod_{i=1}^{n_i}\prod_{a=1}^{n_u}[\frac{1}{k_{ia}^{*ru}}, k_{ia}^{*ru}] \times [\frac{1}{k_{ia}^{*ru}}, k_{ia}^{*ru}] \times [\frac{1}{k_{ia}^{*ru}}, k_{ia}^{*ru}]$. From (6.97), let $k_0 \equiv k_{00\,\min}^{*00}$.

Let $k_{00}^{00} \in \mathbb{R}_+^{n^2}$ be an arbitrary vector whose components k_{ia}^{ru} satisfy $k_{ia}^{ru} \geq k_{ia}^{*ru}$, $i \in I(1,n_i)$, $a \in I(1,n_u)$, $r,u \in I(1,M)$. Let the local solution $x_{00}^{00}(t) \in \prod_{r=1}^{M}\prod_{u=1}^{M}\prod_{i=1}^{n_i} \times \prod_{a=1}^{n_u}[\frac{1}{k_{ia}^{ru}}, k_{ia}^{ru}] \times [\frac{1}{k_{ia}^{ru}}, k_{ia}^{ru}] \times [\frac{1}{k_{ia}^{ru}}, k_{ia}^{ru}]$ for $t \in (0,t_e)$, where t_e is the first hitting time of the solution process. For $t \leq t_e$, it follows that $S_{ia}^{ru}(t), I_{ia}^{ru}(t), R_{ia}^{ru}(t) \in [\frac{1}{\|k_{00}^{00}\|_\infty}, k_{00\min}^{00}]$ for all $i \in I(1,n_r)$, $a \in I(1,n_u)$, $r,u \in I(1,M)$. Using (6.97), define the stopping time for the process

$$\tau_k = \inf\left\{t \in (0,t_e) : \min_{1\leq r,u\leq M, 1\leq i\leq n_r, 1\leq a\leq n_u} (S_{ia}^{ru}(t), I_{ia}^{ru}(t), R_{ia}^{ru}(t)) \leq \frac{1}{\|k_{00}^{00}\|_\infty},\right.$$

$$\left.\max_{1\leq r,u\leq M, 1\leq i\leq n_r, 1\leq a\leq n_u} (S_{ia}^{ru}(t), I_{ia}^{ru}(t), R_{ia}^{ru}(t)) \geq k\right\}, \quad \text{and} \tag{6.98}$$

$$\tau_k(t) = \min\{t, \tau_k\} \quad \text{for } t \geq t_0, \tag{6.99}$$

where k is defined in (6.97). Furthermore, we set $\inf \varnothing = \infty$. It follows from (6.99) that τ_k increases as $k \to \infty$. Let $\tau_\infty = \lim_{k\to\infty} \tau_k$. By (6.99) we have that

$$\tau_\infty \leq t_e \quad \text{a.s.} \tag{6.100}$$

Therefore to show $t_e = \infty$, we only need to show that $\tau_\infty = \infty$ a. s.

On the contrary, suppose $\tau_\infty < \infty$. Then there is $T > 0$ such that for a given $0 < \epsilon < 1$, $P(\tau_\infty \leq T) > \epsilon$. This means that $\{\tau_k\}$ is a finite sequence. Moreover, by the definition of a finite sequence there exists a vector $k_{00}^{100} \in \mathbb{R}^{n^2}$ with $k_{00\,\min}^{100} \equiv k_1 \geq k_0$ (where $k_1 \equiv k_{00\,\min}^{100}$ is defined by (6.97) and (6.96)) and

$$P(\tau_k \leq T) \geq \epsilon \tag{6.101}$$

for $k \geq k_1$. From (6.74), (6.73) can be rewritten as

$$V(x_{00}^{00}) = \sum_{r=1}^{M}\sum_{i=1}^{n_r}\sum_{u=1}^{M}\sum_{a=1}^{n_u} [(S_{ia}^{ru} - 1 - \log S_{ia}^{ru}) + (I_{ia}^{ru} - 1 - \log I_{ia}^{ru}) + (R_{ia}^{ru} - 1 - \log R_{ia}^{ru})]. \tag{6.102}$$

By Lemmas 6.6.2 and 6.6.3 the stopped solution process (6.53)–(6.55) satisfies the following stochastic inequality for some $\tilde{K} > 0$:

$$dV(x_{00}^{00}(t)) \leq \tilde{K}dt + \sum_{r=1}^{M}\sum_{i=1}^{n_r}\sum_{u=1}^{M}\sum_{a=1}^{n_u}\sum_{v=1}^{M}\sum_{b=1}^{n_v}\left(1 - \frac{S_{ia}^{ru}}{I_{ia}^{ru}}\right)v_{aib}^{urv} I_{ba}^{vu} dw_{aib}^{urv}. \tag{6.103}$$

Furthermore, for all $t_1 \le T$, integrating both sides of (6.103) on $[t_0, t_1 \wedge \tau_k]$ and taking the expected values of both sides, we get that

$$\begin{aligned} E(V(x_{00}^{00}(t_1 \wedge \tau_k))) &\le V(x_{00}^{00}(t_0)) + \tilde{K}(t_1 \wedge \tau_k) \\ &\le V(x_{00}^{00}(t_0)) + \tilde{K}T. \end{aligned} \tag{6.104}$$

For $k \ge k_1$, we set $E_k = \{\tau_k \le T\}$. Then from (6.101) we see that $P(E_k) \ge \epsilon$. If $\omega \in E_k$, then ω is an event at the stopping time where at least one of $S_{ia}^{ru}(\tau_k, \omega)$, $I_{ia}^{ru}(\tau_k, \omega)$, or $R_{ia}^{ru}(\tau_k, \omega)$ for $r, u \in I(1, M)$, $i \in I(1, n_r)$, and $a \in I(1, n_u)$ is $\frac{1}{\|k_{00}^{00}\|_\infty}$ or $k \equiv k_{\min}$. This implies by (6.102) that

$$V(x_{00}^{00}(\tau_k, \omega)) \ge [k_{\min} - 1 - \log k_{\min}] \wedge \left[\frac{1}{\|k_{00}^{00}\|_\infty} - 1 - \log\|k_{00}^{00}\|_\infty\right], \quad \omega \in E_k. \tag{6.105}$$

It follows from (6.104) and (6.105) that

$$\begin{aligned} V(x_{i0}^{00}(t_0)) + \tilde{K}T &\ge E(I_{E_k(\omega)} V(x_{00}^{00}(\tau_k, \omega))) \\ &\ge \epsilon\left\{[k_{\min} - 1 - \log k_{\min}] \wedge \left[\frac{1}{\|k_{00}^{00}\|_\infty} - 1 - \log\|k_{00}^{00}\|_\infty\right]\right\}, \end{aligned} \tag{6.106}$$

where $I_{E_k(\omega)}$ is the indicator function of E_k.

Hence as $k = k_{\min} \to \infty$, (6.106) implies that $V(x_{00}^{00}(t_0)) + \tilde{K}T \to \infty$, which leads to a contradiction to the existence of a local solution. Therefore we have $\tau_\infty = \infty$, and the rest of the proof follows. □

Remark 6.6.3. For any $r \in I(1, M)$ and $i \in I(1, n_r)$, Lemmas 6.6.1, 6.6.2, and 6.6.3 and Theorem 6.6.4 show that there exists a positive self-invariant set for system (6.53)–(6.55) given by

$$A = \left\{(S_{ia}^{ru}, I_{ia}^{ru}, R_{ia}^{ru}) : y_{ia}^{ru}(t) \ge 0 \quad \text{and} \quad \sum_{r=1}^{M}\sum_{u=1}^{M}\sum_{i=1}^{n_r}\sum_{a=1}^{n_u} y_{ia}^{ru}(t) \le \frac{1}{\mu}\sum_{r=1}^{M}\sum_{i=1}^{n_r} B_i^r\right\} \tag{6.107}$$

for $u \in I^r(1, M)$ and $a \in I_i^r(1, n_u)$. We denote

$$\bar{B} \equiv \frac{1}{\mu}\sum_{r=1}^{M}\sum_{i=1}^{n_r} B_i^r. \tag{6.108}$$

6.7 Disease eradication in the two-scale network dynamic SDE model

In this section, we investigate the SDE model in (6.53)–(6.55) for disease eradication conditions for the outbreak in the two-scale network population $C = C_{00}^{00}$. Note that the conditions for disease-eradication in a differential equation disease epidemic dynamic model are obtained by investigating system (6.53)–(6.55) for (i) the existence of a disease-free equilibrium E_0 and (ii) computing threshold parameters in the epidemic model, such as the *basic reproduction number* $\mathfrak{R}_0$ for the disease, where for $\mathfrak{R}_0 < 1$, the disease-free equilibrium E_0 is *asymptotically stochastically stable*, and hence the disease is eradicated; otherwise, the disease persists in the population (see [29]).

6.7.1 Existence of disease-free equilibrium

For an SDE (6.65), that is,

$$dx(t) = f(x(t), t)dt + g(x(t), t)d\omega, \tag{6.109}$$

the equilibria $E = x_0 \in \mathbb{R}_+^{3n}$ of are obtained by solving the system for the drift and diffusion components given by

$$f(x(t), t) = 0, \quad g(x(t), t) = 0. \tag{6.110}$$

The disease-free equilibrium for (6.53)–(6.55) is obtained by solving the system of algebraic equations (6.110) obtained by setting the drift and diffusion parts of the system of stochastic differential equations to zero. In addition, the conditions $I = R = 0$ in the event when there is no disease in the population must be applied. We summarize the results as follows. For $r, u \in I(1, M)$, $i \in I(1, n_r)$, and $a \in I(1, n_u)$, define the constant

$$D_i^r = \gamma_i^r + \sigma_i^r + \delta_i^r - \sum_{a=1}^{n_r} \frac{\rho_{ia}^{rr}\sigma_{ia}^{rr}}{\rho_{ia}^{rr} + \delta_a^r} - \sum_{u \neq r}^{M} \sum_{a=1}^{n_u} \frac{\rho_{ia}^{rr}\gamma_{ia}^{ru}}{\rho_{ia}^{ru} + \delta_a^u} > 0. \tag{6.111}$$

Note that based on the definitions of the parameters in model (6.53)–(6.55), $D_i^r > 0$ for all $r \in I(1, M)$ and $i \in I(1, n_r)$.

Furthermore, let $(S_{ia}^{ru*}, I_{ia}^{ru*}, R_{ia}^{ru*})$ denote the equilibrium state of system (6.53)–(6.55). We can see that the disease-free equilibrium state is given by $E_{ia}^{ru} = (S_{ia}^{ru*}, 0, 0)$, where

$$S_{ia}^{ru*} = \begin{cases} \frac{B_i^r}{D_i^r}, & u = r, a = i, \\ \frac{B_i^r}{D_i^r} \frac{\sigma_{ij}^{rr}}{\rho_{ij}^{rr} + \delta_j^r}, & u = r, a \neq i, \\ \frac{B_i^r}{D_i^r} \frac{\gamma_{ia}^{ru}}{\rho_{ia}^{ru} + \delta_a^u}, & u \neq r. \end{cases} \tag{6.112}$$

6.7.2 Stochastic stability of the disease-free equilibrium

According to the Lyapunov stability results in Section 1.4 and Theorem 1.4.1 applying the *Lyapunov second method*, the equilibrium $E = x_0 \in \mathbb{R}_+^{3n}$ of the SDE system (6.109) is said to be *stable in probability* if every path of the process $\{x(t), t \geq 0\}$ that starts in the neighborhood of E has a *high probability* to remain near E for all times. For the epidemic dynamics in (6.53)–(6.55), if the disease-free equilibrium is stable in probability, it suggests that the disease epidemic is under control as the population stays over time near the disease-free equilibrium.

The methods and results for stochastic stability in Section 1.4 and Theorem 1.4.1 are applied subsequently. The asymptotic stability property of E_{ia}^{ru} will be established by verifying the conditions of Theorem 1.4.1. First, system (6.53)–(6.55) is transformed with respect to the equilibrium state $(S_{ia}^{ru*}, 0, 0)$ using the following:

$$\begin{cases} U_{ia}^{ru} = S_{ia}^{ru} - S_{ia}^{ru*}, \\ V_{ia}^{ru} = I_{ia}^{ru}, \\ W_{ia}^{ru} = R_{ia}^{ru}. \end{cases} \tag{6.113}$$

By employing (6.113) in (6.53)–(6.55) we obtain the following system:

$$dU_{il}^{rq} = \begin{cases} [\sum_{q\neq r}^{M} \sum_{a=1}^{n_q} \rho_{ia}^{rq} U_{ia}^{rq} + \eta_i^r V_{ii}^{rr} + \alpha_i^r W_{ii}^{rr} \\ \quad - (\gamma_i^r + \sigma_i^r + \delta_i^r) U_{ii}^{rr} - \sum_{u=1}^{M} \sum_{a=1}^{n_u} \beta_{iia}^{rru} (S_{ii}^{rr*} + U_{ii}^{rr}) V_{ai}^{ur}]dt \\ \quad - [\sum_{u=1}^{M} \sum_{a=1}^{n_u} \nu_{iia}^{rru} (S_{ii}^{rr*} + U_{ii}^{rr}) V_{ai}^{ur} dw_{iia}^{rru}(t)], \quad q = r, l = i, \\ [\sigma_{ij}^{rr} U_{ii}^{rr} + \eta_j^r V_{ij}^{rr} + \alpha_j^r W_{ij}^{rr} - (\rho_{ij}^{rr} + \delta_j^r) U_{ij}^{rr} - \sum_{u=1}^{M} \sum_{a=1}^{n_u} \beta_{jia}^{rru} (S_{ij}^{rr*} + U_{ij}^{rr}) V_{aj}^{ur}]dt \\ \quad - [\sum_{u=1}^{M} \sum_{a=1}^{n_u} \nu_{jia}^{rru} (S_{ij}^{rr*} + U_{ij}^{rr}) V_{aj}^{ur} dw_{jia}^{rru}(t)], \quad q = r, l = j, j \neq i, \\ [\gamma_{il}^{rq} U_{ii}^{rr} + \eta_l^q V_{il}^{rq} + \alpha_l^q W_{il}^{rq} - (\rho_{il}^{rq} + \delta_l^q) U_{il}^{rq} \\ \quad - \sum_{u=1}^{M} \sum_{a=1}^{n_u} \beta_{lia}^{qru} S_{il}^{rq} I_{al}^{uq}]dt - [\sum_{u=1}^{M} \sum_{a=1}^{n_u} \nu_{lia}^{qru} (S_{il}^{rq*} + U_{il}^{rq}) V_{al}^{uq} dw_{lia}^{qru}(t)], \quad q \neq r, \end{cases} \tag{6.114}$$

$$dV_{il}^{rq} = \begin{cases} [\sum_{q=1}^{M} \sum_{a=1}^{n_q} \rho_{ia}^{rq} V_{ia}^{rq} - (\eta_i^r + \varrho_i^r + \gamma_i^r + \sigma_i^r + \delta_i^r + d_i^r) W_{ii}^{rr} \\ \quad + \sum_{u=1}^{M} \sum_{a=1}^{n_u} \beta_{iia}^{rru} (S_{ii}^{rr*} + U_{ii}^{rr}) V_{ai}^{ur}]dt \\ \quad + [\sum_{u=1}^{M} \sum_{a=1}^{n_u} \nu_{iia}^{rru} (S_{ii}^{rr*} + U_{ii}^{rr}) V_{ai}^{ur} dw_{iia}^{rru}(t)], \quad q = r, l = i, \\ [\sigma_{ij}^{rr} V_{ii}^{rr} - (\eta_j^r + \varrho_j^r + \rho_{ij}^{rr} + \delta_j^r + d_j^r) V_{ij}^{rr} + \sum_{u=1}^{M} \sum_{a=1}^{n_u} \beta_{jia}^{rru} (S_{ij}^{rr*} + U_{ij}^{rr}) V_{aj}^{ur}]dt \\ \quad + [\sum_{u=1}^{M} \sum_{a=1}^{n_u} \nu_{jia}^{rru} (S_{ij}^{rr*} + U_{ij}^{rr}) V_{aj}^{ur} dw_{jia}^{rru}(t)], \quad q = r, l = j, j \neq i, \\ [\gamma_{il}^{rq} V_{ii}^{rr} - (\eta_l^q + \varrho_l^q + \rho_{il}^{rq} + \delta_l^q + d_l^q) V_{il}^{rq} \\ \sum_{u=1}^{M} \sum_{a=1}^{n_u} \beta_{lia}^{qru} (S_{il}^{rq*} + U_{il}^{rq}) V_{al}^{uq}]dt + [\sum_{u=1}^{M} \sum_{a=1}^{n_u} \nu_{lia}^{qru} (S_{il}^{rq*} + U_{il}^{rq}) V_{al}^{uq} dw_{lia}^{qru}(t)], \\ \quad q \neq r, \end{cases} \tag{6.115}$$

and

$$dW_{il}^{rq} = \begin{cases} [\sum_{q\neq r}^{M} \sum_{l=1}^{n_q} \rho_{il}^{rq} W_{il}^{rq} + \varrho_i^r V_{ii}^{rr} - (\gamma_i^r + \sigma_i^r + \alpha_i^r + \delta_i^r) W_{ii}^{rr}]dt, & q = r, l = i, \\ [\sigma_{ij}^{rr} W_{ii}^{rr} + \varrho_j^r V_{ij}^{rr} - (\rho_{ij}^{rr} + \alpha_j^r + \delta_j^r) W_{ij}^{rr}]dt, & q = r, l = j, j \neq i, \\ [\gamma_{il}^{rq} W_{ii}^{rr} + \varrho_l^q V_{il}^{rq} - (\rho_{il}^{rq} + \alpha_l^q + \delta_l^q) W_{il}^{rq}]dt, & q \neq r. \end{cases} \tag{6.116}$$

We state and prove the following lemmas that would be useful in the proofs of the stability results.

Lemma 6.7.1. *Let* $V : \mathbb{R}^{3n^2} \times \mathbb{R}_+ \to \mathbb{R}_+$ *be a function defined by*

$$V(\tilde{x}_{00}^{00}) = \sum_{r=1}^{M}\sum_{u=1}^{M}\sum_{i=1}^{n_r}\sum_{a=1}^{n_u} V(\tilde{x}_{ia}^{ru}), \tag{6.117}$$

where

$$\begin{aligned} V(\tilde{x}_{ia}^{ru}) &= (S_{ia}^{ru} - S_{ia}^{ru*} + I_{ia}^{ru})^2 + c_{ia}^{ru}(I_{ia}^{ru})^2 + (R_{ia}^{ru})^2, \\ \tilde{x}_{00}^{00} &= (U_{ia}^{ru}, V_{ia}^{ru}, W_{ia}^{ru})^T, \quad \text{and} \quad c_{ia}^{ru} \geq 0. \end{aligned} \tag{6.118}$$

Then $V \in C^{2,1}(\mathbb{R}^{3n^2} \times \mathbb{R}_+, \mathbb{R}_+)$, *and it satisfies*

$$b(\|\tilde{x}_{00}^{00}\|) \leq V(\tilde{x}_{00}^{00}(t)) \leq a(\|\tilde{x}_{00}^{00}\|), \tag{6.119}$$

where

$$\begin{aligned} b(\|\tilde{x}_{00}^{00}\|) &= \min_{1\leq r,u\leq M, 1\leq i\leq n_r, 1\leq a\leq n_u} \left\{\frac{c_{ia}^{ru}}{2 + c_{ia}^{ru}}\right\} \sum_{r=1}^{M}\sum_{u=1}^{M}\sum_{i=1}^{n_r}\sum_{a=1}^{n_u} [(U_{ia}^{ru})^2 + (V_{ia}^{ru})^2 + (W_{ia}^{ru})^2], \\ a(\|\tilde{x}_{00}^{00}\|) &= \max_{1\leq r,u\leq M, 1\leq i\leq n_r, 1\leq a\leq n_u} \{c_{ia}^{ru} + 2\} \sum_{r=1}^{M}\sum_{u=1}^{M}\sum_{i=1}^{n_r}\sum_{a=1}^{n_u} [(U_{ia}^{ru})^2 + (V_{ia}^{ru})^2 + (W_{ia}^{ru})^2]. \end{aligned} \tag{6.120}$$

Proof. From (6.114)–(6.116), (6.117) can be written as

$$\begin{aligned} V(x_{ia}^{ru}) &= (U_{ia}^{ru} + V_{ia}^{ru})^2 + c_{ia}^{ru}(V_{ia}^{ru})^2 + (W_{ia}^{ru})^2 \\ &= (U_{ia}^{ru})^2 + 2U_{ia}^{ru}V_{ia}^{ru} + (c_{ia}^{ru} + 1)(V_{ia}^{ru})^2 + (W_{ia}^{ru})^2 \\ &= (U_{ia}^{ru})^2 + (c_{ia}^{ru} + 1)(V_{ia}^{ru})^2 + 2\Bigg(\frac{1}{\sqrt{1 + \frac{c_{ia}^{ru}}{2}}} U_{ia}^{ru}\Bigg)\Bigg(\sqrt{1 + \frac{c_{ia}^{ru}}{2}} V_{ia}^{ru}\Bigg) + (W_{ia}^{ru})^2 \\ &= \Bigg(-\frac{1}{1 + \frac{c_{ia}^{ru}}{2}} + 1\Bigg)(U_{ia}^{ru})^2 + \Bigg(-\Big(1 + \frac{c_{ia}^{ru}}{2}\Big) + c_{ia}^{ru} + 1\Bigg)(V_{ia}^{ru})^2 + (W_{ia}^{ru})^2 \\ &\quad + \Bigg[\Bigg(\frac{1}{\sqrt{1 + \frac{c_{ia}^{ru}}{2}}} U_{ia}^{ru}\Bigg) + \Bigg(\sqrt{1 + \frac{c_{ia}^{ru}}{2}} V_{ia}^{ru}\Bigg)\Bigg]^2. \end{aligned}$$

Therefore, by noting the fact that $\min\{1 - \frac{1}{1+\frac{c_{ia}^{ru}}{2}}, \frac{c_{ia}^{ru}}{2}, 1\}$ we have

$$V(x_{ia}^{ru}) \geq \frac{c_{ia}^{ru}}{2 + c_{ia}^{ru}}[(U_{ia}^{ru})^2 + (V_{ia}^{ru})^2 + (W_{ia}^{ru})^2]. \tag{6.121}$$

Hence from (6.121) we have

$$\begin{aligned} V(\tilde{x}_{00}^{00}) &\geq \sum_{r=1}^{M}\sum_{u=1}^{M}\sum_{i=1}^{n_r}\sum_{a=1}^{n_u} \frac{c_{ia}^{ru}}{2+c_{ia}^{ru}}[(U_{ia}^{ru})^2+(V_{ia}^{ru})^2+(W_{ia}^{ru})^2] \\ &\geq b(\|\tilde{x}_{00}^{00}\|). \end{aligned} \tag{6.122}$$

On the other hand, it follows from (6.117) that

$$\begin{aligned} V(x_{ia}^{ru}) &= (U_{ia}^{ru})^2 + 2U_{ia}^{ru}V_{ia}^{ru} + (c_{ia}^{ru}+1)(V_{ia}^{ru})^2 + (W_{ia}^{ru})^2 \\ &\leq 2(U_{ia}^{ru})^2 + (c_{ia}^{ru}+2)(V_{ia}^{ru})^2 + (W_{ia}^{ru})^2 \\ &\leq (c_{ia}^{ru}+2)[(U_{ia}^{ru})^2+(V_{ia}^{ru})^2+(W_{ia}^{ru})^2]. \end{aligned} \tag{6.123}$$

Thus from (6.121) and (6.123) we have

$$\begin{aligned} V(x_{00}^{00}(t)) &\leq \sum_{r=1}^{M}\sum_{u=1}^{M}\sum_{i=1}^{n_r}\sum_{a=1}^{n_u}(c_{ia}^{ru}+2)[(U_{ia}^{ru})^2+(V_{ia}^{ru})^2+(W_{ia}^{ru})^2] \\ &\leq a(\|\tilde{x}_{00}^{00}\|). \end{aligned} \tag{6.124}$$

Therefore from (6.117), (6.122), and (6.124) we establish the desired inequality. □

Remark 6.7.1. Lemma 6.7.1 shows that the Lyapunov function V defined in (6.117) is positive definite (6.122), decrescent, and radially unbounded (6.124) function [53, 24].

We now state the following lemma.

Lemma 6.7.2. *Assume that the hypothesis of Lemma* 6.7.1 *be satisfied. For* $r, u, v \in I(1, M)$, $i \in I(1, n_r)$, $a \in I(1, n_u)$, *and* $b \in I(1, n_v)$, *let*

$$d_{ai}^{ur} = \sum_{v=1}^{M}\sum_{b=1}^{n_r} c_{ba}^{vu}\left[\beta_{abi}^{uvr}\left(\frac{S_{ba}^{vu*}}{\mu_{ba}^{vu}}+\frac{\bar{B}^2}{\mu_{ba}^{vu}}\right)+(\nu_{abi}^{uvr})^2(S_{ba}^{vu*}+\bar{B})^2\right] \tag{6.125}$$

for some positive numbers c_{ia}^{ru}. *Furthermore, let*

$$\mathfrak{U}_{ia}^{ru} = \begin{cases} \dfrac{[2\sum_{u=1}^{M}\sum_{a=1}^{n_u}\mu_{ia}^{ru}+\sum_{u\neq r}^{M}\sum_{a=1}^{n_r}\frac{(\gamma_{ia}^{ru})^2}{\mu_{ii}^{rr}}+\sum_{a\neq i}^{n_r}\frac{(\sigma_{ia}^{rr})^2}{\mu_{ii}^{rr}}+\frac{3}{2}\mu_{ii}^{rr}]}{(\gamma_i^r+\sigma_i^r+\delta_i^r)}, & u=r, i=a, \\ \dfrac{[\frac{(\rho_{ia}^{rr})^2}{\mu_{ia}^{rr}}+\mu_{ii}^{rr}+\frac{3}{2}\mu_{ia}^{rr}]}{(\rho_{ia}^{rr}+\delta_a^r)}, & u=r, a\neq i, \\ \dfrac{[\frac{(\rho_{ia}^{ru})^2}{\mu_{ia}^{ru}}+\mu_{ii}^{rr}+\frac{3}{2}\mu_{ia}^{ru}]}{(\rho_{ia}^{ru}+\delta_a^u)}, & u\neq r, \end{cases} \tag{6.126}$$

$$\mathfrak{R}_{ia_0}^{ru} = \begin{cases} \dfrac{\sum_{u=1}^{M}\sum_{a=1}^{n_u}\frac{1}{2}\mu_{ia}^{ru}+\sum_{v=1}^{M}\sum_{b=1}^{n_v}\frac{1}{2}\beta_{iib}^{rrv}(S_{ii}^{rr*}+\mu_{ii}^{rr})+\frac{1}{2}d_{ii}^{rr}}{\eta_i^r+\varrho_i^r+\gamma_i^r+\sigma_i^r+\delta_i^r+d_i^r}, & a=i, u=r, \\ \dfrac{\frac{1}{2}\mu_{ii}^{rr}+\sum_{v=1}^{M}\sum_{b=1}^{n_v}\frac{1}{2}\beta_{aib}^{rrv}(S_{ia}^{rr*}+\mu_{ia}^{rr})+\frac{1}{2}d_{ai}^{rr}}{\eta_a^r+\varrho_a^r+\rho_{ia}^{rr}+\delta_a^r+d_a^r}, & a\neq i, u=r, \\ \dfrac{\frac{1}{2}\mu_{ii}^{rr}+\sum_{v=1}^{M}\sum_{b=1}^{n_v}\frac{1}{2}\beta_{aib}^{urv}(S_{ii}^{ru*}+\mu_{ia}^{ru})+\frac{1}{2}d_{ai}^{ur}}{\eta_a^u+\varrho_a^u+\rho_{ia}^{ru}+\delta_a^u+d_a^u}, & u\neq r, \end{cases} \tag{6.127}$$

and

$$\mathfrak{W}_{ia}^{ru} = \begin{cases} \dfrac{[\frac{1}{2}\sum_{u=1}^{M}\sum_{a=1}^{n_u}\mu_{ia}^{ru} + \frac{1}{2}\mu_{ii}^{rr} + \frac{1}{2}\sum_{a\neq i}^{n_r}\frac{(\sigma_{ia}^{rr})^2}{\mu_{ii}^{rr}} + \frac{1}{2}\sum_{u\neq r}^{M}\sum_{a=1}^{n_r}\frac{(\gamma_{ia}^{ru})^2}{\mu_{ii}^{rr}} + \frac{(\alpha_i^r)^2}{\mu_{ii}^{rr}}]}{(\gamma_i^r + \sigma_i^r + \alpha_i^r + \delta_i^r)}, & u = r, a = i, \\ \dfrac{[\frac{1}{2}\frac{(\rho_{ia}^{rr})^2}{\mu_{ia}^{rr}} + \frac{1}{2}\mu_{ii}^{rr} + \frac{1}{2}\mu_{ia}^{rr} + \frac{(\alpha_a^r)^2}{\mu_{ia}^{rr}}]}{(\rho_{ia}^{rr} + \alpha_a^r + \delta_a^r)}, & u = r, a \neq i, \\ \dfrac{[\frac{1}{2}\frac{(\rho_{ia}^{ru})^2}{\mu_{ia}^{ru}} + \frac{1}{2}\mu_{ii}^{rr} + \frac{1}{2}\mu_{ia}^{ru} + \frac{(\alpha_a^u)^2}{\mu_{ia}^{ru}}]}{(\rho_{ia}^{ru} + \alpha_a^u + \delta_a^u)}, & u \neq r, \end{cases} \tag{6.128}$$

for some suitably defined positive number μ_{ia}^{ru}, depending on δ_a^u, for all $r, u \in I^r(1, M)$, $i \in I(1, n)$, and $a \in I_i^r(1, n_r)$. Assume that $\mathfrak{U}_{ia}^{ru} \leq 1$, $\mathfrak{R}_{ia_0}^{ru} < 1$, and $\mathfrak{W}_{ia}^{ru} \leq 1$. Then there exist positive numbers ϕ_{ia}^{ru}, ψ_{ia}^{ru}, and φ_{ia}^{ru} such that the differential operator LV associated with Itô–Doob-type stochastic system (6.53)–(6.55) satisfies the following inequality:

$$\begin{aligned} LV(\tilde{x}_{00}^{00}) \leq \sum_{r=1}^{M}\sum_{i=1}^{n_r}\Big[&-[\phi_{ii}^{rr}(U_{ii}^{rr})^2 + \psi_{ii}^{rr}(V_{ii}^{rr})^2 + \varphi_{ii}^{rr}(W_{ii}^{rr})^2] \\ &- \sum_{a\neq i}^{n_r}[\phi_{ia}^{rr}(U_{ia}^{rr})^2 + \psi_{ia}^{rr}(V_{ia}^{rr})^2 + \varphi_{ia}^{rr}(W_{ia}^{rr})^2] \\ &- \sum_{u\neq r}^{M}\sum_{a=1}^{n_u}[\phi_{ia}^{ru}(U_{ia}^{rr})^2 + \psi_{ia}^{ru}(V_{ia}^{ru})^2 + \varphi_{ia}^{ru}(W_{ia}^{ru})^2]\Big]. \end{aligned} \tag{6.129}$$

Moreover,

$$LV(\tilde{x}_{00}^{00}) \leq -cV(\tilde{x}_{00}^{00}), \tag{6.130}$$

where a positive constant c is defined by

$$c = \frac{\min_{1\leq r,u\leq M, 1\leq i\leq n_r, 1\leq a\leq n_u}(\phi_{ia}^{ru}, \psi_{ia}^{ru}, \varphi_{ia}^{ru})}{\max_{1\leq r,u\leq M, 1\leq i\leq n_r, 1\leq a\leq n_u}\{C_{ia}^{ru} + 2\}}. \tag{6.131}$$

Proof. We compute the differential operator [24, 53] applied to the Lyapunov function V in (6.117) with respect to the large-scale system of Itô–Doob-type stochastic differential equation (6.53)–(6.55) as follows:

$$\begin{aligned} LV(\tilde{x}_{ii}^{rr}) = 2\sum_{u=1}^{M}\sum_{a=1}^{n_u}&[(1 + C_{ii}^{rr})\rho_{ia}^{ru}V_{ia}^{ru}V_{ii}^{rr} + \rho_{ia}^{ru}U_{ia}^{ru}U_{ii}^{rr} + \rho_{ia}^{ru}V_{ia}^{ru}U_{ii}^{rr} + \rho_{ia}^{ru}U_{ia}^{ru}V_{ii}^{rr} \\ &+ \rho_{ia}^{ru}W_{ia}^{ru}W_{ii}^{rr}] + 2\alpha_i^r U_{ii}^{rr}W_{ii}^{rr} + 2(\alpha_i^r + \varrho_i^r)V_{ii}^{rr}W_{ii}^{rr} \\ &- 2[\varrho_i^r + d_i^r + 2(\gamma_i^r + \sigma_i^r + \delta_i^r)]V_{ii}^{rr}U_{ii}^{rr} - 2(\gamma_i^r + \sigma_i^r + \delta_i^r)(U_{ii}^{rr})^2 \\ &- 2[c_{ii}^{rr}\eta_i^r + 2(c_{ii}^{rr} + 1)(\varrho_i^r + \gamma_i^r + \sigma_i^r + \delta_i^r + d_i^r)](V_{ii}^{rr})^2 \\ &- 2(\gamma_i^r + \sigma_i^r + \alpha_i^r + \delta_i^r)(W_{ii}^{rr})^2 \end{aligned}$$

$$
\begin{aligned}
&+2\sum_{u=1}^{M}\sum_{a=1}^{n_u} c_{ii}^{rr}\beta_{iia}^{rru}(S_{ii}^{rr*}+U_{ii}^{rr})V_{ai}^{ur}V_{ii}^{rr}\\
&+c_{ii}^{rr}\sum_{u=1}^{M}\sum_{a=1}^{n_u}(v_{iia}^{rru})^2(S_{ii}^{rr*}+U_{ii}^{rr})^2(V_{ai}^{ur})^2, \quad u=r,\ a=i,
\end{aligned}
\tag{6.132}
$$

$$
\begin{aligned}
\sum_{a\neq i}^{n_r} LV(\tilde{x}_{ia}^{rr}) = \sum_{a\neq r}^{n_r}\{&2(1+c_{ia}^{rr})\sigma_{ia}^{rr}V_{ia}^{rr}V_{ii}^{rr}+2\sigma_{ia}^{rr}U_{ia}^{rr}U_{ii}^{rr}+2\sigma_{ia}^{rr}V_{ia}^{rr}U_{ii}^{rr}+2\sigma_{ia}^{rr}U_{ia}^{rr}V_{ii}^{rr}\\
&+2\sigma_{ia}^{rr}W_{ia}^{rr}W_{ii}^{rr}\\
&-2[c_{ia}^{rr}\eta_a^r+2(c_{ia}^{rr}+1)(\varrho_a^r+\rho_{ia}^{rr}+\delta_a^r)](V_{ia}^{rr})^2-2(\rho_{ia}^{rr}+\delta_a^r)(U_{ia}^{rr})^2\\
&-2(\rho_{ia}^{rr}+\alpha_a^r+\delta_a^r)(W_{ia}^{rr})^2+2\alpha_a^rW_{ia}^{rr}U_{ia}^{rr}+2(\alpha_a^r+\varrho_a^r)V_{ia}^{rr}W_{ia}^{rr}\\
&-2[\varrho_a^r+d_a^r+2(\rho_{ia}^{rr}+\delta_a^r)]V_{ia}^{rr}U_{ia}^{rr}\}\\
&+2\sum_{a\neq r}^{n_r}\sum_{v=1}^{M}\sum_{b=1}^{n_v} c_{ia}^{rr}\beta_{aib}^{rrv}(S_{ia}^{rr*}+U_{ia}^{rr})V_{ba}^{vr}V_{ia}^{rr}\\
&+\sum_{a\neq r}^{n_r} c_{ia}^{rr}\sum_{v=1}^{M}\sum_{b=1}^{n_v}(v_{aib}^{rrv})^2(S_{ia}^{rr*}+U_{ia}^{rr})^2(V_{ba}^{vr})^2, \quad u=r,\ a\neq i,
\end{aligned}
\tag{6.133}
$$

$$
\begin{aligned}
\sum_{u=1}^{M}\sum_{a=1}^{n_r} LV(\tilde{x}_{ia}^{ru}) = \sum_{u=1}^{M}\sum_{a=1}^{n_u}\{&2(1+c_{ia}^{ru})\gamma_{ia}^{ru}V_{ia}^{ru}V_{ii}^{rr}+2\gamma_{ia}^{ru}U_{ia}^{ru}U_{ii}^{rr}+2\gamma_{ia}^{ru}V_{ia}^{ru}U_{ii}^{rr}+2\gamma_{ia}^{ru}U_{ia}^{ru}V_{ii}^{rr}\\
&+2\gamma_{ia}^{ru}W_{ia}^{ru}W_{ii}^{rr}-2[c_{ia}^{ru}\eta_a^u+2(c_{ia}^{ru}+1)(\varrho_a^u+\rho_{ia}^{ru}+\delta_a^u+d_a^u)](V_{ia}^{ru})^2\\
&-2(\rho_{ia}^{ru}+\delta_a^u)(U_{ia}^{ru})^2\\
&-2(\rho_{ia}^{ru}+\alpha_a^u+\delta_a^u)(W_{ia}^{rr})^2+2\alpha_a^uW_{ia}^{ru}U_{ia}^{ru}+2(\alpha_a^u+\varrho_a^u)V_{ia}^{ru}W_{ia}^{ru}\\
&-2[\varrho_a^u+d_a^u+2(\rho_{ia}^{ru}+\delta_a^u)]V_{ia}^{ru}U_{ia}^{ru}\}\\
&+2\sum_{u=1}^{M}\sum_{a=1}^{n_u}\sum_{v=1}^{M}\sum_{b=1}^{n_v} c_{ia}^{ru}\beta_{aib}^{urv}(S_{ia}^{ru*}+U_{ia}^{ru})V_{ba}^{vu}V_{ia}^{ru}\\
&+\sum_{u=1}^{M}\sum_{a=1}^{n_r} c_{ia}^{ru}\sum_{v=1}^{M}\sum_{b=1}^{n_v}(v_{aib}^{urv})^2(S_{ia}^{ru*}+U_{ia}^{ru})^2(V_{ba}^{vu})^2, \quad u\neq r.
\end{aligned}
\tag{6.134}
$$

By using (6.107) and the algebraic inequality

$$
2ab \le \frac{a^2}{g(c)} + b^2 g(c), \quad a,b,c\in\mathbb{R},
\tag{6.135}
$$

where the function g is such that $g(c)\ge 0$, we estimate the sixth term in (6.132), (6.133), and (6.134) as follows:

$$
\begin{aligned}
&2\sum_{v=1}^{M}\sum_{b=1}^{n_v} c_{ii}^{rr}\beta_{iib}^{rrv}(S_{ii}^{rr*}+U_{ii}^{rr})V_{bi}^{vr}V_{ii}^{rr}\\
&\le \sum_{v=1}^{M}\sum_{b=1}^{n_v} c_{ii}^{rr}\beta_{iib}^{rrv}(S_{ii}^{rr*}g_i^r(\delta_i^r)+g_i^r(\delta_i^r))(V_{ii}^{rr})^2
\end{aligned}
$$

$$+\sum_{v=1}^{M}\sum_{b=1}^{n_v} c_{ii}^{rr}\beta_{iib}^{rrv}\left(\frac{S_{ii}^{rr*}}{g_i^r(\delta_i^r}\right)+\frac{\bar{B}^2}{g_i^r(\delta_i^r)})(V_{bi}^{vr})^2,$$

$$\sum_{a\neq r}^{n_r} 2\sum_{v=1}^{M}\sum_{b=1}^{n_v} c_{ia}^{rr}\beta_{aib}^{rrv}(S_{ia}^{rr*}+U_{ia}^{rr})V_{ba}^{vr}V_{ia}^{rr}$$
$$\leq \sum_{a\neq r}^{n_r}\sum_{v=1}^{M}\sum_{b=1}^{n_v} c_{ia}^{rr}\beta_{aib}^{rrv}(S_{ia}^{rr*}g_i^r(\delta_a^r)+g_i^r(\delta_a^r))(V_{ia}^{rr})^2$$
$$+\sum_{a\neq r}^{n_r}\sum_{v=1}^{M}\sum_{b=1}^{n_v} c_{ia}^{rr}\beta_{aib}^{rrv}\left(\frac{S_{ia}^{rr*}}{g_i^r(\delta_a^r}\right)+\frac{\bar{B}^2}{g_i^r(\delta_a^r)})(V_{bi}^{vr})^2$$

and

$$2\sum_{u=1}^{M}\sum_{a=1}^{n_u}\sum_{v=1}^{M}\sum_{b=1}^{n_v} c_{ia}^{ru}\beta_{aib}^{urv}(S_{ia}^{ru*}+U_{ia}^{ru})V_{ba}^{vu}V_{ia}^{ru}$$
$$\leq \sum_{u\neq r}^{M}\sum_{a=1}^{n_u}\sum_{v=1}^{M}\sum_{b=1}^{n_v} c_{ia}^{ru}\beta_{aib}^{urv}(S_{ia}^{ru*}g_i^r(\delta_a^u)+g_i^r(\delta_a^u))(V_{ia}^{ru})^2$$
$$+\sum_{u\neq r}^{M}\sum_{a=1}^{n_u}\sum_{v=1}^{M}\sum_{b=1}^{n_v} c_{ia}^{ru}\beta_{aib}^{urv}\left(\frac{S_{ia}^{ru*}}{g_i^r(\delta_a^u}\right)+\frac{\bar{B}^2}{g_i^r(\delta_a^u)})(V_{ba}^{vu})^2. \tag{6.136}$$

From (6.132) and (6.133), by repeated usage of inequalities (6.135) and (6.136), coupled with algebraic manipulations and simplifications, we have the inequality

$$LV(\tilde{x}_{00}^{00}) \leq \sum_{r=1}^{M}\sum_{i=1}^{n_r}\left\{\left[2\sum_{u=1}^{M}\sum_{a=1}^{n_u}\mu_{ia}^{ru}+3\mu_{ii}^{rr}+2\sum_{a\neq i}^{n_r}\frac{(\sigma_{ia}^{rr})^2}{\mu_{ii}^{rr}}+2\sum_{a\neq r}^{M}\sum_{a=1}^{n_r}\frac{(\gamma_{ia}^{ru})^2}{\mu_{ii}^{rr}}\right.\right.$$
$$\left.-2(\gamma_i^r+\sigma_i^r+\delta_i^r)\right](U_{ii}^{rr})^2$$
$$+\left[\sum_{u=1}^{M}\sum_{a=1}^{n_u}[(2+c_{ii}^{rr})\mu_{ia}^{ru}]+\mu_{ii}^{rr}+\frac{(\varrho_i^r)^2}{\mu_{ii}^{rr}}+\frac{(\varrho_i^r+d_i^r)^2}{\mu_{ii}^{rr}}+4\frac{(\gamma_i^r+\sigma_i^r+\delta_i^r)^2}{\mu_{ii}^{rr}}\right.$$
$$-2[c_{ii}^{rr}\eta_i^r+(c_{ii}^{rr}+1)(\varrho_i^r+\gamma_i^r+\sigma_i^r+\delta_i^r+d_i^r)]+\sum_{a\neq r}^{n_r}\frac{(2+c_{ia}^{rr})(\sigma_{ia}^{rr})^2}{\mu_{ii}^{rr}}$$
$$\left.+\sum_{u\neq r}^{M}\sum_{a=1}^{n_r}\frac{(2+c_{ia}^{ru})(\gamma_{ia}^{ru})^2}{\mu_{ii}^{rr}}+c_{ii}^{rr}\sum_{v=1}^{M}\sum_{b=1}^{n_r}\beta_{iib}^{rrv}(S_{ii}^{rr*}\mu_{ii}^{rr}+\mu_{ii}^{rr})\right](V_{ii}^{rr})^2$$
$$+\left[\sum_{u=1}^{M}\sum_{a=1}^{n_u}\mu_{ia}^{ru}+2\frac{(\alpha_i^r)^2}{\mu_{ii}^{rr}}+\mu_{ii}^{rr}+\sum_{a\neq i}^{n_r}\frac{(\sigma_{ia}^{rr})^2}{\mu_{ii}^{rr}}+\sum_{a\neq r}^{M}\sum_{a=1i}^{n_r}\frac{(\gamma_{ia}^{ru})^2}{\mu_{ii}^{rr}}\right.$$
$$\left.-2(\gamma_i^r+\sigma_i^r+\alpha_i^r+\delta_i^r)\right](W_{ii}^{rr})^2$$
$$+\sum_{a\neq i}^{n_r}\left\{\left[2\frac{(\rho_{ia}^{rr})^2}{\mu_{ia}^{rr}}+2\mu_{ii}^{rr}+3\mu_{ia}^{rr}-2(\rho_{ia}^{rr}+\delta_a^r)\right](U_{ia}^{rr})^2+\left[(2+c_{ii}^{rr})\frac{(\rho_{ia}^{rr})^2}{\mu_{ia}^{rr}}\right.\right.$$
$$\times(2+c_{ia}^{rr})\mu_{ii}^{rr}-2[c_{ia}^{rr}\eta_a^r+(1+c_{ia}^{rr})(\eta_a^r+\rho_{ia}^{rr}+\delta_a^r+d_a^r)]+\frac{(\varrho_a^r+d_a^r)^2}{\mu_{ia}^{rr}}$$

$$
\begin{aligned}
&+ 4\frac{(\rho^{rr}_{ia}+\delta^{r}_{a})^2}{\mu^{rr}_{ia}} + \mu^{rr}_{ia} + c^{rr}_{ia}\sum_{v=1}^{M}\sum_{b=1}^{n_r}\beta^{rrv}_{aib}(S^{rr}_{ia}\mu^{rr}_{ia}+\mu^{rr}_{ia})\Bigg](V^{rr}_{ia})^2\\
&+\Bigg[\frac{(\rho^{rr}_{ia})^2}{\mu^{rr}_{ia}} + \mu^{rr}_{ii} + \mu^{rr}_{ia} + \frac{2(\alpha^{r}_{a})^2}{\mu^{rr}_{ia}} - 2(\rho^{rr}_{ia}+\alpha^{r}_{a}+\delta^{r}_{a})\Bigg](W^{rr}_{ia})^2\Bigg\}\\
&+\sum_{u\neq r}^{M}\sum_{a=1}^{n_u}\Bigg\{\Bigg[2\frac{(\rho^{ru}_{ia})^2}{\mu^{ru}_{ia}} + 2\mu^{rr}_{ii} + 3\mu^{ru}_{ia} - 2(\rho^{ru}_{ia}+\delta^{u}_{a})\Bigg](U^{ru}_{ia})^2\\
&+\Bigg[(2+c^{rr}_{ii})\frac{(\rho^{ru}_{ia})^2}{\mu^{ru}_{ia}} + (2+c^{ru}_{ia})\mu^{rr}_{ii} - 2[c^{ru}_{ia}\eta^{u}_{a} + (1+c^{ru}_{ia})(\eta^{u}_{a}+\rho^{ru}_{ia}+\delta^{u}_{a}+d^{u}_{a})]\\
&+\frac{(\varrho^{u}_{a}+d^{u}_{a})^2}{\mu^{ru}_{ia}} + 4\frac{(\rho^{ru}_{ia}+\delta^{u}_{a})^2}{\mu^{ru}_{ia}} + \mu^{ru}_{ia} + c^{ru}_{ia}\sum_{v=1}^{M}\sum_{b=1}^{n_r}\beta^{urv}_{aib}(S^{ru*}_{ia}\mu^{ru}_{ia}+\mu^{ru}_{ia})\Bigg](V^{ru}_{ia})^2\\
&+\Bigg[\frac{(\rho^{ru}_{ia})^2}{\mu^{ru}_{ia}} + \mu^{rr}_{ii} + \mu^{ru}_{ia} + \frac{2(\alpha^{u}_{a})^2}{\mu^{ru}_{ia}} - 2(\rho^{ru}_{ia}\alpha^{u}_{a}+\delta^{u}_{a})\Bigg](W^{ru}_{ia})^2\Bigg\}\Bigg\}\\
&+\sum_{r=1}^{M}\sum_{i=1}^{n_r}c^{rr}_{ii}\sum_{v=1}^{M}\sum_{b=1}^{n_r}\Bigg[\beta^{rrv}_{iib}\Bigg(\frac{S^{rr*}_{ii}}{\mu^{rr}_{ii}}+\frac{\bar{B}^2}{\mu^{rr}_{ii}}\Bigg) + (\nu^{rrv}_{iib})^2(S^{rr*}_{ii}+\bar{B})^2\Bigg](V^{vr}_{bi})^2\\
&+\sum_{r=1}^{M}\sum_{i=1}^{n_r}\sum_{a\neq i}^{n_r}c^{rr}_{ia}\sum_{v=1}^{M}\sum_{b=1}^{n_r}\Bigg[\beta^{rrv}_{aib}\Bigg(\frac{S^{rr*}_{ia}}{\mu^{rr}_{ia}}+\frac{\bar{B}^2}{\mu^{rr}_{ia}}\Bigg) + (\nu^{rrv}_{aib})^2(S^{rr*}_{ia}+\bar{B})^2\Bigg](V^{vr}_{ba})^2\\
&+\sum_{r=1}^{M}\sum_{i=1}^{n_r}\sum_{u\neq r}^{M}\sum_{a=1}^{n_r}c^{ru}_{ia}\sum_{v=1}^{M}\sum_{b=1}^{n_r}\Bigg[\beta^{urv}_{aib}\Bigg(\frac{S^{ru*}_{ia}}{\mu^{ru}_{ia}}+\frac{\bar{B}^2}{\mu^{ru}_{ia}}\Bigg) + (\nu^{urv}_{aib})^2(S^{ru*}_{ia}+\bar{B})^2\Bigg](V^{vu}_{ba})^2,
\end{aligned}
\tag{6.137}
$$

where $\mu^{ru}_{ia} = g^{r}_{i}(\delta^{u}_{a})$ with g^{r}_{i} appropriately defined by (6.135).

For all $r,u \in I(1,M)$, $i \in I(1,n_r)$, and $a \in I(1,n_u)$, using algebraic manipulations and (6.126), (6.127), and (6.128), we determine the coefficients ϕ^{ru}_{ia}, ψ^{ru}_{ia}, and φ^{ru}_{ia} of $(U^{ru}_{ia})^2$, $(V^{ru}_{ia})^2$, and $(W^{ru}_{ia})^2$, respectively:

$$
\phi^{ru}_{ia} = \begin{cases}
2(\gamma^{r}_{i}+\sigma^{r}_{i}+\delta^{r}_{i})(1-\mathfrak{U}^{ru}_{ia}), & u=r, a=i,\\
2(\rho^{rr}_{ia}+\delta^{r}_{a}+\delta^{r}_{a})(1-\mathfrak{U}^{ru}_{ia}), & u=r, a\neq i,\\
2(\rho^{ru}_{ia}+\delta^{u}_{a}+\delta^{u}_{a})(1-\mathfrak{U}^{ru}_{ia}), & u\neq r,
\end{cases}
\tag{6.138}
$$

$$
\psi^{ru}_{ia} = \begin{cases}
[2c^{rr}_{ii}(1-\mathfrak{R}^{rr}_{ii_0})(\eta^{r}_{i}+\varrho^{r}_{i}+\gamma^{r}_{i}+\sigma^{r}_{i}+\delta^{r}_{i}+d^{r}_{i}) - \mathfrak{C}^{rr}_{ii}]\\
\quad + 2(\varrho^{r}_{i}+\gamma^{r}_{i}+\sigma^{r}_{i}+\delta^{r}_{i}+d^{r}_{i}), \quad u=r,\ a=i,\\
[2c^{rr}_{ia}(1-\mathfrak{R}^{rr}_{ia_0})(\eta^{r}_{a}+\varrho^{r}_{a}+\rho^{rr}_{ia}+\delta^{r}_{a}+d^{r}_{a}) - \mathfrak{C}^{rr}_{ia}]\\
\quad + 2(\varrho^{r}_{a}+\rho^{rr}_{ia}+\delta^{r}_{a}+d^{r}_{a}), \quad u=r,\ a\neq i,\\
[2c^{ru}_{ia}(1-\mathfrak{R}^{ru}_{ia_0})(\eta^{u}_{a}+\varrho^{u}_{a}+\rho^{ru}_{ia}+\delta^{u}_{a}+d^{u}_{a}) - \mathfrak{C}^{ru}_{ia}]\\
\quad + 2(\varrho^{u}_{a}+\rho^{ru}_{ia}+\delta^{u}_{a}+d^{u}_{a}), \quad u\neq r,
\end{cases}
\tag{6.139}
$$

and

$$\varphi_{ia}^{ru} = \begin{cases} 2(\gamma_i^r + \sigma_i^r + \alpha_i^r + \delta_i^r)(1 - \mathfrak{W}_{ia}^{ru}), & u = r, a = i, \\ 2(\rho_{ia}^{rr} + \delta_a^r)(1 - \mathfrak{W}_{ia}^{ru}), & u = r,\ a \neq i, \\ 2(\rho_{ia}^{ru} + \delta_a^u)(1 - \mathfrak{W}_{ia}^{ru}), & u \neq r, \end{cases} \tag{6.140}$$

where

$$\mathfrak{C}_{ia}^{ru} = \begin{cases} \Big[2\sum_{u=1}^{M}\sum_{a=1}^{n_u}\mu_{ia}^{ru} + \mu_{ii}^{rr} + \frac{(\varrho_i^r)^2}{\mu_{ii}^{rr}} + \frac{(\varrho_i^r+d_i^r)^2}{\mu_{ii}^{rr}} + 4\frac{(\gamma_i^r+\sigma_i^r+\delta_i^r}{\mu_{ii}^{rr}} + \sum_{a\neq r}^{n_r}\frac{(2+c_{ia}^{rr})(\sigma_{ia}^{rr})^2}{\mu_{ii}^{rr}} \\ \quad + \sum_{u\neq r}^{M}\sum_{a=1}^{n_r}(\frac{(2+c_{ia}^{ru})(\gamma_{ia}^{ru})^2}{\mu_{ii}^{rr}}) + \sum_{b\neq i}^{n_r}c_{bi}^{rr}d_{ii}^{rr} + \sum_{v\neq r}^{M}\sum_{b=1}^{n_r}c_{bi}^{vr}d_{ii}^{rr}\Big], & u = r,\ a = i, \\ (2 + c_{ii}^{rr})\frac{(\rho_{ia}^{rr})^2}{\mu_{ia}^{rr}} + \frac{(\varrho_a^r)^2}{\mu_{ia}^{rr}} + 2\mu_{ii}^{rr} + \frac{(\varrho_a^r+d_a^r)^2}{\mu_{ia}^{rr}} \\ \quad + 4\frac{(\rho_{ia}^{rr}+\delta_a^r)^2}{\mu_{ia}^{rr}} + \mu_{ia}^{rr} + \sum_{b\neq i}^{n_r}c_{ba}^{rr}d_{ia}^{rr} + \sum_{v\neq r}^{M}\sum_{b=1}^{n_r}c_{ba}^{vr}d_{ia}^{rr}, & u = r, a \neq i, \\ (2 + c_{ii}^{rr})\frac{(\rho_{ia}^{ru})^2}{\mu_{ia}^{ru}} + \frac{(\varrho_a^u)^2}{\mu_{ia}^{ru}} + 2\mu_{ii}^{rr} + \frac{(\varrho_a^u+d_a^u)^2}{\mu_{ia}^{ru}} \\ \quad + 4\frac{(\rho_{ia}^{ru}+\delta_a^u)^2}{\mu_{ia}^{ru}} + \mu_{ia}^{ru} + \sum_{b\neq i}^{n_r}c_{ba}^{ru}d_{ia}^{ru} + \sum_{v\neq r}^{M}\sum_{b=1}^{n_r}c_{ba}^{vu}d_{ia}^{ru}, & u \neq r. \end{cases} \tag{6.141}$$

Under the assumptions on $\mathfrak{U}_{ia}^{ru}$, $\mathfrak{R}_{ia_0}^{ru}$, and $\mathfrak{W}_{ia}^{ru}$, it is clear that ϕ_{ia}^{ru}, ψ_{ia}^{ru}, and φ_{ia}^{ru} are positive for suitable choice of c_{ia}^{ru} defined in (6.118). We substitute (6.125), (6.138), (6.139), and (6.141) into (6.137). Thus inequality (6.137) can be rewritten as

$$\begin{aligned} LV(\tilde{x}_{00}^{00}) \leq \sum_{r=1}^{M}\sum_{i=1}^{n_r} -\Big\{[\phi_{ii}^{rr}(U_{ii}^{rr})^2 + \psi_{ii}^{rr})(V_{ii}^{rr})^2\ \varphi_{ii}^{rr}(W_{ii}^{rr})^2] \\ + \sum_{a\neq r}^{n_r}[\phi_{ia}^{rr}(U_{ia}^{rr})^2 + \psi_{ia}^{rr}(V_{ia}^{rr})^2 + \varphi_{ia}^{rr}(W_{ia}^{rr})^2] \\ + \sum_{u\neq r}^{M}\sum_{a=1}^{n_u}[\phi_{ia}^{ru}(U_{ia}^{ru})^2 + \psi_{ia}^{ru}(V_{ia}^{ru})^2 + \varphi_{ia}^{ru}(W_{ia}^{ru})^2]\Big\}. \end{aligned} \tag{6.142}$$

This proves inequality (6.129). Now (6.130) follows from (6.129), that is,

$$LV(\tilde{x}_{00}^{00}) \leq -cV(\tilde{x}_{00}^{00}),$$

where $c = \frac{\min_{1\leq r,u\leq M,1\leq i\leq n_r,1\leq a\leq n_u}(\phi_{ia}^{ru},\psi_{ia}^{ru},\varphi_{ia}^{ru})}{\max_{1\leq r,u\leq M,1\leq i\leq n_r,1\leq a\leq n_u}\{C_{ia}^{ru}+2\}}$. This establishes the result. □

We now formally state the stochastic stability theorems for the disease-free equilibria.

Theorem 6.7.3. *Let $r, u \in I(1, M)$, $i \in I(1, n_r)$, and $a \in I(1, n_u)$. Let the hypotheses of Lemma 6.7.2 be satisfied. Then the disease-free solutions E_{ia}^{ru} are asymptotically stable in the large. Moreover, the solutions E_{ia}^{ru} are exponentially mean square stable.*

Proof. Applying the comparison result [53, 24], the proof of stochastic asymptotic stability immediately follows. Moreover, the disease-free equilibrium state is exponentially mean square stable. □

Remark 6.7.2 (Interpretation of the disease-eradication conditions). Note that the threshold parameter (6.127) contains the basic reproduction number for the disease. Indeed, it is easy to see that

$$\mathfrak{R}^{ru}_{ia_0} = \begin{cases} \frac{\sum_{u=1}^{M}\sum_{a=1}^{n_u}\frac{1}{2}\mu^{ru}_{ia}+\frac{1}{2}d^{rr}_{ii}}{\eta^r_i+\varrho^r_i+\gamma^r_i+\sigma^r_i+\delta^r_i+d^r_i} + \frac{\sum_{v=1}^{M}\sum_{b=1}^{n_v}\frac{1}{2}\beta^{rrv}_{iib}(S^{rr*}_{ii})}{\eta^r_i+\varrho^r_i+\gamma^r_i+\sigma^r_i+\delta^r_i+d^r_i} + \frac{\sum_{v=1}^{M}\sum_{b=1}^{n_v}\frac{1}{2}\beta^{rrv}_{iib}(\mu^{rr}_{ii})}{\eta^r_i+\varrho^r_i+\gamma^r_i+\sigma^r_i+\delta^r_i+d^r_i}, & a = i, u = r, \\ \frac{\frac{1}{2}\mu^{rr}_{ii}+\frac{1}{2}d^{rr}_{ai}}{\eta^r_a+\varrho^r_a+\rho^{rr}_{ia}+\delta^r_a+d^r_a} + \frac{\sum_{v=1}^{M}\sum_{b=1}^{n_v}\frac{1}{2}\beta^{rrv}_{aib}(S^{rr*}_{ia})}{\eta^r_a+\varrho^r_a+\rho^{rr}_{ia}+\delta^r_a+d^r_a} + \frac{\sum_{v=1}^{M}\sum_{b=1}^{n_v}\frac{1}{2}\beta^{rrv}_{aib}(\mu^{rr}_{ia})}{\eta^r_a+\varrho^r_a+\rho^{rr}_{ia}+\delta^r_a+d^r_a}, & a \neq i, u = r, \\ \frac{\frac{1}{2}\mu^{rr}_{ii}+\frac{1}{2}d^{ur}_{ai}}{\eta^u_a+\varrho^u_a+\rho^{ru}_{ia}+\delta^u_a+d^u_a} + \frac{\sum_{v=1}^{M}\sum_{b=1}^{n_v}\frac{1}{2}\beta^{urv}_{aib}(S^{ru*}_{ii})}{\eta^u_a+\varrho^u_a+\rho^{ru}_{ia}+\delta^u_a} + \frac{\sum_{v=1}^{M}\sum_{b=1}^{n_v}\frac{1}{2}\beta^{urv}_{aib}(\mu^{ru}_{ia})}{\eta^u_a+\varrho^u_a+\rho^{ru}_{ia}+\delta^u_a+d^u_a}, & u \neq r. \end{cases} \tag{6.143}$$

Note that the terms

$$\bar{\mathfrak{R}}^{ru}_{ia_0} = \begin{cases} \frac{\sum_{v=1}^{M}\sum_{b=1}^{n_v}\beta^{rrv}_{iib}(S^{rr*}_{ii})}{\eta^r_i+\varrho^r_i+\gamma^r_i+\sigma^r_i+\delta^r_i+d^r_i}, & a = i, u = r, \\ \frac{\sum_{v=1}^{M}\sum_{b=1}^{n_v}\beta^{rrv}_{aib}(S^{rr*}_{ia})}{\eta^r_a+\varrho^r_a+\rho^{rr}_{ia}+\delta^r_a+d^r_a}, & a \neq i, u = r, \\ \frac{\sum_{v=1}^{M}\sum_{b=1}^{n_v}\beta^{urv}_{aib}(S^{ru*}_{ii})}{\eta^u_a+\varrho^u_a+\rho^{ru}_{ia}+\delta^u_a}, & u \neq r, \end{cases} \tag{6.144}$$

represent the average number of new infections over the average lifespan $\frac{1}{\eta^u_a+\varrho^u_a+\rho^{ru}_{ia}+\delta^u_a}$, $u \in I(1,M)$, $a \in I(1,n_u)$, at the site level s^r_i given by $\bar{\mathfrak{R}}^{rr}_{ii_0} = \frac{\sum_{v=1}^{M}\sum_{b=1}^{n_v}\beta^{rrv}_{iib}(S^{rr*}_{ii})}{\eta^r_i+\varrho^r_i+\gamma^r_i+\sigma^r_i+\delta^r_i+d^r_i}$, at the intraregional site s^r_a given by $\bar{\mathfrak{R}}^{rr}_{ia_0} = \frac{\sum_{v=1}^{M}\sum_{b=1}^{n_v}\beta^{rrv}_{aib}(S^{rr*}_{ia})}{\eta^r_a+\varrho^r_a+\rho^{rr}_{ia}+\delta^r_a+d^r_a}$, $a \neq i$, and at the interregional site level s^u_a given by $\bar{\mathfrak{R}}^{ru}_{ia_0} = \frac{\sum_{v=1}^{M}\sum_{b=1}^{n_v}\beta^{urv}_{aib}(S^{ru*}_{ii})}{\eta^u_a+\varrho^u_a+\rho^{ru}_{ia}+\delta^u_a}$, $u \neq r$. These parameters must be the **basic reproduction numbers (BRNs)** at the different sites in the two-scale network population.

Furthermore, the threshold condition for $\mathfrak{R}^{ru}_{ia_0} < 1$, $u \in I(1,M)$, $a \in I(1,n_u)$, in Lemma 6.7.2 and Theorem 6.7.3 implies the BRNs $\bar{\mathfrak{R}}^{rr}_{ii_0} < 1$, $\bar{\mathfrak{R}}^{rr}_{ia_0} < 1$ and $\bar{\mathfrak{R}}^{ru}_{ia_0} < 1$, $u \in I(1,M)$, $a \in I(1,n_u)$. In other words, the disease is eradicated completely from the two-scale network population C, provided that the disease is eradicated at every site interconnected with site s^r_i through the human mobility network.

We now consider the following corollary to Theorem 6.7.3.

Corollary 6.7.4. *Let $r \in I(1,M)$ and $i \in I(1,n_r)$. Assume that $\sigma^r_i = \gamma^r_i = 0$ for all $r \in I(1,M)$ and $i \in I(1,n_r)$. Denote*

$$\mathfrak{U}^{ru}_{ia} = \begin{cases} \frac{\frac{1}{(\delta^r_i)}}{[\frac{1}{2\sum_{u=1}^{M}\sum_{a=1}^{n_u}\mu^{ru}_{ia}+\frac{3}{2}\mu^{rr}_{ii}}]}, & u = r, i = a, \\ \frac{\frac{1}{(\delta^r_a)}}{\frac{1}{[\mu^{rr}_{ii}+\mu^{rr}_{ia}]}}, & u = r, a \neq i, \\ \frac{\frac{1}{(\delta^u_a)}}{\frac{1}{[\mu^{rr}_{ii}+\mu^{ru}_{ia}]}}, & u \neq r, \end{cases} \tag{6.145}$$

$$\mathfrak{R}_{ia_0}^{ru} = \begin{cases} \frac{\sum_{u=1}^{M}\sum_{a=1}^{n_u}\frac{1}{2}\mu_{ia}^{ru}+\sum_{v=1}^{M}\sum_{b=1}^{n_v}\frac{1}{2}\beta_{iib}^{rrv}(S_{ii}^{rr*}+\mu_{ii}^{rr})+\frac{1}{2}d_{ii}^{rr}}{\eta_i^r+\varrho_i^r+\delta_i^r+d_i^r}, & a=i, u=r, \\ \frac{\frac{1}{2}\mu_{ii}^{rr}+\sum_{v=1}^{M}\sum_{b=1}^{n_v}\frac{1}{2}\beta_{aib}^{rrv}(S_{ia}^{rr*}+\mu_{ia}^{rr})+\frac{1}{2}d_{ia}^{rr}}{\eta_a^r+\varrho_a^r+\delta_a^r+d_a^r}, & a \neq i, u=r, \\ \frac{\frac{1}{2}\mu_{ii}^{rr}+\sum_{v=1}^{M}\sum_{b=1}^{n_v}\frac{1}{2}\beta_{aib}^{urv}(S_{ii}^{ru*}+\mu_{ia}^{ru})+\frac{1}{2}d_{ia}^{ru}}{\eta_a^u+\varrho_a^u+\delta_a^u+d_a^u}, & u \neq r, \end{cases} \tag{6.146}$$

and

$$\mathfrak{W}_{ia}^{ru} = \begin{cases} \frac{[\frac{1}{2}\sum_{u=1}^{M}\sum_{a=1}^{n_u}\mu_{ia}^{ru}+\frac{(\alpha_i^r)^2}{\mu_{ii}^{rr}}+\frac{1}{2}\mu_{ii}^{rr}]}{(\alpha_i^r+\delta_i^r)}, & u=r, a=i, \\ \frac{\frac{1}{2}\mu_{ii}^{rr}+\frac{1}{2}\mu_{ia}^{rr}+\frac{(\alpha_a^r)^2}{\mu_{ia}^{rr}}}{(\alpha_a^r+\delta_a^r)}, & u=r, a \neq i, \\ \frac{[\frac{1}{2}\mu_{ii}^{rr}+\frac{1}{2}\mu_{ia}^{ru}+\frac{(\alpha_a^u)^2}{\mu_{ia}^{ru}}]}{\alpha_a^u+\delta_a^u}, & u \neq r. \end{cases} \tag{6.147}$$

Then the equilibrium state E_{ii}^{rr} is stochastically asymptotically stable if $\mathfrak{U}_{ia}^{ru}, \mathfrak{W}_{ia}^{ru} \leq 1$ and $\mathfrak{R}_{ia_0}^{ru} < 1$ for all $u \in I^r(1, M)$ and $a \in I_i^r(1, n_u)$.

Proof. Follows immediately from the hypotheses of Lemma 6.7.2 (letting $\sigma_i^r = \gamma_i^r = 0$), Theorem 6.7.3, and some algebraic manipulations. □

Remark 6.7.3. The presented results about the two-level large-scale SIRS disease dynamic model depend on the underlying system parameters. In particular, the sufficient conditions are algebraically simple, computationally attractive, and explicit in terms of the rate parameters. As a result, several scenarios can be discussed and exhibit practical course of action to control the disease. For simplicity, we present the following illustration: the conditions $\sigma_i^r = \gamma_i^r = 0$ for all r, i in Corollary 6.7.4 signify that the arbitrary site s_i^r is a sink [21, 20] for all other sites in the inter- and intraregional accessible domain. This scenario is displayed in Figure 6.3. The condition $\mathfrak{U}_{ia}^{ru} \leq 1$ exhibits that the average infectious period is smaller than the joint average lifespan of individuals in the intra- and interregional accessible domain of site s_i^r. Furthermore, the condition $\mathfrak{R}_{ia_0}^{ru} < 1$ signifies that the magnitude of disease inhibitory processes, for example, the magnitude of the recovery process, is greater than the disease transmission process. A future detailed study of the disease dynamics in the two-scale network dynamic structure for many real-life scenarios using the presented two-level large-scale SIRS disease dynamic model will appear elsewhere.

In the next section, we present simulation results to demonstrate the results of this section.

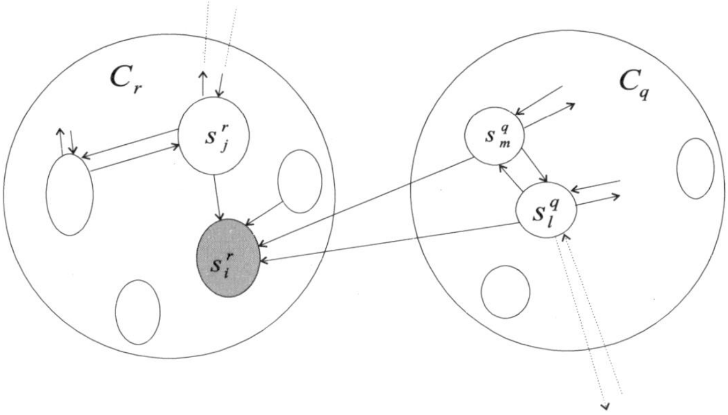

Figure 6.3: Shows that residents of site s_i^r are present only at their home site s_i^r. Hence they isolate every site from their inter- and intraregional accessible domain $C(s_i^r)$. Site s_i^r is a "sink" in the context of the compartmental system [21, 20]. The arrows represent a transport network between any two sites and regions, and the dotted lines and arrows indicate connection with other sites and regions.

6.8 Simulation results for a two-scale stochastic network SIR influenza epidemic model

The simulations given in this section extend the example in Section 5.7 and utilizes the data in Tables 5.1 and 5.2 to define the two-scale network population structure C and the intra- and interregional rates for the human mobility network. It is assumed that there is an outbreak of influenza in the two-scale network population, and the disease related exponential rates are obtained in the literature [16]. The simulations for the influenza disease epidemic dynamics are given for the SIR compartmental model by restricting the parameters $\eta_i^r = \alpha_i^r = 0, r \in I(1, M), i \in I(1, n_r)$, in the SDE model (6.53)–(6.55). The compartmental framework for the SIR epidemic model is exhibited in Figure 6.2 with the restrictions $\eta_i^r = \alpha_i^r = 0, r \in I(1, M), i \in I(1, n_r)$. Furthermore, the diagram illustrating the interpatch connections in the example for two-scale dynamic epidemic model represented in this example is shown in Figure 5.10. In the absence of real data for the intra- and interregional human mobility return rates $\rho_{ia}^{ru}, r, u \in I(1, M), i \in I(1, n_r)$, $a \in I(1, n_u)$, we simulate the intra- and interregional human mobility return rates. We display the intra- and interregional human mobility return rates in Tables 6.1 and 6.2, respectively. We make the following assumptions concerning the influenza epidemic process represented in this example:

(a_1) The population structure and influenza transmission process at every site $s_i^r, r = 1, 2, 3, i = 1, 2, 3$, in the region $C_r, r = 1, 2, 3$, is similar to the population structure and the influenza transmission process represented in the simulation model of [16]. That is, we assume that every person in site s_i^r belongs to one age-dependent stratum (ages ≥ 0). In addition, each individual belongs to three mixing or contact groups $z_j, j = 1, 2, 3$, for example, the household, marketplace, and community. Every day, a

Table 6.1: The intraregional return rates of residents of sites in the two-scale network of spatial patches illustrated in Figure 5.10 are simulated based on the special human mobility pattern and the probabilistic formulation for the mobility process (see Chapter 5 or [45]).

$(\rho_{12}^{11}, \rho_{13}^{11}, \rho_{21}^{11}, \rho_{23}^{11})$	**(0.000092504,0.000177496,0.164327,0.0001173)**
$(\rho_{31}^{11}, \rho_{32}^{11})$	(0.013230408, 0.001305838)
$(\rho_{12}^{22}, \rho_{13}^{22}, \rho_{21}^{22}, \rho_{23}^{22})$	(0.000092504, 0.000177496, 0.164327, 0.0001173)
$(\rho_{31}^{22}, \rho_{32}^{22})$	(0.013230408, 0.001305838)
$(\rho_{12}^{33}, \rho_{13}^{33}, \rho_{21}^{33}, \rho_{23}^{33})$	(0.000092504, 0.000177496, 0.164327, 0.0001173)
$(\rho_{31}^{33}, \rho_{32}^{33})$	(0.013230408, 0.001305838)

Table 6.2: The interregional return rates of residents of sites in the two-scale network of spatial patches illustrated in Figure 5.10 are simulated based on the mobility structure and the probabilistic formulation for the mobility process (see Chapter 5 or [45]).

$(\rho_{11}^{12}, \rho_{12}^{12}, \rho12_{13}, \rho12_{21}, \rho12_{22})$	**(0.1995,0.035,0.0985,0.007892,0.02748)**
$(\rho12_{23}, \rho12_{31}, \rho12_{32}, \rho12_{33})$	(0.075824, 0.04256, 0.009616, 0.028628)
$(\rho_{11}^{21}, \rho_{12}^{21}, \rho21_{13}, \rho21_{21}, \rho21_{22})$	(0.002096896, 0.00175424, 0.003460864, 0.00043856, 0.0001664)
$(\rho21_{23}, \rho21_{31}, \rho21_{32}, \rho21_{33})$	(0.00071504, 0.001944052, 0.00119788, 0.0001713912)
$(\rho_{11}^{23}, \rho_{12}^{23}, \rho23_{13}, \rho23_{21}, \rho23_{22})$	(0.018512, 0.03290368, 0.0272192, 0.04883712, 0.00151648)
$(\rho23_{23}, \rho23_{31}, \rho23_{32}, \rho23_{33})$	(0.0219232, 0.00383316, 0.0025404, 0.000414644)
$(\rho_{11}^{31}, \rho_{12}^{31}, \rho31_{13}, \rho31_{21}, \rho31_{22})$	(0.001285712, 0.00085328, 0.001725008, 0.0004380944, 0.000379536)
$(\rho31_{23}, \rho31_{31}, \rho31_{32}, \rho31_{33})$	(0.0005991696, 0.000000371428, 0.00000026332, 0.000000281252)
$(\rho_{11}^{32}, \rho_{12}^{32}, \rho32_{13}, \rho32_{21}, \rho31_{22})$	(0.0003230096, 0.00036224, 0.0004619664, 0.00043146104, 0.0003741576)
$(\rho32_{23}, \rho32_{31}, \rho32_{32}, \rho32_{33})$	(0.00059126136, 0.000498339428, 0.00042838332, 0.000070993252)

susceptible person A has contacts with other individuals in his or her contact zones. The probability of acquiring infection depends on (a) the number of different persons A has contacts within the contact group, (b) the time duration, in minutes, of all contacts, and (c) the rate of infection transmission per minute if the contacted person is infectious (see [16]). We assume that in a given day, a susceptible person makes three contacts in the mixing group z_1, ten contacts in the mixing group z_2, and three contacts in the mixing group z_3. In addition, each contacted person is infectious. Furthermore, the time duration d and the per minute influenza transmission rate λ per contact in all contact zones are [zone z_1: $d \approx 92$ minutes, $\lambda = 0.00062$], [zone z_2: $d \approx 120$ minutes, $\lambda = 0.00061$], and [zone z_3: $d \approx 51$ minutes, $\lambda = 0.00061$]. Furthermore, we assume that the number and duration of contacts are the same on weekdays and weekend days. We utilize the probability model $1 - \exp(-\lambda d)$ for the influenza transmission occurring during a contact of d minutes and a transmission rate λ (see [16]) to find the infection probability β_{aib}^{urv} of the two-scale SIRS epidemic dynamic model. It is easy to see that the infection probability per day for a susceptible person at site s_i^r is $\beta_{aib}^{urv} = 1 - \Pr$ (no disease transmission in zones z_1, z_2, z_3) $= 1 - \exp(-3(92)(0.00062) - 10(120)(0.00061) - 3(51)(0.00061)) \approx 0.6277$.

(a_2) In the absence of data for the recovery and disease-related death processes, we take the recovery and disease mortality rate to be $\varrho_a^u = 0.05067$ and $d_a^u = 0.01838$, $u, a = 1, 2, 3$.

(a_3) The population in this example is assumed to be remote and lacking the high technological facilities found in the developed world. Furthermore, we assume that influenza is highly endemic in this population. As a result, we can assume that the time duration of the epidemic is comparable with the average lifespan of individuals in the population. In the absence of data concerning average birth rates, we use the yearly birth rate data from [8] for the people of the Dominican Republic, $B = \frac{\text{births}}{1000} = \frac{22.39}{1000}$ as an estimate. Furthermore, we assume that this birth rate is the same for all residents of sites in the population, that is, $B_a^u = \frac{\text{births}}{1000} = \frac{22.39}{1000}$ per year for $u, a = 1, 2, 3$.

(a_4) In addition, using the average lifespan of the people of Dominican Republic [9], the natural death rate of the residents at all sites and regions is the same and is calculated as the reciprocal of the average lifespan of individuals in the population, that is, $\delta_a^u = \frac{1}{77.15 \times 365}$, $u, a = 1, 2, 3$, per day.

(a_5) The effects of the fluctuating environment on the dynamics of the influenza epidemic is assumed to be the same at all sites and regions. We take the standard deviation of the environmental fluctuations to be $\nu_{aib}^{urv} = 0.5$, $r, u, v, a, b, i = 1, 2, 3$.

6.8.1 Mathematical algorithm and simulation results

We use the standard Euler–Maruyama stochastic approximation scheme [19] to generate the trajectories for the residents of sites s_1^1, s_1^2, and s_1^3 in the regions C_1, C_2, and C_3, respectively, for different population disease classifications (S, I, R) and current locations at some sites in the intra- and interregional accessible domain of the sites. Given a scalar autonomous stochastic differential equation

$$dX(t) = f(X(t))dt + g(X(t))dW(t), \quad X(0) = X_0, \quad T_0 \le t \le T, \tag{6.148}$$

let $T_0 = \tau_0 \le \tau_1 \le \tau_2 \le \cdots \le \tau_n = T$ be a regular partition of $[T_0, T]$, where $\Delta\tau = \tau_j - \tau_{j-1} = \frac{T-T_0}{L}$, $\tau_j = T_0 + j\Delta\tau$, $j = 1, \ldots, L$, and L is a positive integer. The Euler–Maruyama method takes the form

$$X(\tau_j) = X(\tau_{j-1}) + f(X(\tau_{j-1}))\Delta\tau + g(X(\tau_{j-1}))(W(\tau_j) - W(\tau_{j-1})), \quad j = 1, \ldots, L. \tag{6.149}$$

Using (6.148) as a general representation of system (6.53)–(6.55) in the context of the scenario considered in this example, the algorithm to execute the Euler–Maruyama method to finding the solution process of (6.148) consists of the following steps,

Step 1: parameter specification. The system rate parameters for the epidemic model (6.53)–(6.55) represented in this example are specified in **(a_1)–(a_5)** above. Further the following convenient initial conditions are used for the simulation process: for $r, u, i, a \in I(1,3)$,

$$S_{ia}^{ru}(0) = \begin{cases} 9, & r = u, i = a, \\ 8, & r = u, i \neq a, \\ 7, & r \neq u, \end{cases}$$

$$I_{ia}^{ru}(0) = \begin{cases} 6, & r = u, i = a, \\ 4, & r = u, i \neq a, \\ 3, & r \neq u, \end{cases}$$

and $R_{ia}^{ru}(0) = 2, r, u, i, a \in I(1,3)$. Furthermore, the trajectories were generated over the time interval $t \in [0,1]$.

Step 2: generate Brownian motion path: Given that $W(t)$ is a standard Brownian motion or Wiener process over the time interval $[T_0, T]$, then (1) $W(0) = 0$, (2) for all $T_0 \leq s < t \leq T$, the increments $W(t) - W(s)$ are independent and have normal distribution with mean 0 and variance $t - s$. In other words, $W(t) - W(s) \sim \sqrt{t-s}N(0,1)$, where $N(0,1)$ represents a normally distributed random variable with zero mean and unit variance. In view of properties (1) and (2), we discretize the Brownian motion as follows: we let $T_0 = t_0 \leq t_1 \leq t_2 \leq \cdots \leq t_n = T$ be a regular partition of $[T_0, T]$, where $\delta t = t_j - t_{j-1} = \frac{T-T_0}{N}$, $t_j = T_0 + j\delta t, j = 1, \ldots, N$, and N is a positive integer. The Brownian path is generated as the solution to the g difference equation

$$\begin{cases} W(0) = 0, \\ W(t_j) = W(t_{j-1}) + dW(t_j), \quad j = 1, \ldots, N. \end{cases} \tag{6.150}$$

We simulated 1000 sample points for the Brownian motion over the interval $[0,1]$.

Step 3: generate solution path for the susceptible, infectious, and removal populations. Using (6.148) as a general representation of each equation in system (6.53)–(6.55), we use discretization (6.149) to find solution paths for each equation in the system. For convenience, we choose $\Delta\tau = R\delta t$ [18], where the integer $R \geq 1$. Moreover, from (6.149) it follows that

$$W(\tau_j) - W(\tau_{j-1}) = W(jR\delta t) - W((j-1)R\delta t) = \sum_{k=jR-R+1}^{jR} dW_k, \tag{6.151}$$

where dW_k is given by the Brownian path (6.150). We choose $R = 1$ for this example. Moreover, from (6.149), (6.150), and (6.151) we obtain trajectories for susceptible, infectious, and removal populations of residents of sites s_1^1, s_1^2, and s_1^3 in the regions C_1, C_2, and C_3 over the time interval $[0, 1]$. The trajectories for the residents of sites s_1^1, s_1^2, and s_1^3 in the regions C_1, C_2, and C_3 are exhibited in Figures 6.4, 6.5, and 6.6, respectively.

Remark 6.8.1. From Figure 6.4 we observe that Figures (a), (b), and (c) represent the trajectories of the different disease classes of residents of site s_1^1 at home. Figures (d), (e), and (f) represent the trajectories of the different disease classes of residents of site s_1^1 visiting site s_2^1 in the home region C_1. These two groups of figures are representatives of the disease dynamics of influenza affecting the residents of site s_1^1 at the intraregional

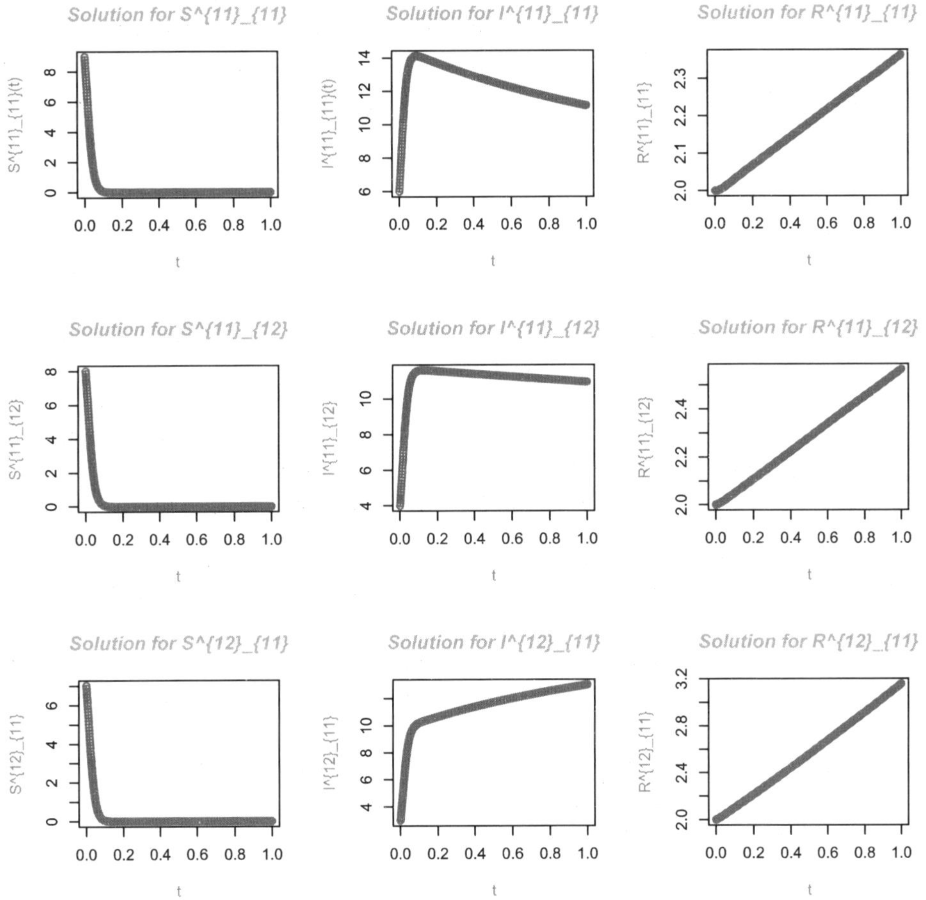

Figure 6.4: Trajectories of the disease classification (S, I, R) for residents of site s_1^1 in the region C_1 at their home site and the current location in the two-scale spatial patch dynamic structure. See Remark 6.8.1 for more comments on this figure.

level. Figures (g), (h), and (i) represent the trajectories of the different disease classes of residents of site s_1^1 visiting site s_1^2 in the region C_2. These figures reflect the behavior of the disease affecting the residents of site s_1^1 at the interregional level. Furthermore, we observe that the trajectories of the susceptible (S) and infectious (I) populations saturate to their equilibrium states. We further note that the trajectories are random in character, but because of the scale of the pictures presented in the figure, they seem to be smooth.

Remark 6.8.2. From Figure 6.5 we observe that Figures (a), (b), and (c) represent the trajectories of the different disease classes of residents of site s_1^2 at home. Figures (d), (e), and (f) represent the trajectories of the different disease classes of residents of site s_1^2

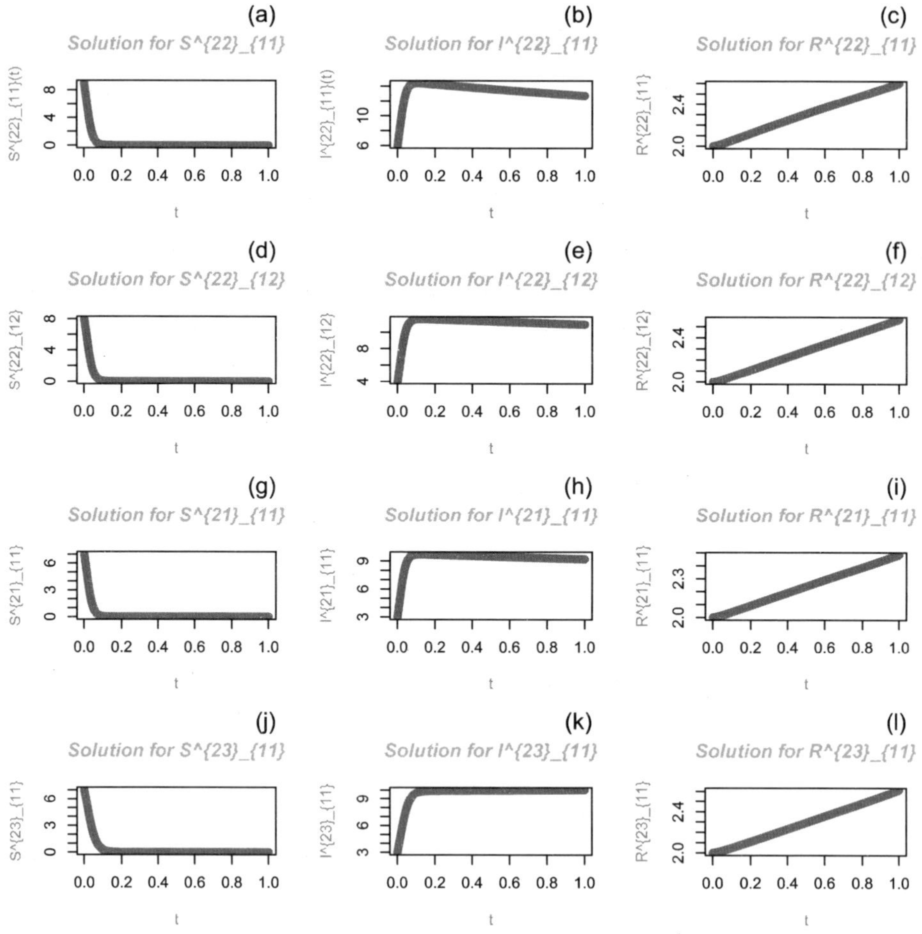

Figure 6.5: Trajectories of the disease classification (S, I, R) for residents of site s_1^2 in region C_2 at their home site and at their current locations in the two-scale spatial patch dynamic structure. See Remark 6.8.2 for more comments on this figure.

visiting site s_2^2 in the home region C_2. These two groups of figures are representatives of the disease dynamics of influenza affecting the residents of site s_1^2 at the intraregional level. Figures (g), (h), and (i) represent the trajectories of the different disease classes of residents of site s_1^2 visiting site s_1^1 in the region C_1. Figures (j), (k), and (l) represent the trajectories of the different disease classes of residents of site s_1^2 visiting site s_1^3 in the region C_3. These last two groups of figures reflect the behavior of the disease affecting the residence of site s_1^2 at the interregional level. Furthermore, we observe that the trajectories of the susceptible (S) and infectious (I) populations saturate to their equilibrium states. We further note that the trajectory paths are random in character, but because of the scale of the pictures presented in this figure, they seem to be smooth.

Remark 6.8.3. From Figure 6.6 we observe that Figures (a), (b), and (c) represent the trajectories of the different disease classes of residents of site s_1^3 at home. Figures (d), (e), and (f) represent the trajectories of the different disease classes of residents of site s_1^3 visiting site s_2^3 in the home region C_3. These two groups of figures are representatives of the disease dynamics of influenza affecting the residents of site s_1^3 at the intraregional level. Figures (g), (h), and (i) represent the trajectories of the different disease classes of residents of site s_1^3 visiting site s_1^1 in the region C_1. Figures (j), (k), and (l) represent the trajectories of the different disease classes of residents of site s_1^3 visiting site s_1^2 in the region C_2. The last two groups of figures reflect the behavior of the disease affecting the residence of site s_1^3 at the interregional level. Furthermore, we observe that the trajectories of the susceptible (S) and infectious (I) populations saturate to their equilibrium states. We further note that the trajectory paths are random in character, but because of the scale of the pictures presented in this figure, they seem to be smooth.

6.9 Discussion of the results and other types of disease epidemic dynamic models

The results presented in this chapter, including Theorem 6.7.3, provide a clear indication that effectively controlling disease spread within a two-scale network population, denoted as C, requires targeted efforts at all three primary interaction sites: the home site s_i^r, intraregional sites s_j^r where $j \neq i$, and interregional sites s_a^u with $u \neq r$ and $a \in I(1, n_u)$. Successful disease eradication is contingent on ensuring that the basic reproduction number at each of these interaction levels is below one. This highlights the need for multilevel interventions that address the complexities of the network structure and population dynamics.

The model in (6.53)–(6.55) offers a simplified representation of random environmental fluctuations influencing disease transmission rates. It is important to acknowledge that other sources of random fluctuations—such as those stemming from socioeconomic factors, behavioral changes, or local environmental conditions—are also likely to affect the epidemic trajectory. These factors were not explicitly incorporated in the current

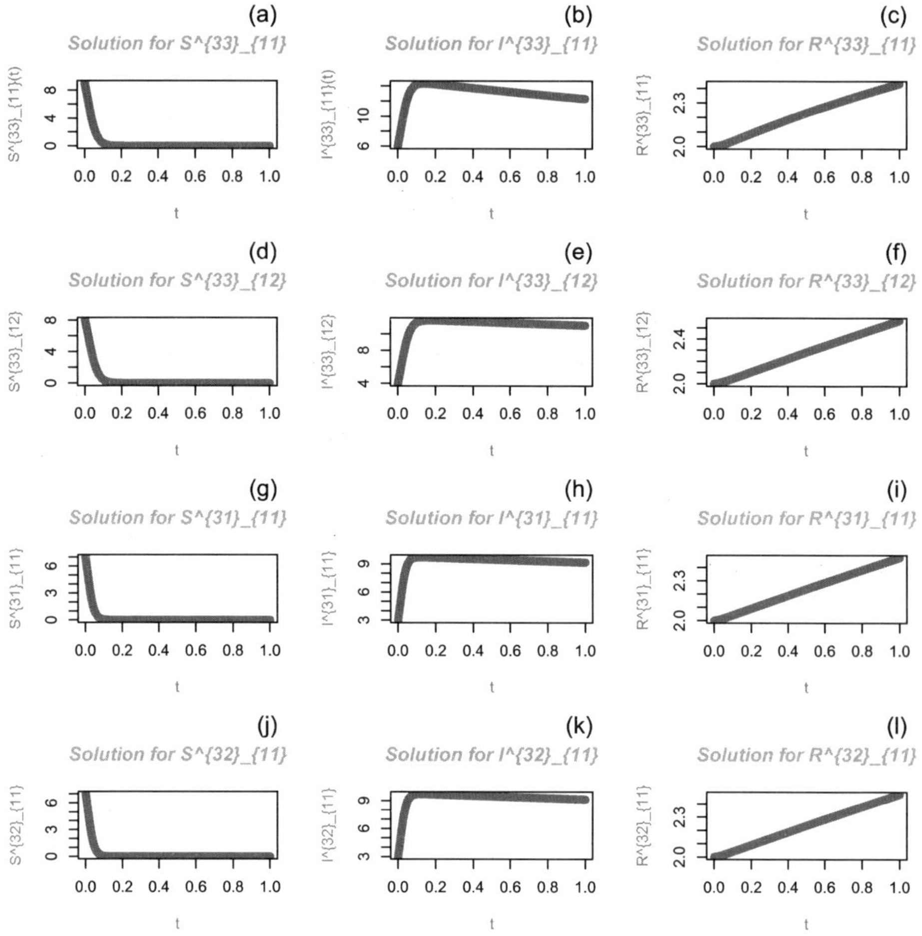

Figure 6.6: Trajectories of the disease classification (S, I, R) for residents of site s_1^3 in the region C_3 at their current location in the two-scale spatial patch dynamic structure. See Remark 6.8.3 for more comments on this figure.

model, but future extensions could incorporate them to provide a more comprehensive understanding of the dynamics. For instance, accounting for more granular sources of stochasticity could lead to insights into the resilience of control strategies under varying real-world conditions.

Corollary 6.7.4 further emphasizes the significance of human mobility network patterns in shaping disease dynamics within the two-scale network population. The mobility patterns of individuals, such as travel frequency and the connectivity between different regions or sites, play a crucial role in determining the vulnerability of particular areas to disease outbreaks. The findings suggest that regions with higher interregional connectivity or those that function as “sinks” in the mobility network are more prone to outbreaks. These regions may therefore require more intensive disease control efforts to mitigate the spread of infections. Additionally, regions exhibiting uneven population distributions, as influenced by human mobility patterns, could either facilitate or hinder the spread of diseases depending on local network characteristics.

Despite these insights, this chapter did not account for delays or other critical aspects of disease dynamics, such as the incubation period of the disease and the delays associated with acquired immunity. These factors can introduce periodic behavior in the epidemic dynamics, resulting in cycles of disease resurgence and decline. The incorporation of time delays is expected to provide a more accurate representation of the disease spread, particularly in complex, heterogeneous populations where the timing of interventions and the natural progression of the disease can significantly alter the course of an outbreak. The upcoming chapters will explore these delays in greater detail, investigating their impact on disease dynamics within the two-scale network population. This will allow for a more nuanced understanding of the mechanisms that drive epidemic trajectories, ultimately leading to more robust and adaptive strategies for disease prevention and control.

In summary, the model developed in this chapter lays the groundwork for understanding disease dynamics in finite-scale network populations. However, the extension of this work to include delays and other forms of randomness will provide a more holistic view of epidemic behavior, paving the way for improved forecasting and more effective intervention strategies.

7 Modeling the stochastic dynamics of communicable infectious disease epidemics with delay SDEs in finite-scale network populations

7.1 Introduction

Delay differential equation (DDE) models, whether in stochastic or deterministic frameworks, are powerful tools for capturing complex features of epidemic dynamics that nondelay models often overlook. These delays can represent critical biological processes, such as the incubation period—the time between exposure to infection and the onset of symptoms—or the temporary immunity conferred by recovery, as seen in diseases like *chickenpox* and *measles*. The latter, commonly referred to as immunity delay, accounts for the period during which individuals are protected against reinfection through either natural or artificial immunity.

In short, a time delay can be used to represent a time lag for a latent behavior or hidden activity in a dynamic process that implicitly affects the system by accounting only for the statistical properties of the survival distribution over the time lag. For example, Wanduku [35] has investigated delays associated with HIV infection, particularly the intervals between infection and the initiation of antiretroviral therapy (ART) and between infection and the progression to AIDS for untreated individuals. The literature is saturated with papers addressing time delays in a variety of dynamic processes not particularly related to disease epidemic dynamics (see Wanduku et al. [42, 43, 13] and Ladde et al. [25]). However, in this monograph the subject of time-delays is addressed in the context of disease epidemic dynamics in finite-scale network populations. While this monograph discusses only time delays for communicable infectious diseases in a general context, there are also vector-borne diseases with delays accounting for the life cycle of the infectious agent, such as parasites in humans as well as in vectors. For instance, in the case of *malaria* transmissions, delays capture the time taken for parasites to mature within mosquitoes and humans (see [36, 33, 34]). Additional examples, such as *dengue fever* or *Lyme disease*, could further illustrate these mechanisms.

The inclusion of time delays in epidemic models has significant implications for understanding disease dynamics. Delays can result in oscillatory patterns in disease prevalence, with alternating periods of low and high transmission. For instance, immunity delays can lead to phases where the disease appears to be under control due to high immunity levels, followed by resurgent outbreaks when immunity wanes. Similarly, delays in vector-borne diseases can influence seasonal transmission patterns. Without accounting for these delays, control strategies risk being less effective, underscoring the importance of delay modeling in epidemic analysis. Moreover, these delays often necessitate the implementation of time-sensitive interventions, such as booster vaccination campaigns, to sustain population immunity.

https://doi.org/10.1515/9783111613215-007

This chapter explores the impacts of naturally acquired immunity on the dynamics of SIRS disease epidemic dynamics. Regardless of the type of immunity conferred, whether naturally after recovery from infection or artificially through vaccination, immunity effects wane over time due to declining exposure to the pathogen, necessitating booster interventions. For example, vaccinated individuals typically exhibit lower immunity levels than those who recover naturally from measles [28]. Similarly, immunity against pertussis diminishes 6–12 years after the last infection or booster dose [14]. The literature includes a significant number of papers addressing acquired immunity delays to capture these nuances (see [13]).

In this chapter, we build upon the stochastic epidemic dynamic model introduced in Chapter 6 by incorporating the effects of a temporary immunity delay period while accounting for the survival probability distribution over this period. Specific techniques, such as survival probability functions and delay kernels, will be utilized to model these delays rigorously. We focus on a communicable infectious disease that provides natural immunity to all recovered individuals immediately following the infectious stage.

The steps of Algorithm 6.2.1 are applied to derive the epidemic model. Specifically, in Step 1 the mean-field dynamic equations are derived. In Steps 2 and 3 the stochastic model is developed by introducing white noise perturbations and incorporating the survival probability distributions to account for the effects of temporary immunity delays.

7.2 Assumptions for a disease epidemic dynamics with delays

We consider an SIRS (Susceptible–Infective–Removal–Susceptible) epidemic model that adheres to the assumptions regarding population structure, human mobility processes, and disease dynamics outlined in Chapter 6. In this model the removal compartment R exclusively comprises individuals who have acquired natural immunity to the disease. For a strictly SIRS process, where no instantaneous transitions occur from the infectious to the susceptible state, the recovery rates η_a^u and α_a^u in the SIRS epidemic framework (6.42)–(6.44) and (6.53)–(6.55) are zero, that is, $\eta_a^u = \alpha_a^u = 0$ for all $u \in I(1, M)$ and $a \in I(1, n_u)$. The transition from the recovered state R back to the susceptible state S occurs after a delay period, during which acquired immunity wanes. Once this immunity is lost, the individual reenters the susceptible compartment, that is, we assume that for all $r \in I(1, M)$ and $i \in I(1, n_r)$, an infectious individual I_{ia}^{ru} residing at site $s^r i$ in the region C_r and visiting site s_a^u in the region Cu recovers from the disease and acquires temporary immunity, either natural or infection-induced, immediately upon recovery. This immunity wanes after a fixed period T_i^r, at which point the individual transitions back to the susceptible state. The immunity period T_i^r is assumed to be constant for all naturally immune residents of site s_i^r, whether they are at their home site or visiting any site s_a^u in the region C_u, where $u \in I(1, M)$ and $a \in I(1, n_u)$, within the two-scale network population.

To incorporate temporary immunity into the epidemic dynamics described by (6.42)–(6.44) and (6.53)–(6.55), we introduce the term

$$\varrho_a^u I_{ia}^{ru}(t - T_i^r)e^{-\delta_a^u T_i^r}, \tag{7.1}$$

where $e^{-\delta_a^u T_i^r}$ denotes the probability of survival from natural death during the immunity period T_i^r. This ensures that the transition back to the susceptible class occurs only after both immunity loss and survival. Derivation of the term (7.2) is shown in the following example that extends Example 6.2.1.

Example 7.2.1. Deriving a delay SDE system. Recall the dimensionless version (6.11) of the SIRS epidemic model studied in Example 6.2.1 given by

$$\begin{aligned} \frac{di(t)}{dt} &= \beta_{IS} i(t)[1 - i(t)] - (\alpha_{IS} + \mu)i(t), \\ s(t) &= 1 - i(t), \quad s(t) + i(t) = 1, \end{aligned} \tag{7.2}$$

further denoted by

$$di(t) = f(i(t), t)dt, \tag{7.3}$$

where $f(i(t), t)$ is the right-hand-side of (7.2), $s(t) = \frac{s(t)}{N}$, and $i(t) = \frac{I(t)}{N}$.

Suppose instead of an instantaneous transition from the infectious state I back to the susceptible state S given by the rate α_{IS} as in (7.2) with the term $\alpha_{IS} i(t)$, it is assumed that individuals recover from the disease at the exponential rate α_{IR} and acquire natural immunity that lasts over the period $T > 0$. When the immunity wanes, the individual returns to the susceptible state S. Thus a new state $R(t)$ denotes the number of recovered individuals with acquired immunity at time t. Recall Example 6.2.1, where the states of system (7.2) are averages. The derivation of the delay term in (7.2) is obtained through integral equations that represent averages over time as shown below.

Indeed, the average of $i(t)$ at time t is the sum of the averages of all new infections over time $[t_0, t]$, where t_0 is the initial time, and all infectious individuals that survive natural death and are still in the state I. Recall Example 6.2.1, where all survivals through compartments in the underlying random jump process follow the **exponential distribution**. That is, suppose τ_{ND} is the random time until an individual (of the S, I, or R status) dies naturally in the population; then based on the model (7.2), the distribution of τ_{ND} is exponential with mean $\frac{1}{\mu}$. This is denoted by $\tau_{IR} \sim Exp(\alpha_{IR})$. Furthermore, by the **no-memory property** of the exponential distribution it follows that the conditional survival function over the time interval $[t_0, t]$ is given by

$$\mathbb{P}_{\tau_{ND}}(t|t_0) = \mathbb{P}(\tau_{ND} > t|\tau_{ND} > t_0) = e^{-\mu(t-t_0)}. \tag{7.4}$$

Define another lifetime random variable τ_{IR} representing time until an individual in the infectious state I leaves the state through recovery at the rate α_{IR}, where $\tau_{IS} \sim Exp(\alpha_{IR})$.

Thus this sum for the average number of infectious individuals over the interval $[t_0, t]$ is

$$
\begin{aligned}
i(t) &= i(t_0)\mathbb{P}_{\tau_{ND}}(t|t_0)\mathbb{P}_{\tau_{IS}}(t|t_0) + \int_{t_0}^{t} \beta_{IS}s(\theta)i(\theta)\mathbb{P}_{\tau_{ND}}(t|\theta)\mathbb{P}_{\tau_{IS}}(t|\theta)d\theta \\
&= i(t_0)e^{-\mu(t-t_0)}e^{-\alpha_{IR}(t-t_0)} + \int_{t_0}^{t} \beta_{IS}s(\theta)i(\theta)e^{-\mu(t-\theta)}e^{-\alpha_{IR}(t-\theta)}d\theta. \qquad (7.5)
\end{aligned}
$$

Differentiating (7.5) with respect to time leads to

$$\frac{di(t)}{dt} = \beta_{IS}s(t)i(t) - \alpha_{IR}i(t) - \mu i(t). \qquad (7.6)$$

By the same approach in (7.5)–(7.6) it is easy to see that the mean number of recovered individuals at time t is as given below. First, define τ_{RS} be the time until a naturally immune individual loses the immunity as it wanes over time. Then with the assumption of a discrete delay time $T > 0$ of natural immunity, the distribution of τ_{RS} is best represented by

$$\mathbb{P}_{\tau_{RS}}(t|t_0) = \begin{cases} 1, & T \leq t - t_0, \\ 0, & T > t - t_0, \end{cases} \qquad (7.7)$$

$$r(t) = r(t_0)\mathbb{P}_{\tau_{ND}}(t|t_0)\mathbb{P}_{\tau_{RS}}(t|t_0) + \int_{t_0}^{t} \alpha_{IR}I(s)\mathbb{P}_{\tau_{ND}}(t|\theta)\mathbb{P}_{\tau_{RS}}(t|\theta)d\theta. \qquad (7.8)$$

Assuming that the disease epidemic has been in the population for a significantly large amount of time so that $t \gg (t_0 - T)$ and all initial perturbations have died out, then it follows from (7.7)–(7.8) that

$$r(t) = \int_{t-T}^{t} \alpha_{IR}i(s)e^{-\mu(t-\theta)}d\theta, \qquad (7.9)$$

and differentiating (7.9) leads to

$$\frac{dr(t)}{dt} = \alpha_{IR}i(t) - \alpha_{IR}i(t-T)e^{-\mu T} - \mu r(t). \qquad (7.10)$$

Thus the delay term $\alpha_{IR}I(t-T)e^{-\mu T}$ represents new susceptible individuals who were previously infectious at time $t - T$, recovered at rate α_{IR}, survived natural death with probability $e^{-\mu T}$ over the period of their active natural immunity T, and are back in the susceptible state S.

Therefore the complete SIRS epidemic model with natural immunity delay is given as follows:

$$\begin{aligned} \frac{ds(t)}{dt} &= \mu s(t) - \beta_{IS}s(t)i(t) + \alpha_{IR}i(t-T)e^{-\mu T} - \mu s(t), \\ \frac{di(t)}{dt} &= \beta_{IS}s(t)i(t) - \alpha_{IR}i(t) - \mu i(t), \\ \frac{dr(t)}{dt} &= \alpha_{IR}i(t) - \alpha_{IR}I(t-T)e^{-\mu T} - \mu r(t), \\ s(t) &+ i(t) + r(t) = 1, \end{aligned} \tag{7.11}$$

with initial conditions

$$(s(a), i(a), r(a)) = (\phi_1(a), \phi_2(a), \phi_3(a)), \quad a \in [t_0 - T, t_0], \tag{7.12}$$

where $\phi_k \in C([t_0 - T, t_0])$, $k = 1, 2, 3$, are continuous bounded positive real-valued functions, and $C([t_0 - T, t_0], \mathbb{R}_+)$ is the Banach space equipped with the supremum norm

$$\|\phi_k\|_\infty = \sup_{t_0 - T \le t \le t_0} |\phi_k(t)|. \tag{7.13}$$

In the complete probability space $(\Omega, \{\mathfrak{F}_t\}_{t\ge 0}, \mathbb{P})$, by introducing the noise terms in (6.29) back into the delay mean-field differential equation system in (7.11), the following Itô–Doob-type stochastic differential equation SIRS model is obtained with the temporary time delay T:

$$\begin{aligned} ds(t) &= [\mu s(t) - \beta_{IS}s(t)i(t) + \alpha_{IR}i(t-T)e^{-\mu T} - \mu s(t)]dt + \sigma_S s(t)dw_S(t), \\ di(t) &= [\beta_{IS}s(t)i(t) - \alpha_{IR}i(t) - \mu i(t)]dt + \sigma_I i(t)dw_I(t), \\ dr(t) &= [\alpha_{IR}i(t) - \alpha_{IR}I(t-T)e^{-\mu T} - \mu r(t)]dt + \sigma_R r(t)dw_R(t), \end{aligned} \tag{7.14}$$

where $\phi_k \in C([t_0 - T, t_0])$, $k = 1, 2, 3$, are $\mathfrak{F}_0$-measurable continuous bounded positive real-valued functions, and $C([t_0 - T, t_0], \mathbb{R}_+)$ is the Banach space equipped with the supremum norm

$$\|\phi_k\|_\infty = \sup_{t_0 - T \le t \le t_0} |\phi_k(t)|. \tag{7.15}$$

7.2.1 The stochastic delay two-scale SIRS disease epidemic dynamic model

Hence, applying the method similar to that in Example 7.2.1, from (6.53)–(6.55) we obtain the following stochastic two-scale SIRS disease epidemic dynamic model with the immunity delay term in (7.1):

$$dS_{ia}^{ru} = \begin{cases} [B_i^r + \sum_{k=1}^{n_r} \rho_{ik}^{rr} S_{ik}^{rr} + \sum_{q \neq r}^{M} \sum_{a=1}^{n_q} \rho_{ia}^{rq} S_{ia}^{rq} + \varrho_i^r I_{ii}^{rr}(t - T_i^r) e^{-\delta_i^r T_i^r} \\ \quad - (\gamma_i^r + \sigma_i^r + \delta_i^r) S_{ii}^{rr} - \sum_{u=1}^{M} \sum_{a=1}^{n_u} \beta_{iia}^{rru} S_{ii}^{rr} I_{ai}^{ur}] dt \\ \quad - [\sum_{u=1}^{M} \sum_{a=1}^{n_u} \nu_{iia}^{rru} S_{ii}^{rr} I_{ai}^{ur} dw_{iia}^{rru}(t)], \quad u = r,\ a = i, \\ [\sigma_{ij}^{rr} S_{ii}^{rr} + \varrho_j^r I_{ij}^{rr}(t - T_i^r) e^{-\delta_j^r T_i^r} - (\rho_{ij}^{rr} + \delta_j^r) S_{ij}^{rr} - \sum_{u=1}^{M} \sum_{a=1}^{n_u} \beta_{jia}^{rru} S_{ij}^{rr} I_{aj}^{ur}] dt \\ \quad - [\sum_{u=1}^{M} \sum_{a=1}^{n_u} \nu_{jia}^{rru} S_{ij}^{rr} I_{aj}^{ur} dw_{jia}^{rru}(t)], \quad u = r,\ a = j,\ j \neq i, \\ [\gamma_{il}^{rq} S_{ii}^{rr} + \varrho_l^q I_{il}^{rq}(t - T_i^r) e^{-\delta_l^q T_i^r} - (\rho_{il}^{rq} + \delta_l^q) S_{il}^{rq} - \sum_{u=1}^{M} \sum_{a=1}^{n_u} \beta_{lia}^{qru} S_{il}^{rq} I_{al}^{uq}] dt \\ \quad - [\sum_{u=1}^{M} \sum_{a=1}^{n_u} \nu_{lia}^{qru} S_{il}^{rq} I_{al}^{uq} dw_{lia}^{qru}(t)], \quad u = q,\ a = l,\ q \neq r, \end{cases} \tag{7.16}$$

$$dI_{ia}^{ru} = \begin{cases} [\sum_{k=1}^{n_r} \rho_{ik}^{rr} I_{ik}^{rr} + \sum_{q \neq r}^{M} \sum_{a=1}^{n_q} \rho_{ia}^{rq} I_{ia}^{rq} - \varrho_i^r I_{ii}^{rr} \\ \quad - (\gamma_i^r + \sigma_i^r + \delta_i^r + d_i^r) I_{ii}^{rr} + \sum_{u=1}^{M} \sum_{a=1}^{n_u} \beta_{iia}^{rru} S_{ii}^{rr} I_{ai}^{ur}] dt \\ \quad + [\sum_{u=1}^{M} \sum_{a=1}^{n_u} \nu_{iia}^{rru} S_{ii}^{rr} I_{ai}^{ur} dw_{iia}^{rru}(t)], \quad u = r,\ a = i, \\ [\sigma_{ij}^{rr} I_{ii}^{rr} - \varrho_j^r I_{ij}^{rr} - (\rho_{ij}^{rr} + \delta_j^r + d_j^r) I_{ij}^{rr} + \sum_{u=1}^{M} \sum_{a=1}^{n_u} \beta_{jia}^{rru} S_{ij}^{rr} I_{aj}^{ur}] dt \\ \quad + [\sum_{u=1}^{M} \sum_{a=1}^{n_u} \nu_{jia}^{rru} S_{ij}^{rr} I_{aj}^{ur} dw_{jia}^{rru}(t)], \quad u = r,\ a = j,\ j \neq i, \\ [\gamma_{il}^{rq} I_{ii}^{rr} - \varrho_l^q I_{il}^{rq} - (\rho_{il}^{rq} + \delta_l^q + d_l^q) I_{il}^{rq} + \sum_{u=1}^{M} \sum_{a=1}^{n_u} \beta_{lia}^{qru} S_{il}^{rq} I_{al}^{uq}] dt \\ \quad + [\sum_{u=1}^{M} \sum_{a=1}^{n_u} \nu_{lia}^{qru} S_{il}^{rq} I_{al}^{uq} dw_{lia}^{qru}(t)], \quad u = q,\ a = l,\ q \neq r, \end{cases} \tag{7.17}$$

$$R_{ia}^{ru} = \begin{cases} [\sum_{k=1}^{n_r} \rho_{ik}^{rr} R_{ik}^{rr} + \sum_{q \neq r}^{M} \sum_{l=1}^{n_q} \rho_{il}^{rq} R_{il}^{rq} + \varrho_i^r I_{ii}^{rr} - \varrho_i^r I_{ii}^{rr}(t - T_i^r) e^{-\delta_i^r T_i^r} \\ \quad - (\gamma_i^r + \sigma_i^r + \delta_i^r) R_{ii}^{rr}] dt, \quad u = r, a = i, \\ [\sigma_{ij}^{rr} R_{ii}^{rr} + \varrho_j^r I_{ij}^{rr} - \varrho_j^r I_{ij}^{rr}(t - T_i^r) e^{-\delta_j^r T_i^r} - (\rho_{ij}^{rr} + \delta_j^r) R_{ij}^{rr}] dt, \\ \quad u = r, a = j, j \neq i, \\ [\gamma_{il}^{rq} R_{ii}^{rr} + \varrho_l^q I_{il}^{rq} - \varrho_l^q I_{il}^{rq}(t - T_i^r) e^{-\delta_l^q T_i^r} - (\rho_{il}^{rq} + \delta_l^q) R_{il}^{rq}] dt, \\ \quad u = q, a = l, q \neq r, \end{cases} \tag{7.18}$$

where all parameters are previously defined. Furthermore, for all $r \in I(1, M)$ and $i \in I(1, n_r)$, we have the following initial conditions:

$$\begin{gathered} (S_{ia}^{ru}(t, w), I_{ia}^{ru}(t, w), R_{ia}^{ru}(t, w)) = (\varphi_{ia1}^{ru}(t), \varphi_{ia2}^{ru}(t), \varphi_{ia3}^{ru}(t)), \quad t \in [t_0 - T_i^r, t_0], \\ \varphi_{iak}^{ru} \in \mathcal{C}([t_0 - T_i^r, t_0], \mathbb{R}_+), \quad k = 1, 2, 3,\ r, q \in I(1, M),\ a \in I(1, n_u),\ i \in I(1, n_r), \\ \varphi_{iak}^{ru}(t_0) > 0, \quad k = 1, 2, 3, \end{gathered} \tag{7.19}$$

where $\mathcal{C}([t_0 - T_i^r, t_0], \mathbb{R}_+)$ is the space of continuous functions with the supremum norm

$$\|\varphi\|_\infty = \sup_{t_0 - T_i^r \le t \le t_0} |\varphi(t)|, \tag{7.20}$$

and w is a Wiener process. Furthermore, the random continuous functions φ_{iak}^{ru}, $k = 1, 2, 3$, are $\mathfrak{F}_0$-measurable or independent of $w(t)$ for all $t \ge 0$.

Similarly, using (6.45), we express the state of system (7.16)–(7.18) in vector form. Furthermore, using expression (6.57), it follows from (7.16)–(7.18) that for all $i \in I(1, n_r)$, $l \in I_i^r(1, n_q)$, $r \in I(1, M)$, and $q \in I^r(1, M)$,

$$dy_{il}^{rq} = \begin{cases} [B_i^r + \sum_{k\neq i}^{n_r} \rho_{ik}^{rr} y_{ik}^{rr} + \sum_{q\neq r}^{M} \sum_{a=1}^{n_q} \rho_{ia}^{rq} y_{ia}^{rq} - (\gamma_i^r + \sigma_i^r + \delta_i^r) y_{ii}^{rr} - d_i^r I_{ii}^{rr}]dt, & q = r, l = i, \\ [\sigma_{ij}^{rr} y_{ii}^{rr} - (\rho_{ij}^{rr} + \delta_j^r) y_{ij}^{rr} - d_j^r I_{ij}^{rr}]dt, & q = r,\ a = j,\ i \neq j, \\ [\gamma_{il}^{rq} y_{ii}^{rr} - (\rho_{il}^{rq} + \delta_l^q) y_{il}^{rq} - d_l^q I_{il}^{rq}]dt, & q \neq r,\ y_{il}^{rq}(t_0) \geq 0. \end{cases} \tag{7.21}$$

7.3 Model validation results for the SIRS delay stochastic dynamic model

Similarly to (6.53)–(6.55), it is necessary to establish the existence of a positive solution process for the delay system (7.16)–(7.18). Similar methods applying the Lyapunov energy functions utilized in Theorem 6.6.4 are employed to establish the existence result (see [53, 24, 17]).

We observe from (7.16)–(7.18) that (7.18) decouples from the first two equations in the system. Therefore it suffices to prove the existence of a positive solution process for $(S_{ia}^{ru}, I_{ia}^{ru})$. We utilize notations (6.45) and keep in mind that $X_{ia}^{ru} = (S_{ia}^{ru}, I_{ia}^{ru})^T$ for the decoupled system.

Theorem 7.3.1. *Let $r, u \in I(1, M)$, $i \in I(1, n_r)$, and $a \in I(1, n_u)$. Given any initial conditions (7.19) and (7.20), there exists a unique solution process $X_{ia}^{ru}(t, w) = (S_{ia}^{ru}(t, w), I_{ia}^{ru}(t, w))^T$ satisfying (7.16) and (7.17) for all $t \geq t_0$. Moreover, the solution process is positive for all $t \geq t_0$ a. s., that is, $S_{ia}^{ru}(t, w) > 0$ and $I_{ia(t,w)}^{ru} > 0$ for all $t \geq t_0$ a. s.*

Proof. It is easy to see that the coefficients of (7.16) and (7.17) satisfy the local Lipschitz condition for the given initial data (7.19). Therefore there exists a unique maximal local solution $X_{ia}^{ru}(t, w)$ on $t \in [-T_i^r, \tau_e(w)]$, where $\tau_e(w)$ is the first hitting time or the explosion time [53]. We show subsequently that $S_{ia}^{ru}(t, w), I_{ia}^{ru}(t, w) > 0$ for all $t \in [-T_i^r, \tau_e(w)]$ almost surely. We define the stopping time

$$\begin{cases} \tau_+ = \sup\{t \in (t_0, \tau_e(w)) : S_{ia}^{ru}|_{[t_0,t]} > 0 \text{ and } I_{ia}^{ru}|_{[t_0,t]} > 0\}, \\ \tau_+(t) = \min(t, \tau_+), \quad t \geq t_0, \end{cases} \tag{7.22}$$

and we show that $\tau_+(t) = \tau_e(w)$ a. s. Suppose on the contrary that $P(\tau_+(t) < \tau_e(w)) > 0$. Let $w \in \{\tau_+(t) < \tau_e(w)\}$ and $t \in [t_0, \tau_+(t))$. Define

$$\begin{cases} V(X_{00}^{00}) = \sum_{r=1}^{M} \sum_{i=1}^{n_r} \sum_{u=1}^{M} \sum_{a=1}^{n_u} V(X_{ia}^{ru}), \\ V(X_{ia}^{ru}) = \ln(S_{ia}^{ru}) + \ln(I_{ia}^{ru}), \quad t \leq \tau_+(t). \end{cases} \tag{7.23}$$

We rewrite (7.23) as

$$V(X_{00}^{00}) = \sum_{r=1}^{M} \sum_{i=1}^{n_r} \left[V(X_{ii}^{rr}) + \sum_{j\neq i}^{n_r} V(X_{ij}^{rr}) + \sum_{q\neq r}^{M} \sum_{l=1}^{n_q} V(X_{il}^{rq}) \right], \tag{7.24}$$

and (7.24) further implies that

$$dV(X_{00}^{00}) = \sum_{r=1}^{M}\sum_{i=1}^{n_r}\left[dV(X_{ii}^{rr}) + \sum_{j\neq i}^{n_r} dV(X_{ij}^{rr}) + \sum_{q\neq r}^{M}\sum_{l=1}^{n_q} dV(X_{il}^{rq})\right], \tag{7.25}$$

where dV is the Itô–Doob differential operator with respect to system (7.16)–(7.18). We express the terms of the right-hand-side of (7.25) as follows.

Site level: From (7.23) the terms of the right-hand side of (7.25) for the case of $u = r$, $a = i$ are

$$\begin{aligned} dV(X_{ii}^{rr}) = &\left[\frac{B_i^r}{S_{ii}^{rr}} + \sum_{k\neq i}^{n_r} \rho_{ik}^{rr}\frac{S_{ik}^{rr}}{S_{ii}^{rr}} + \sum_{q\neq r}^{M}\sum_{l=1}^{n_q}\rho_{ia}^{rq}\frac{S_{ia}^{rq}}{S_{ii}^{rr}} + \frac{\varrho_i^r I_{ii}^{rr}(t-T_i^r)e^{-\delta_i^r T_i^r}}{S_{ii}^{rr}}\right.\\ &\left. -(\gamma_i^r+\sigma_i^r+\delta_i^r) - \sum_{u=1}^{M}\sum_{a=1}^{n_u}\beta_{iia}^{rru} I_{ai}^{ur} - \frac{1}{2}\sum_{u=1}^{M}\sum_{a=1}^{n_u}(\nu_{iia}^{rru})^2 (I_{ai}^{ur})^2\right] dt\\ &\times\left[\sum_{k\neq i}^{n_r}\rho_{ik}^{rr}\frac{I_{ik}^{rr}}{S_{ii}^{rr}} + \sum_{q\neq r}^{M}\sum_{l=1}^{n_q}\rho_{ia}^{rq}\frac{I_{ia}^{rq}}{S_{ii}^{rr}} - \varrho_i^r - (\gamma_i^r+\sigma_i^r+\delta_i^r+d_i^r)\right.\\ &\left. -\sum_{u=1}^{M}\sum_{a=1}^{n_u}\beta_{iia}^{rru}\frac{S_{ii}^{rr}}{I_{ii}^{rr}} I_{ai}^{ur} - \frac{1}{2}\sum_{u=1}^{M}\sum_{a=1}^{n_u}(\nu_{iia}^{rru})^2\frac{(S_{ii}^{rr})^2}{(I_{ii}^{rr})^2}(I_{ai}^{ur})^2\right] dt\\ &-\sum_{u=1}^{M}\sum_{a=1}^{n_u}\nu_{iia}^{rru} I_{ai}^{ur}\,dw_{iia}^{rru}(t) + \sum_{u=1}^{M}\sum_{a=1}^{n_u}\nu_{iia}^{rru}\frac{S_{ii}^{rr}}{I_{ii}^{rr}} I_{ai}^{ur}\,dw_{iia}^{rru}(t). \end{aligned} \tag{7.26}$$

Intraregional level: From (7.23) the terms of the right-hand side of (7.25) for the case of $u = r$, $a = j, j \neq i$ are

$$\begin{aligned} dV(X_{ij}^{rr}) = &\left[\sigma_{ij}^{rr}\frac{S_{ii}^{rr}}{S_{ij}^{rr}} + \frac{\varrho_j^r I_{ij}^{rr}(t-T_i^r)e^{-\delta_j^r T_i^r}}{S_{ij}^{rr}}\right.\\ &\left. -(\rho_{ij}^{rr}+\delta_j^r) - \sum_{u=1}^{M}\sum_{a=1}^{n_u}\beta_{jia}^{rru} I_{aj}^{ur} - \frac{1}{2}\sum_{u=1}^{M}\sum_{a=1}^{n_u}(\nu_{jia}^{rru})^2 (I_{aj}^{ur})^2\right] dt\\ &+\left[\sigma_{ij}^{rr}\frac{I_{ii}^{rr}}{I_{ij}^{rr}} - \varrho_j^r - (\rho_{ij}^{rr}+\delta_j^r+d_j^r)\right.\\ &\left. +\sum_{u=1}^{M}\sum_{a=1}^{n_u}\beta_{jia}^{rru}\frac{S_{ij}^{rr}}{I_{ij}^{rr}} I_{aj}^{ur} - \frac{1}{2}\sum_{u=1}^{M}\sum_{a=1}^{n_u}(\nu_{jia}^{rru})^2\frac{(S_{ij}^{rr})^2}{(I_{ij}^{rr})^2}(I_{aj}^{ur})^2\right] dt\\ &-\sum_{u=1}^{M}\sum_{a=1}^{n_u}\nu_{jia}^{rru} I_{aj}^{ur}\,dw_{jia}^{rru}(t) + \sum_{u=1}^{M}\sum_{a=1}^{n_u}\nu_{jia}^{rru}\frac{S_{ij}^{rr}}{I_{ij}^{rr}} I_{aj}^{ur}\,dw_{jia}^{rru}(t). \end{aligned} \tag{7.27}$$

Regional level: From (7.23) the terms of the right-hand side of (7.25) for the case of $u = q, q \neq r, a = l$ are

$$
\begin{aligned}
dV(X_{il}^{rq}) = {} & \left[\gamma_{il}^{rq} \frac{S_{ii}^{rr}}{S_{iq}^{rq}} + \frac{\varrho_l^q I_{il}^{rq}(t - T_i^r) e^{-\delta_l^q T_i^r}}{S_{il}^{rq}} \right. \\
& \left. - (\rho_{il}^{rq} + \delta_l^q) - \sum_{u=1}^{M} \sum_{a=1}^{n_u} \beta_{lia}^{qru} I_{al}^{uq} - \frac{1}{2} \sum_{u=1}^{M} \sum_{a=1}^{n_u} (v_{lia}^{qru})^2 (I_{al}^{uq})^2 \right] dt \\
& + \left[\gamma_{il}^{rq} \frac{I_{ii}^{rr}}{I_{il}^{rq}} - \varrho_l^q - (\rho_{il}^{rq} + \delta_l^q + d_l^q) \right. \\
& \left. + \sum_{u=1}^{M} \sum_{a=1}^{n_u} \beta_{lia}^{qru} \frac{S_{il}^{rq}}{I_{il}^{rq}} I_{al}^{uq} - \frac{1}{2} \sum_{u=1}^{M} \sum_{a=1}^{n_u} (v_{lia}^{qru})^2 \frac{(S_{il}^{rq})^2}{(I_{il}^{rq})^2} (I_{al}^{uq})^2 \right] dt \\
& - \sum_{u=1}^{M} \sum_{a=1}^{n_u} v_{lia}^{qru} I_{al}^{uq} dw_{lia}^{qru}(t) + \sum_{u=1}^{M} \sum_{a=1}^{n_u} v_{lia}^{qru} \frac{S_{il}^{rq}}{I_{il}^{rq}} I_{al}^{uq} dw_{lia}^{qru}(t).
\end{aligned}
\tag{7.28}
$$

It follows from (7.26)–(7.28), (7.25), and (7.22) that for $t < \tau_+(t)$,

$$
\begin{aligned}
& V(X_{00}^{00}(t)) - V(X_{00}^{00}(t_0)) \\
& \geq \sum_{r=1}^{M} \sum_{i=1}^{n_r} \int_{t_0}^{t} \left[\frac{\varrho_i^r I_{ii}^{rr}(t - T_i^r) e^{-\delta_i^r T_i^r}}{S_{ii}^{rr}} - (\gamma_i^r + \sigma_i^r + \delta_i^r) - \sum_{u=1}^{M} \sum_{a=1}^{n_u} \beta_{iia}^{rru} I_{ai}^{ur} \right. \\
& \left. - \frac{1}{2} \sum_{u=1}^{M} \sum_{a=1}^{n_u} (v_{iia}^{rru})^2 (I_{ai}^{ur})^2 \right] ds + \sum_{r=1}^{M} \sum_{i=1}^{n_r} \int_{t_0}^{t} \left[-\varrho_i^r - (\gamma_i^r + \sigma_i^r + \delta_i^r + d_i^r) \right. \\
& \left. - \sum_{u=1}^{M} \sum_{a=1}^{n_u} \beta_{iia}^{rru} \frac{S_{ii}^{rr}}{I_{ii}^{rr}} I_{ai}^{ur} - \frac{1}{2} \sum_{u=1}^{M} \sum_{a=1}^{n_u} (v_{iia}^{rru})^2 \frac{(S_{ii}^{rr})^2}{(I_{ii}^{rr})^2} (I_{ai}^{ur})^2 \right] ds \\
& - \sum_{r=1}^{M} \sum_{i=1}^{n_r} \sum_{u=1}^{M} \sum_{a=1}^{n_u} \int_{t_0}^{t} v_{iia}^{rru} I_{ai}^{ur} dw_{iia}^{rru}(s) + \sum_{r=1}^{M} \sum_{i=1}^{n_r} \sum_{u=1}^{M} \sum_{a=1}^{n_u} \int_{t_0}^{t} v_{iia}^{rru} \frac{S_{ii}^{rr}}{I_{ii}^{rr}} I_{ai}^{ur} dw_{iia}^{rru}(s) \\
& \times \sum_{r=1}^{M} \sum_{i=1}^{n_r} \sum_{j \neq i}^{n_r} \int_{t_0}^{t} \left[\frac{\varrho_j^r I_{ij}^{rr}(t - T_i^r) e^{-\delta_j^r T_i^r}}{S_{ij}^{rr}} \right. \\
& \left. - (\rho_{ij}^{rr} + \delta_j^r) - \sum_{u=1}^{M} \sum_{a=1}^{n_u} \beta_{jia}^{rru} I_{aj}^{ur} - \frac{1}{2} \sum_{u=1}^{M} \sum_{a=1}^{n_u} (v_{jia}^{rru})^2 (I_{aj}^{ur})^2 \right] ds \\
& + \sum_{r=1}^{M} \sum_{i=1}^{n_r} \sum_{j \neq i}^{n_r} \int_{t_0}^{t} \left[-\varrho_j^r - (\rho_{ij}^{rr} + \delta_j^r + d_j^r) - \frac{1}{2} \sum_{u=1}^{M} \sum_{a=1}^{n_u} (v_{jia}^{rru})^2 \frac{(S_{ij}^{rr})^2}{(I_{ij}^{rr})^2} (I_{aj}^{ur})^2 \right] ds \\
& - \sum_{r=1}^{M} \sum_{i=1}^{n_r} \sum_{j \neq i}^{n_r} \sum_{u=1}^{M} \sum_{a=1}^{n_u} \int_{t_0}^{t} v_{jia}^{rru} I_{aj}^{ur} dw_{jia}^{rru}(s) + \sum_{r=1}^{M} \sum_{i=1}^{n_r} \sum_{j \neq i}^{n_r} \sum_{u=1}^{M} \sum_{a=1}^{n_u} \int_{t_0}^{t} v_{jia}^{rru} \frac{S_{ij}^{rr}}{I_{ij}^{rr}} I_{aj}^{ur} dw_{jia}^{rru}(s) \\
& + \sum_{r=1}^{M} \sum_{i=1}^{n_r} \sum_{q \neq r}^{M} \sum_{l=1}^{n_q} \int_{t_0}^{t} \left[\frac{\varrho_l^q I_{il}^{rq}(t - T_i^r) e^{-\delta_l^q T_i^r}}{S_{il}^{rq}} - (\rho_{il}^{rq} + \delta_l^q) - \sum_{u=1}^{M} \sum_{a=1}^{n_u} \beta_{lia}^{qru} I_{al}^{uq} \right.
\end{aligned}
$$

$$-\frac{1}{2}\sum_{u=1}^{M}\sum_{a=1}^{n_u}(v_{lia}^{qru})^2(I_{al}^{uq})^2\Bigg]ds+\sum_{r=1}^{M}\sum_{i=1}^{n_r}\sum_{q\neq r}^{M}\sum_{l=1}^{n_q}\int_{t_0}^{t}\Bigg[-\varrho_l^q-(\rho_{il}^{rq}+\delta_l^q+d_l^q)$$

$$-\frac{1}{2}\sum_{u=1}^{M}\sum_{a=1}^{n_u}(v_{lia}^{qru})^2\frac{(S_{il}^{rq})^2}{(I_{il}^{rq})^2}(I_{al}^{uq})^2\Bigg]ds-\sum_{r=1}^{M}\sum_{i=1}^{n_r}\sum_{q\neq r}^{M}\sum_{l=1}^{n_q}\sum_{u=1}^{M}\sum_{a=1}^{n_u}\int_{t_0}^{t}v_{lia}^{qru}I_{al}^{uq}dw_{lia}^{qru}(s)$$

$$+\sum_{r=1}^{M}\sum_{i=1}^{n_r}\sum_{q\neq r}^{M}\sum_{l=1}^{n_q}\sum_{u=1}^{M}\sum_{a=1}^{n_u}\int_{t_0}^{t}v_{lia}^{qru}\frac{S_{il}^{rq}}{I_{il}^{rq}}I_{al}^{uq}dw_{lia}^{qru}(s). \tag{7.29}$$

Taking the limit on (7.29) as $t \to \tau_+(t)$, it follows from (7.23) and (7.22) that $V(X_{00}^{00}(t)) - V(X_{00}^{00}(t_0)) \leq -\infty$. This contradicts the finiteness of the right-hand-side of inequality (7.29). Hence $\tau_+(t) = \tau_e(w)$ a. s. We show subsequently that $\tau_e(w) = \infty$. Let $k > 0$ be a positive integer such that $\|\varphi_{00}^{00}\|_1 \leq k$, where the vector of initial values $\varphi_{00}^{00} = (\varphi_{ia}^{ru})_{1\leq r,u\leq M,1\leq i\leq n_r,1\leq a\leq n_u} \in \mathbb{R}^{2n^2}$ is defined in (7.19). Furthermore, $\|\cdot\|_1$ is the p-sum norm (6.46) for the case of $p = 1$. We define the stopping time

$$\begin{cases}\tau_k = \sup\{t \in [t_0, \tau_e) : \|X_{00}^{00}(s)\|_1 \leq k, s \in [0,t]\},\\ \tau_k(t) = \min(t, \tau_k),\end{cases} \tag{7.30}$$

where, by (6.46),

$$\|X_{00}^{00}(s)\|_1 = \sum_{r=1}^{M}\sum_{u=1}^{M}\sum_{i=1}^{n_r}\sum_{a=1}^{n_u}(S_{ia}^{ru}(s) + I_{ia}^{ru}(s)). \tag{7.31}$$

It is easy to see that as $k \to \infty$, $\tau_k(t)$ increases. Set $\lim_{k\to\infty}\tau_k(t) = \tau_\infty$. Then $\tau_\infty \leq \tau_e$ a. s. We show in the following that (1) $\tau_e = \tau_\infty$ a. s. $\Leftrightarrow P(\tau_e \neq \tau_\infty) = 0$ and (2) $\tau_\infty = \infty$ a. s. $\Leftrightarrow P(\tau_\infty = \infty) = 1$.

Suppose on the contrary that $P(\tau_\infty < \tau_e) > 0$. Let $w \in \{\tau_\infty < \tau_e\}$ and $t \leq \tau_\infty$. In the same line as (7.23) and (7.25), define

$$\begin{cases}V_1(X_{00}^{00}) = \sum_{r=1}^{M}\sum_{i=1}^{n_r}\sum_{u=1}^{M}\sum_{a=1}^{n_u}V(X_{ia}^{ru}),\\ V_1(X_{ia}^{ru}) = e^{\delta_a^u t}(S_{ia}^{ru} + I_{ia}^{ru}), \quad t \leq \tau_k(t).\end{cases} \tag{7.32}$$

From (7.32), using expression (7.25), the Itô–Doob differential dV_1 with respect to system (7.16)–(7.18) is given as follows.

Site level: From (7.32) the terms of the right-hand side of (7.25) for the case of $u = r$, $a = i$ are

$$dV_1(X_{ii}^{rr}) = e^{\delta_i^r t}\Bigg[B_i^r + \sum_{k\neq i}^{n_r}\rho_{ik}^{rr}S_{ik}^{rr} + \sum_{q\neq r}^{M}\sum_{l=1}^{n_q}\rho_{ia}^{rq}S_{ia}^{rq} + \varrho_i^r I_{ii}^{rr}(t - T_i^r)e^{-\delta_i^r T_i^r} - (\gamma_i^r + \sigma_i^r)S_{ii}^{rr}\Bigg]dt$$

$$+ e^{\delta_i^r t}\Bigg[\sum_{k\neq i}^{n_r}\rho_{ik}^{rr}I_{ik}^{rr} + \sum_{q\neq r}^{M}\sum_{l=1}^{n_q}\rho_{ia}^{rq}I_{ia}^{rq} - \varrho_i^r I_{ii}^{rr} - (\gamma_i^r + \sigma_i^r + d_i^r)I_{ii}^{rr}\Bigg]dt. \tag{7.33}$$

Intraregional level: From (7.32) the terms of the right-hand side of (7.25) for the case of $u = r$, $a = j, j \neq i$ are

$$dV_1(X_{ij}^{rr}) = e^{\delta_i^r t}[\sigma_{ij}^{rr} S_{ii}^{rr} + \varrho_j^r I_{ij}^{rr}(t - T_i^r)e^{-\delta_j^r T_i^r} - \rho_{ij}^{rr} S_{ij}^{rr}]dt$$
$$+ e^{\delta_j^r t}[\sigma_{ij}^{rr} I_{ii}^{rr} + \varrho_j^r I_{ij}^{rr} - (\rho_{ij}^{rr} + d_j^r) I_{ij}^{rr}]dt. \quad (7.34)$$

Regional level: From (7.32) the terms of the right-hand side of (7.25) for the case of $u = q, q \neq r, a = l$ are

$$dV_1(X_{il}^{rq}) = e^{\delta_l^q t}[\gamma_{il}^{rq} S_{ii}^{rr} + \varrho_l^q I_{il}^{rq}(t - T_i^r)e^{-\delta_l^q T_i^r} - \rho_{il}^{rq} S_{il}^{rq}]dt$$
$$+ e^{\delta_l^q t}[\gamma_{il}^{rq} I_{ii}^{rr} + \varrho_l^q I_{il}^{rq} - (\rho_{il}^{rq} + d_l^q) I_{il}^{rq}]dt. \quad (7.35)$$

By (7.33)–(7.35), integrating (7.25) over $[t_0, \tau]$ leads to

$$\begin{aligned}
&V_1(X_{00}^{00}(\tau)) \\
&\quad = V_1(X_{00}^{00}(t_0)) \\
&\quad + \sum_{r=1}^{M}\sum_{i=1}^{n_r}\int_{t_0}^{\tau} e^{\delta_i^r s}\left[B_i^r + \sum_{k\neq i}^{n_r} \rho_{ik}^{rr} S_{ik}^{rr} + \sum_{q\neq r}^{M}\sum_{l=1}^{n_q} \rho_{ia}^{rq} S_{ia}^{rq} + \varrho_i^r I_{ii}^{rr}(t - T_i^r)e^{-\delta_i^r T_i^r} - (\gamma_i^r + \sigma_i^r) S_{ii}^{rr}\right] ds \\
&\quad + \sum_{r=1}^{M}\sum_{i=1}^{n_r}\int_{t_0}^{\tau} e^{\delta_i^r s}\left[\sum_{k\neq i}^{n_r} \rho_{ik}^{rr} I_{ik}^{rr} + \sum_{q\neq r}^{M}\sum_{l=1}^{n_q} \rho_{ia}^{rq} I_{ia}^{rq} - \varrho_i^r I_{ii}^{rr} - (\gamma_i^r + \sigma_i^r + d_i^r) I_{ii}^{rr}\right] ds \\
&\quad + \sum_{r=1}^{M}\sum_{i=1}^{n_r}\sum_{j\neq i}^{n_r}\int_{t_0}^{\tau} e^{\delta_i^r s}[\sigma_{ij}^{rr} S_{ii}^{rr} + \varrho_j^r I_{ij}^{rr}(t - T_i^r)e^{-\delta_j^r T_i^r} - \rho_{ij}^{rr} S_{ij}^{rr}] ds \\
&\quad + \sum_{r=1}^{M}\sum_{i=1}^{n_r}\sum_{j\neq i}^{n_r}\int_{t_0}^{\tau} e^{\delta_j^r s}[\sigma_{ij}^{rr} I_{ii}^{rr} - \varrho_j^r I_{ij}^{rr} - (\rho_{ij}^{rr} + d_j^r) I_{ij}^{rr}] ds \\
&\quad + \sum_{r=1}^{M}\sum_{i=1}^{n_r}\sum_{q\neq r}^{n_r}\sum_{l=1}^{n_q}\int_{t_0}^{\tau} e^{\delta_l^q s}[\gamma_{il}^{rq} S_{ii}^{rr} + \varrho_l^q I_{il}^{rq}(t - T_i^r)e^{-\delta_l^q T_i^r} - \rho_{il}^{rq} S_{il}^{rq}] ds \\
&\quad + \sum_{r=1}^{M}\sum_{i=1}^{n_r}\sum_{q\neq r}^{n_r}\sum_{l=1}^{n_q}\int_{t_0}^{\tau} e^{\delta_l^q s}[\gamma_{il}^{rq} I_{ii}^{rr} - \varrho_l^q I_{il}^{rq} - (\rho_{il}^{rq} + d_l^q) I_{il}^{rq}] ds. \qquad (7.36)
\end{aligned}$$

From (7.36) we let $\tau = \tau_k(t)$, where $\tau_k(t)$ is defined in (7.30). It is easy to see from (7.36), (7.30), (7.31), and (7.32) that

$$k = \|X_{00}^{00}(\tau_k(t))\|_1 \leq V_1(X_{00}^{00}(\tau_k(t))). \quad (7.37)$$

Taking the limit on (7.37) as $k \to \infty$ leads to a contradiction because the left-hand-side of inequality (7.37) is infinite and the right-hand side is finite. Hence $\tau_e = \tau_\infty$ a. s. In the following, we show that $\tau_e = \tau_\infty = \infty$ a. s.

Let $w \in \{\tau_e < \infty\}$. Applying some algebraic manipulations and simplifications to (7.36), we have

$$\begin{aligned}
&I_{\{\tau_e<\infty\}}V_1(X_{00}^{00}(\tau))\\
&= I_{\{\tau_e<\infty\}}V_1(X_{00}^{00}(t_0)) + I_{\{\tau_e<\infty\}}\sum_{r=1}^{M}\sum_{i=1}^{n_r}\frac{B_i^r}{\delta_i^r}(e^{\delta_i^r\tau}-1)\\
&+ I_{\{\tau_e<\infty\}}\sum_{r=1}^{M}\sum_{i=1}^{n_r}\sum_{q=1}^{M}\sum_{l=1}^{n_q}\left[\varrho_l^q\int_{-T_i^r}^{t_0} I_{il}^{rq}(s)e^{\delta_l^q s}ds - \varrho_l^q\int_{\tau-T_i^r}^{\tau} I_{il}^{rq}(s)e^{\delta_l^q s}ds\right]\\
&- I_{\{\tau_e<\infty\}}\sum_{r=1}^{M}\sum_{i=1}^{n_r}\int_{t_0}^{\tau}\left[\sigma_i^r e^{\delta_i^r s} - \sum_{j\neq i}^{n_r}\sigma_{ij}^{rr}e^{\delta_j^r s}\right](S_{ii}^{rr}+I_{ii}^{rr})ds\\
&- I_{\{\tau_e<\infty\}}\sum_{r=1}^{M}\sum_{i=1}^{n_r}\int_{t_0}^{\tau}\left[\gamma_i^r e^{\delta_i^r s} - \sum_{q=1}^{M}\sum_{l=1}^{n_q}\gamma_{il}^{rq}e^{\delta_l^q s}\right](S_{ii}^{rr}+I_{ii}^{rr})ds\\
&- I_{\{\tau_e<\infty\}}\sum_{r=1}^{M}\sum_{i=1}^{n_r}d_i^r\int_{t_0}^{\tau} I_{ii}^{rr}e^{\delta_i^r s}ds\\
&- I_{\{\tau_e<\infty\}}\sum_{r=1}^{M}\sum_{i=1}^{n_r}\sum_{j\neq i}^{n_r}d_j^r\int_{t_0}^{\tau} I_{ij}^{rr}e^{\delta_j^r s}ds\\
&- I_{\{\tau_e<\infty\}}\sum_{r=1}^{M}\sum_{i=1}^{n_r}\sum_{q\neq r}^{M}\sum_{l=1}^{n_q}d_l^q\int_{t_0}^{\tau} I_{il}^{rq}e^{\delta_l^q s}ds,
\end{aligned}\tag{7.38}$$

where I_A is the indicator function of a set A.

We recall [45] that $\sigma_i^r = \sum_{j\neq i}^{n_r}\sigma_{ij}^{rr}$ and $\gamma_i^r = \sum_{q\neq r}^{M}\sum_{l=1}^{n_q}\gamma_{il}^{rq}$. Hence the fourth and fifth terms on the right-hand side of (7.38) are such that $[\sigma_i^r e^{\delta_i^r s} - \sum_{j\neq i}^{n_r}\sigma_{ij}^{rr}e^{\delta_j^r s}] \geq 0$ for all $\delta_i^r \geq \delta_j^r, j \neq i$, and $[\gamma_i^r e^{\delta_i^r s} - \sum_{q=1}^{M}\sum_{l=1}^{n_q}\gamma_{il}^{rq}e^{\delta_l^q s}] \geq 0$ for all $\delta_i^r \geq \delta_l^q, q \neq r, l \in I(1, n_q)$. We now let $\tau = \tau_k(t) \wedge T$ in (7.38), $T > 0$, where $\tau_k(t)$ is defined in (7.30). The expected value of (7.38) is estimated as follows:

$$\begin{aligned}
&E[I_{\{\tau_e<\infty\}}V_1(X_{00}^{00}(\tau_k(t)\wedge T))]\\
&\leq V_1(X_{00}^{00}(t_0)) + \sum_{i=1}^{n_r}\frac{B_i^r}{\delta_i^r}e^{\delta_i^r\tau_k(t)\wedge T} + \sum_{r=1}^{M}\sum_{i=1}^{n_r}\sum_{q=1}^{M}\sum_{l=1}^{n_q}\left[\varrho_l^q\int_{-T_i^r}^{t_0}\varphi_{il2}^{rq}(s)e^{\delta_l^q s}ds\right].
\end{aligned}\tag{7.39}$$

Furthermore, from (7.31), (7.32), and the definition of the indicator function I_A it follows that

$$I_{\{\tau_e<\infty,\tau_k(t)\le T\}}\|X_{00}^{00}(\tau_k(t))\|_1 \le I_{\{\tau_e<\infty\}}V_1(X_{00}^{00}(\tau_k(t)\wedge T)). \tag{7.40}$$

It follows from (7.39), (7.40), and (7.30) that

$$\begin{aligned}
&P(\{\tau_e<\infty,\tau_k(t)\le T\})k \\
&= E[I_{\{\tau_e<\infty,\tau_k(t)\le T\}}\|X_{00}^{00}(\tau_k(t))\|_1] \\
&\le E[I_{\{\tau_e<\infty\}}V(X_{00}^{00}(\tau_k(t)\wedge T))] \\
&\le V_1(X_{00}^{00}(t_0)) + \sum_{i=1}^{n_r}\frac{B_i^r}{\delta_i^r}e^{\delta_i^r T} + \sum_{r=1}^{M}\sum_{i=1}^{n_r}\sum_{q=1}^{M}\sum_{l=1}^{n_q}\left[\varrho_l^q\int_{-T_i^r}^{t_0}\varphi_{il2}^{rq}(s)e^{\delta_l^q s}ds\right].
\end{aligned}\tag{7.41}$$

It follows immediately from (7.41) that $P(\{\tau_e<\infty,\tau_\infty\le T\})\to 0$ as $k\to\infty$. Furthermore, since $T<\infty$ is arbitrary, we conclude that $P(\{\tau_e<\infty,\tau_\infty<\infty\})=0$.

Finally, by the total probability principle,

$$\begin{aligned}
P(\{\tau_e<\infty\}) &= P(\{\tau_e<\infty,\tau_\infty=\infty\}) + P(\{\tau_e<\infty,\tau_\infty<\infty\}) \\
&\le P(\{\tau_e\ne\tau_\infty\}) + P(\{\tau_e<\infty,\tau_\infty<\infty\}) \\
&= 0.
\end{aligned}\tag{7.42}$$

Thus by (7.42), $\tau_e=\tau_\infty=\infty$ a. s., as required. □

Remark 7.3.1. For any $r\in I(1,M)$ and $i\in I(1,n_r)$, Theorem 7.3.1 signifies that the number of residents of site s_i^r of all categories present at home site s_i^r or visiting intra- and interregional sites s_j^r and s_l^q, respectively, are nonnegative. This implies that the total population of residents of site s_i^r present at home and also visiting sites in regions in their intra- and interregional accessible domains [47], given by the sum $N_{i0}^{rr}(t)=\sum_{u=1}^{M}\sum_{a=1}^{n_u}y_{ia}^{ru}$, is nonnegative. Moreover, the total effective population [47], defined by $eff(N_{i0}^{rr})(t)=\sum_{u=1}^{M}\sum_{a=1}^{n_u}y_{ai}^{ur}$, at any site s_i^r in the region C_r is also nonnegative at all times $t\ge t_0$.

The following result defines an upper bound for the solution process of system (7.16)–(7.18). We use of Theorem 7.3.1 to establish this result.

Theorem 7.3.2. *Suppose the hypotheses of Theorem* 7.3.1 *is satisfied. Let* $\mu=\min_{1\le u\le M,1\le a\le n_u}(\delta_a^u)$. *If*

$$\sum_{r=1}^{M}\sum_{u=1}^{M}\sum_{i=1}^{n_r}\sum_{a=1}^{n_u}y_{ia}^{ru}(t_0)\le\frac{1}{\mu}\sum_{r=1}^{M}\sum_{i=1}^{n_r}B_i^r, \tag{7.43}$$

then

$$\sum_{r=1}^{M}\sum_{u=1}^{M}\sum_{i=1}^{n_r}\sum_{a=1}^{n_u} y_{ia}^{ru}(t) \leq \frac{1}{\mu}\sum_{r=1}^{M}\sum_{i=1}^{n_r} B_i^r \quad for\ t \geq t_0 \quad a.s. \tag{7.44}$$

Proof. The proof of this result is similar to that of Lemma 6.6.2. □

Remark 7.3.2. From Theorems 7.3.1 and 7.3.2 we conclude that a closed ball $\bar{\mathfrak{B}}_{R^{3n^2}}(\vec{0};r)$ in R^{3n^2} under the sum norm $\|\cdot\|_1$ centered at the origin $\vec{0} \in R^{3n^2}$ with radius $r = \frac{1}{\mu}\sum_{r=1}^{M}\sum_{i=1}^{n_r} B_i^r$ is positive self-invariant with regard to the two-scale network SIRS delay SDE system (7.16)–(7.18) subject to the influence of the dynamic human mobility process in Chapter 5, that is,

$$\bar{\mathfrak{B}}_{R^{3n^2}}(\vec{0};r) = \left\{(S_{ia}^{ru}, I_{ia}^{ru}, R_{ia}^{ru}) : y_{ia}^{ru}(t) \geq 0 \text{ and } \|x_{00}^{00}\|_1 = \sum_{r=1}^{M}\sum_{u=1}^{M}\sum_{i=1}^{n_r}\sum_{a=1}^{n_u} y_{ia}^{ru}(t) \leq \frac{1}{\mu}\sum_{r=1}^{M}\sum_{i=1}^{n_r} B_i^r\right\} \tag{7.45}$$

is a positive self-invariant set for system (7.16)–(7.18). We denote

$$\bar{B} \equiv \frac{1}{\mu}\sum_{r=1}^{M}\sum_{i=1}^{n_r} B_i^r. \tag{7.46}$$

In terms of the epidemic dynamics, this result signifies that the maximum spread of the SIRS disease epidemic in the manner described by the stochastic model (7.16)–(7.18) is confined in the space $\bar{\mathfrak{B}}_{R^{3n^2}}(\vec{0};r)$. Thus, if disease eradication or persistence of the disease are investigated, then it is best to conduct these investigations in this space.

7.4 Disease eradication in the delay SIRS stochastic epidemic dynamic model

In this section, we investigate disease eradication in the two-scale delay SIRS disease epidemic model. Although the goals of investigation are similar to those in Section 6.7, the methods for achieving the goals are slightly different because of the presence of a time lag in the delay SDE system (7.16)–(7.18). Two epidemiological results are expected from this investigation: (a) The basic reproduction number for the disease and (b) restriction thresholds based on the discrete delay T for the complete eradication of the disease in the two-scale network population $C = C_{00}^{00}$.

Disease eradication conditions are obtained by examining the stochastic stability of the disease-free equilibrium (DFE). Thus, first, the DFE is obtained, and, second, the stochastic stability of the DFE is obtained.

7.4.1 Existence of a disease-free equilibrium

The disease-free equilibrium is obtained by solving the system of algebraic equations obtained from (7.16)–(7.18) by setting the drift and diffusion parts of the system of stochastic differential equations to zero. In addition, the conditions $I = R = 0$ are applied in the event when there is no disease in the population. We summarize the results as follows. For any $r, u \in I(1, M)$, $i \in I(1, n_r)$, and $a \in I(1, n_u)$, let

$$D_i^r = \gamma_i^r + \sigma_i^r + \delta_i^r - \sum_{a=1}^{n_r} \frac{\rho_{ia}^{rr}\sigma_{ia}^{rr}}{\rho_{ia}^{rr} + \delta_a^r} - \sum_{u\neq r}^{M} \sum_{a=1}^{n_u} \frac{\rho_{ia}^{rr}\gamma_{ia}^{ru}}{\rho_{ia}^{ru} + \delta_a^u} > 0. \tag{7.47}$$

Furthermore, let $(S_{ia}^{ru*}, I_{ia}^{ru*}, R_{ia}^{ru*})$ be the equilibrium state of the delayed system (7.16)–(7.18). The disease-free equilibrium state is given by $E_{ia}^{ru} = (S_{ia}^{ru*}, 0, 0)$, where

$$S_{ia}^{ru*} = \begin{cases} \frac{B_i^r}{D_i^r}, & u = r,\ a = i, \\ \frac{B_i^r}{D_i^r}\frac{\sigma_{ij}^{rr}}{\rho_{ij}^{rr}+\delta_j^r}, & u = r,\ a \neq i, \\ \frac{B_i^r}{D_i^r}\frac{\gamma_{ia}^{ru}}{\rho_{ia}^{ru}+\delta_a^u}, & u \neq r. \end{cases} \tag{7.48}$$

7.4.2 Stochastic stability of the DFE for the delay SDE system

The asymptotic stability property of E_{ia}^{ru} is established by verifying the conditions of the stochastic version of the Lyapunov second theorem in Section 1.4. The qualitative properties of (7.16)–(7.18) near the equilibrium state $(S_{ia}^{ru*}, 0, 0)$ are easily examined in the transformed system obtained by applying the following change of variables that shift the equilibrium to the origin:

$$\begin{cases} U_{ia}^{ru} = S_{ia}^{ru} - S_{ia}^{ru*}, \\ V_{ia}^{ru} = I_{ia}^{ru}, \\ W_{ia}^{ru} = R_{ia}^{ru}. \end{cases} \tag{7.49}$$

By employing this transformation system (7.16)–(7.18) is transformed into the following forms:

$$dU_{il}^{rq} = \begin{cases} [\sum_{q=1}^{M}\sum_{a=1}^{n_q}\rho_{ia}^{rq}U_{ia}^{rq} + \varrho_i^r V_{ii}^{rr}(t - T_i^r)e^{-\delta_i^r T_i^r} \\ \quad - (\gamma_i^r + \sigma_i^r + \delta_i^r)U_{ii}^{rr} - \sum_{u=1}^{M}\sum_{a=1}^{n_u}\beta_{iia}^{rru}(S_{ii}^{rr*} + U_{ii}^{rr})V_{ai}^{ur}]dt \\ \quad - [\sum_{u=1}^{M}\sum_{a=1}^{n_u}\nu_{iia}^{rru}(S_{ii}^{rr*} + U_{ii}^{rr})V_{ai}^{ur}dw_{iia}^{rru}(t)], \quad q = r, l = i, \\ [\sigma_{ij}^{rr}U_{ii}^{rr} + \varrho_j^r V_{ij}^{rr}(t - T_i^r)e^{-\delta_j^r T_i^r} - (\rho_{ij}^{rr} + \delta_j^r)U_{ij}^{rr} - \sum_{u=1}^{M}\sum_{a=1}^{n_u}\beta_{jia}^{rru}(S_{ij}^{rr*} + U_{ij}^{rr})V_{aj}^{ur}]dt \\ \quad - [\sum_{u=1}^{M}\sum_{a=1}^{n_u}\nu_{jia}^{rru}(S_{ij}^{rr*} + U_{ij}^{rr})V_{aj}^{ur}dw_{jia}^{rru}(t)], \quad q = r, l = j, j \neq i, \\ [\gamma_{il}^{rq}U_{ii}^{rr} + \varrho_l^q V_{il}^{rq}(t - T_i^r)e^{-\delta_l^q T_i^r} - (\rho_{il}^{rq} + \delta_l^q)U_{il}^{rq} - \sum_{u=1}^{M}\sum_{a=1}^{n_u}\beta_{lia}^{qru}S_{il}^{rq}I_{al}^{uq}]dt \\ \quad - [\sum_{u=1}^{M}\sum_{a=1}^{n_u}\nu_{lia}^{qru}(S_{il}^{rq*} + U_{il}^{rq})V_{al}^{uq}dw_{lia}^{qru}(t)], \quad q \neq r, \end{cases} \tag{7.50}$$

$$dV_{il}^{rq} = \begin{cases} [\sum_{q=1}^{M}\sum_{a=1}^{n_q}\rho_{ia}^{rq}V_{ia}^{rq} - (\varrho_i^r + \gamma_i^r + \sigma_i^r + \delta_i^r + d_i^r)W_{ii}^{rr} \\ \quad + \sum_{u=1}^{M}\sum_{a=1}^{n_u}\beta_{iia}^{rru}(S_{ii}^{rr*} + U_{ii}^{rr})V_{ai}^{ur}]dt \\ \quad + [\sum_{u=1}^{M}\sum_{a=1}^{n_u}\nu_{iia}^{rru}(S_{ii}^{rr*} + U_{ii}^{rr})V_{ai}^{ur}dw_{iia}^{rru}(t)], \quad q = r,\ l = i, \\ [\sigma_{ij}^{rr}V_{ii}^{rr} - (\varrho_j^r + \rho_{ij}^{rr} + \delta_j^r + d_j^r)V_{ij}^{rr} + \sum_{u=1}^{M}\sum_{a=1}^{n_u}\beta_{jia}^{rru}(S_{ij}^{rr*} + U_{ij}^{rr})V_{aj}^{ur}]dt \\ \quad + [\sum_{u=1}^{M}\sum_{a=1}^{n_u}\nu_{jia}^{rru}(S_{ij}^{rr*} + U_{ij}^{rr})V_{aj}^{ur}dw_{jia}^{rru}(t)], \quad q = r,\ l = j,\ j \neq i, \\ [\gamma_{il}^{rq}V_{ii}^{rr} - (\varrho_l^q + \rho_{il}^{rq} + \delta_l^q + d_l^q)V_{il}^{rq} + \sum_{u=1}^{M}\sum_{a=1}^{n_u}\beta_{lia}^{qru}(S_{il}^{rq*} + U_{il}^{rq})V_{al}^{uq}]dt \\ \quad + [\sum_{u=1}^{M}\sum_{a=1}^{n_u}\nu_{lia}^{qru}(S_{il}^{rq*} + U_{il}^{rq})V_{al}^{uq}dw_{lia}^{qru}(t)], \quad q \neq r, \end{cases} \tag{7.51}$$

and

$$dW_{il}^{rq} = \begin{cases} [\sum_{q=1}^{M}\sum_{l=1}^{n_q}\rho_{il}^{rq}W_{il}^{rq} + \varrho_i^r V_{ii}^{rr} - \varrho_i^r V_{ii}^{rr}(t - T_i^r)e^{-\delta_i^r T_i^r} - (\gamma_i^r + \sigma_i^r + \delta_i^r)W_{ii}^{rr}]dt, \\ \quad q = r,\ l = i \\ [\sigma_{ij}^{rr}W_{ii}^{rr} + \varrho_j^r V_{ij}^{rr} - \varrho_j^r V_{ij}^{rr}(t - T_i^r)e^{-\delta_j^r T_i^r} - (\rho_{ij}^{rr} + \delta_j^r)W_{ij}^{rr}]dt, \quad q = r,\ l = j,\ j \neq i, \\ [\gamma_{il}^{rq}W_{ii}^{rr} + \varrho_l^q V_{il}^{rq} - \varrho_l^q V_{il}^{rq}(t - T_i^r)e^{-\delta_l^q T_i^r} - (\rho_{il}^{rq} + \delta_l^q)W_{il}^{rq}]dt, \quad q \neq r. \end{cases} \tag{7.52}$$

We state and prove the following lemmas that will be useful in the proofs of the stability results.

Lemma 7.4.1. *Let $V_1 : \mathbb{R}^{3n^2} \times \mathbb{R}_+ \to \mathbb{R}_+$ be the function defined by*

$$\begin{cases} V_1(\tilde{x}_{00}^{00}) = \sum_{r=1}^{M}\sum_{u=1}^{M}\sum_{i=1}^{n_r}\sum_{a=1}^{n_u} V(\tilde{x}_{ia}^{ru}), \\ V_1(\tilde{x}_{ia}^{ru}) = (S_{ia}^{ru} - S_{ia}^{ru*} + I_{ia}^{ru})^2 + c_{ia}^{ru}(I_{ia}^{ru})^2 + (R_{ia}^{ru})^2, \\ \tilde{x}_{ia}^{ru} = (U_{ia}^{ru}, V_{ia}^{ru}, W_{ia}^{ru})^T, \text{ and } c_{ia}^{ru} \geq 0. \end{cases} \tag{7.53}$$

Then $V_1 \in C^{2,1}(\mathbb{R}^{3n^2} \times \mathbb{R}_+, \mathbb{R}_+)$, and it satisfies

$$b(\|\tilde{x}_{00}^{00}\|) \leq V_1(\tilde{x}_{00}^{00}(t)) \leq a(\|\tilde{x}_{00}^{00}\|), \tag{7.54}$$

where

$$\begin{aligned} b(\|\tilde{x}_{00}^{00}\|) &= \min_{1\leq r,u\leq M, 1\leq i\leq n_r, 1\leq a\leq n_u} \left\{\frac{c_{ia}^{ru}}{2 + c_{ia}^{ru}}\right\} \sum_{r=1}^{M}\sum_{u=1}^{M}\sum_{i=1}^{n_r}\sum_{a=1}^{n_u}[(U_{ia}^{ru})^2 + (V_{ia}^{ru})^2 + (W_{ia}^{ru})^2], \\ a(\|\tilde{x}_{00}^{00}\|) &= \max_{1\leq r,u\leq M, 1\leq i\leq n_r, 1\leq a\leq n_u} \{c_{ia}^{ru} + 2\} \sum_{r=1}^{M}\sum_{u=1}^{M}\sum_{i=1}^{n_r}\sum_{a=1}^{n_u}[(U_{ia}^{ru})^2 + (V_{ia}^{ru})^2 + (W_{ia}^{ru})^2]. \end{aligned} \tag{7.55}$$

Proof. The proof is similar to that of Lemma 6.7.1. □

Remark 7.4.1. Lemma 7.4.1 shows that the Lyapunov function V defined in (7.53) is positive definite (see (7.54)), decrescent, and radially unbounded (see (7.54)) function [53, 24].

We state the following lemma.

Lemma 7.4.2. *Assume that the hypothesis of Lemma 7.4.1 is satisfied. Define the Lyapunov functional*

$$V = V_1 + V_2, \tag{7.56}$$

where V_1 is defined by (7.53), and

$$V_2 = 3\sum_{r=1}^{M}\sum_{i=1}^{n_r}\sum_{u=1}^{M}\sum_{a=1}^{n_u}\left[\frac{(\varrho_a^u)^2}{\mu_{ia}^{ru}}e^{-\delta_a^u T_i^r}\right]\int\limits_{t-T_i^r}^{t}(V_{ia}^{ru}(\theta))^2\,d\theta. \tag{7.57}$$

Suppose that

$$\begin{cases}(\gamma_i^r+\sigma_i^r+\delta_i^r) > \max\Big((\sum_{u=1}^{M}\sum_{a=1}^{n_u}\mu_{ia}^{ru}+\sum_{a\neq i}^{n_r}\frac{(\sigma_{ia}^{rr})^2}{\mu_{ii}^{rr}}+\sum_{a\neq r}^{M}\sum_{a=1}^{n_r}\frac{(\gamma_{ia}^{ru})^2}{\mu_{ii}^{rr}}+\frac{3}{2}\mu_{ii}^{rr}),\\ \qquad (\frac{1}{2}\sum_{u=1}^{M}\sum_{a=1}^{n_u}\mu_{ia}^{ru}+\frac{1}{2}\sum_{u\neq r}^{M}\sum_{a=1}^{n_r}\frac{(\gamma_{ia}^{ru})^2}{\mu_{ii}^{rr}}+\frac{1}{2}\sum_{a\neq i}^{n_r}\frac{(\sigma_{ia}^{rr})^2}{\mu_{ii}^{rr}}+\frac{1}{2}\mu_{ii}^{rr})\Big),\\ \quad u=r,\ a=i,\\ (\rho_{ia}^{rr}+\delta_a^r) > \max((\frac{(\rho_{ia}^{rr})^2}{\mu_{ia}^{rr}}+\mu_{ii}^{rr}+\mu_{ia}^{rr}),(\frac{1}{2}\frac{(\rho_{ia}^{rr})^2}{\mu_{ia}^{rr}}+\frac{1}{2}\mu_{ii}^{rr}+\frac{1}{2}\mu_{ia}^{rr})), \quad u=r,\ a\neq i,\\ (\rho_{ia}^{ru}+\delta_a^u) > \max((\frac{(\rho_{ia}^{ru})^2}{\mu_{ia}^{ru}}+\mu_{ii}^{rr}+\mu_{ia}^{ru}),(\frac{1}{2}\frac{(\rho_{ia}^{ru})^2}{\mu_{ia}^{ru}}+\frac{1}{2}\mu_{ii}^{rr}+\frac{1}{2}\mu_{ia}^{ru})), \quad u\neq r.\end{cases}$$

Furthermore, let

$$\mathfrak{M}_{ia}^{ru} = \begin{cases}\max\Big(\frac{1}{\delta_i^r}\log\Big(\frac{\mu_{ii}^{rr}}{2((\gamma_i^r+\sigma_i^r+\delta_i^r)-[\sum_{u=1}^{M}\sum_{a=1}^{n_u}\mu_{ia}^{ru}+\sum_{a\neq i}^{n_r}\frac{(\sigma_{ia}^{rr})^2}{\mu_{ii}^{rr}}+\sum_{a\neq r}^{M}\sum_{a=1}^{n_r}\frac{(\gamma_{ia}^{ru})^2}{\mu_{ii}^{rr}}+\frac{3}{2}\mu_{ii}^{rr}])}\Big),\\ \qquad \frac{1}{\delta_i^r}\log(\frac{\mu_{ii}^{rr}}{2((\gamma_i^r+\sigma_i^r+\delta_i^r)-[\frac{1}{2}\sum_{u=1}^{M}\sum_{a=1}^{n_u}\mu_{ia}^{ru}+\frac{1}{2}\sum_{u\neq r}^{M}\sum_{a=1}^{n_r}\frac{(\gamma_{ia}^{ru})^2}{\mu_{ii}^{rr}}+\frac{1}{2}\sum_{a\neq i}^{n_r}\frac{(\sigma_{ia}^{rr})^2}{\mu_{ii}^{rr}}+\frac{1}{2}\mu_{ii}^{rr}])})\Big),\\ \quad u=r,\ i=a,\\ \max\Big(\frac{1}{\delta_a^r}\log\Big(\frac{\mu_{ia}^{rr}}{2((\rho_{ia}^{rr}+\delta_a^r)-[\frac{(\rho_{ia}^{rr})^2}{\mu_{ia}^{rr}}+\mu_{ii}^{rr}+\mu_{ia}^{rr}])}\Big),\frac{1}{\delta_a^r}\log\Big(\frac{\mu_{ia}^{rr}}{2((\rho_{ia}^{rr}+\delta_a^r)-[\frac{1}{2}\frac{(\rho_{ia}^{rr})^2}{\mu_{ia}^{rr}}+\frac{1}{2}\mu_{ii}^{rr}+\frac{1}{2}\mu_{ia}^{rr}])}\Big)\Big),\\ \quad u=r,\ a\neq i\\ \max\Big(\frac{1}{\delta_a^u}\log\Big(\frac{\mu_{ia}^{ru}}{2((\rho_{ia}^{ru}+\delta_a^u)-[\frac{(\rho_{ia}^{ru})^2}{\mu_{ia}^{ru}}+\mu_{ii}^{rr}+\mu_{ia}^{ru}])}\Big),\frac{1}{\delta_a^u}\log\Big(\frac{\mu_{ia}^{ru}}{2((\rho_{ia}^{ru}+\delta_a^u)-[\frac{1}{2}\frac{(\rho_{ia}^{ru})^2}{\mu_{ia}^{ru}}+\frac{1}{2}\mu_{ii}^{rr}+\frac{1}{2}\mu_{ia}^{ru}])}\Big)\Big),\\ \quad u\neq r,\end{cases}$$

and

$$\mathfrak{V}_{ia}^{ru} = \begin{cases}\frac{\frac{1}{2}\sum_{u=1}^{M}\sum_{a=1}^{n_u}\mu_{ia}^{ru}+\frac{1}{2}\sum_{v=1}^{M}\sum_{b=1}^{n_v}\beta_{iib}^{rrv}(S_{ii}^{rr*}\mu_{ii}^{rr}+\mu_{ii}^{rr})+\frac{1}{2}d_{ii}^{rr}}{\varrho_i^r+\gamma_i^r+\sigma_i^r+\delta_i^r+d_i^r}, & a=i,\ u=r,\\ \frac{\frac{1}{2}\mu_{ii}^{rr}+\frac{1}{2}\sum_{v=1}^{M}\sum_{b=1}^{n_v}\beta_{aib}^{rrv}(S_{ia}^{rr*}\mu_{ia}^{rr}+\mu_{ia}^{rr})+\frac{1}{2}d_{ai}^{rr}}{\varrho_a^r+\rho_{ia}^{rr}+\delta_a^r+d_a^r}, & a\neq i,\ u=r,\\ \frac{\frac{1}{2}\mu_{ii}^{rr}+\frac{1}{2}\sum_{v=1}^{M}\sum_{b=1}^{n_v}\beta_{aib}^{urv}(S_{ii}^{ru*}\mu_{ia}^{ru}+\mu_{ia}^{ru})+\frac{1}{2}d_{ai}^{ur}}{\varrho_a^u+\rho_{ia}^{ru}+\delta_a^u+d_a^u}, & u\neq r,\end{cases} \tag{7.58}$$

for some suitably defined positive number μ_{ia}^{ru}, depending on δ_a^u, for all $r,u \in I^r(1,M)$, $i \in I(1,n)$, and $a \in I_i^r(1,n_r)$. Assume that $\mathfrak{V}_{ia}^{ru} < 1$ and $T_i^r \geq \max_{1\leq r,u\leq M,1\leq i\leq n_r,1\leq a\leq n_u}(\mathfrak{M}_{ia}^{ru})$.

Then there exist positive numbers ϕ_{ia}^{ru}, ψ_{ia}^{ru}, and φ_{ia}^{ru} such that the differential operator LV associated with Itô–Doob-type stochastic system (7.16)–(7.18) satisfies the inequality

$$
\begin{aligned}
LV(\tilde{x}_{00}^{00}) \le \sum_{r=1}^{M}\sum_{i=1}^{n_r}\Bigg[&-[\phi_{ii}^{rr}(U_{ii}^{rr})^2+\psi_{ii}^{rr}(V_{ii}^{rr})^2+\varphi_{ii}^{rr}(W_{ii}^{rr})^2]\\
&-\sum_{a\neq i}^{n_r}[\phi_{ia}^{rr}(U_{ia}^{rr})^2+\psi_{ia}^{rr}(V_{ia}^{rr})^2+\varphi_{ia}^{rr}(W_{ia}^{rr})^2]\\
&-\sum_{u\neq r}^{M}\sum_{a=1}^{n_u}[\phi_{ia}^{ru}(U_{ia}^{rr})^2+\psi_{ia}^{ru}(V_{ia}^{ru})^2+\varphi_{ia}^{ru}(W_{ia}^{ru})^2]\Bigg].
\end{aligned}
\tag{7.59}
$$

Moreover,

$$LV(\tilde{x}_{00}^{00}) \le -cV_1(\tilde{x}_{00}^{00}), \tag{7.60}$$

where the positive constant c is defined by

$$c=\frac{\min_{1\le r,u\le M,1\le i\le n_r,1\le a\le n_u}\{\phi_{ia}^{ru},\psi_{ia}^{ru},\varphi_{ia}^{ru}\}}{\max_{1\le r,u\le M,1\le i\le n_r,1\le a\le n_u}\{C_{ia}^{ru}+2\}}. \tag{7.61}$$

Proof. The computation of differential operator [53, 24] applied to the Lyapunov function V_1 in (7.53) with respect to the large-scale system of Itô–Doob-type stochastic differential equation (7.16)–(7.18) is as follows:

$$LV_1(\tilde{x}_{00}^{00})=\sum_{r=1}^{M}\sum_{i=1}^{n_r}\left[LV_1(\tilde{x}_{ii}^{rr})+\sum_{j\neq i}^{n_r}LV_1(\tilde{x}_{ij}^{rr})+\sum_{u\neq r}^{M}\sum_{a=1}^{n_u}LV_1(\tilde{x}_{ia}^{ru})\right], \tag{7.62}$$

where

$$
\begin{aligned}
LV_1(\tilde{x}_{ii}^{rr}) ={}& 2\sum_{u=1}^{M}\sum_{a=1}^{n_u}[(1+C_{ii}^{rr})\rho_{ia}^{ru}V_{ia}^{ru}V_{ii}^{rr}+\rho_{ia}^{ru}U_{ia}^{ru}U_{ii}^{rr}+\rho_{ia}^{ru}V_{ia}^{ru}U_{ii}^{rr}+\rho_{ia}^{ru}U_{ia}^{ru}V_{ii}^{rr}\\
&+\rho_{ia}^{ru}W_{ia}^{ru}W_{ii}^{rr}]+2\varrho_i^rV_{ii}^{rr}(t-T_i^r)U_{ii}^{rr}e^{-\delta_i^rT_i^r}+2\varrho_i^rV_{ii}^{rr}(t-T_i^r)V_{ii}^{rr}e^{-\delta_i^rT_i^r}\\
&+2\varrho_i^rV_{ii}^{rr}(t-T_i^r)W_{ii}^{rr}e^{-\delta_i^rT_i^r}+2\varrho_i^rV_{ii}^{rr}W_{ii}^{rr}\\
&-2[(\varrho_i^r+d_i^r)+2(\gamma_i^r+\sigma_i^r+\delta_i^r)]V_{ii}^{rr}U_{ii}^{rr}-2(\gamma_i^r+\sigma_i^r+\delta_i^r)(U_{ii}^{rr})^2\\
&-2[(c_{ii}^{rr}+1)\varrho_i^r+2(c_{ii}^{rr}+1)(\gamma_i^r+\sigma_i^r+\delta_i^r+d_i^r)](V_{ii}^{rr})^2\\
&-2(\gamma_i^r+\sigma_i^r+\alpha_i^r+\delta_i^r)(W_{ii}^{rr})^2+2c_{ii}^{rr}\sum_{u=1}^{M}\sum_{a=1}^{n_u}\beta_{iia}^{rru}(S_{ii}^{rr*}+U_{ii}^{rr})V_{ai}^{ur}V_{ii}^{rr}\\
&+c_{ii}^{rr}\sum_{u=1}^{M}\sum_{a=1}^{n_u}(v_{iia}^{rru})^2(S_{ii}^{rr*}+U_{ii}^{rr})^2(V_{ai}^{ur})^2,\quad u=r,\ a=i,
\end{aligned}
\tag{7.63}
$$

$$\sum_{a\neq i}^{n_r}LV_1(\tilde{x}_{ia}^{rr})=\sum_{a\neq r}^{n_r}\Big\{2(1+c_{ia}^{rr})\sigma_{ia}^{rr}V_{ia}^{rr}V_{ii}^{rr}+2\sigma_{ia}^{rr}U_{ia}^{rr}U_{ii}^{rr}+2\sigma_{ia}^{rr}V_{ia}^{rr}U_{ii}^{rr}+2\sigma_{ia}^{rr}U_{ia}^{rr}V_{ii}^{rr}$$

$$
\begin{aligned}
&+ 2\sigma_{ia}^{rr} W_{ia}^{rr} W_{ii}^{rr} + 2\varrho_a^r V_{ia}^{rr}(t-T_i^r) U_{ia}^{rr} e^{-\delta_a^r T_i^r} + 2\varrho_a^r V_{ia}^{rr}(t-T_i^r) V_{ia}^{rr} e^{-\delta_a^r T_i^r} \\
&- 2\varrho_a^r V_{ia}^{rr}(t-T_i^r) W_{ia}^{rr} e^{-\delta_a^r T_i^r} - 2[(c_{ia}^{rr}+1)\varrho_a^r + 2(c_{ia}^{rr}+1)(\rho_{ia}^{rr}+\delta_a^r)](V_{ia}^{rr})^2 \\
&- 2(\rho_{ia}^{rr}+\delta_a^r)(U_{ia}^{rr})^2 - 2(\rho_{ia}^{rr}+\delta_a^r)(W_{ia}^{rr})^2 + 2\varrho_a^r V_{ia}^{rr} W_{ia}^{rr} \\
&- 2[(\varrho_a^r + d_a^r) + 2(\rho_{ia}^{rr}+\delta_a^r)] V_{ia}^{rr} U_{ia}^{rr}\Big\} \\
&+ 2\sum_{a\neq i}^{n_r} c_{ia}^{rr} \sum_{v=1}^{M}\sum_{b=1}^{n_v} \beta_{aib}^{rrv}(S_{ia}^{rr*} + U_{ia}^{rr}) V_{ba}^{vr} V_{ia}^{rr} \\
&+ \sum_{a\neq i}^{n_r} c_{ia}^{rr} \sum_{v=1}^{M}\sum_{b=1}^{n_v} (v_{aib}^{rrv})^2 (S_{ia}^{rr*} + U_{ia}^{rr})^2 (V_{ba}^{vr})^2, \quad u = r,\ a \neq i,
\end{aligned} \tag{7.64}
$$

$$
\begin{aligned}
\sum_{u\neq r}^{M}\sum_{a=1}^{n_r} LV_1(\tilde{x}_{ia}^{ru}) = &\sum_{u\neq r}^{M}\sum_{a=1}^{n_u}\Big\{2(1+c_{ia}^{ru})\gamma_{ia}^{ru} V_{ia}^{ru} V_{ii}^{rr} + 2\gamma_{ia}^{ru} U_{ia}^{ru} U_{ii}^{rr} + 2\gamma_{ia}^{ru} V_{ia}^{ru} U_{ii}^{rr} + 2\gamma_{ia}^{ru} U_{ia}^{ru} V_{ii}^{rr} \\
&+ 2\gamma_{ia}^{ru} W_{ia}^{ru} W_{ii}^{rr} + 2\varrho_a^u V_{ia}^{ru}(t-T_i^r) U_{ia}^{ru} e^{-\delta_a^u T_i^r} + 2\varrho_a^u V_{ia}^{ru}(t-T_i^r) V_{ia}^{ru} e^{-\delta_a^u T_i^r} \\
&- 2\varrho_a^u V_{ia}^{ru}(t-T_i^r) W_{ia}^{ru} e^{-\delta_a^u T_i^r} \\
&- 2[(c_{ia}^{ru}+1)\varrho_a^u + 2(c_{ia}^{ru}+1)(\rho_{ia}^{ru}+\delta_a^u+d_a^u)](V_{ia}^{ru})^2 \\
&- 2(\rho_{ia}^{ru}+\delta_a^u)(U_{ia}^{ru})^2 - 2(\rho_{ia}^{ru}+\alpha_a^u+\delta_a^u)(W_{ia}^{rr})^2 \\
&+ 2\varrho_a^u V_{ia}^{ru} W_{ia}^{ru} - 2[(\varrho_a^u + d_a^u) + 2(\rho_{ia}^{ru}+\delta_a^u)] V_{ia}^{ru} U_{ia}^{ru}\Big\} \\
&+ 2\sum_{u\neq r}^{M}\sum_{a=1}^{n_u} c_{ia}^{ru} \sum_{v=1}^{M}\sum_{b=1}^{n_v} \beta_{aib}^{urv}(S_{ia}^{ru*} + U_{ia}^{ru}) V_{ba}^{vu} V_{ia}^{ru} \\
&+ \sum_{u\neq r}^{M}\sum_{a=1}^{n_r} c_{ia}^{ru} \sum_{v=1}^{M}\sum_{b=1}^{n_v} (v_{aib}^{urv})^2 (S_{ia}^{ru*} + U_{ia}^{ru})^2 (V_{ba}^{vu})^2, \quad u \neq r.
\end{aligned} \tag{7.65}
$$

Using (7.46) and the algebraic inequality

$$
2ab \leq \frac{a^2}{g(c)} + b^2 g(c), \tag{7.66}
$$

where $a, b, c \in \mathbb{R}$, and the function g is such that $g(c) > 0$, the sixth terms in (7.63)–(7.65) are estimated as follows:

$$
\begin{aligned}
2\sum_{v=1}^{M}\sum_{b=1}^{n_v} c_{ii}^{rr}\beta_{iib}^{rrv}(S_{ii}^{rr*} + U_{ii}^{rr}) V_{bi}^{vr} V_{ii}^{rr} \leq &\sum_{v=1}^{M}\sum_{b=1}^{n_v} c_{ii}^{rr}\beta_{iib}^{rrv}(S_{ii}^{rr*} g_i^r(\delta_i^r) + g_i^r(\delta_i^r))(V_{ii}^{rr})^2 \\
&+ \sum_{v=1}^{M}\sum_{b=1}^{n_v} c_{ii}^{rr}\beta_{iib}^{rrv}\left(\frac{S_{ii}^{rr*}}{g_i^r(\delta_i^r)} + \frac{\bar{B}^2}{g_i^r(\delta_i^r)}\right)(V_{bi}^{vr})^2, \\
2\sum_{a\neq r}^{n_r}\sum_{v=1}^{M}\sum_{b=1}^{n_v} c_{ia}^{rr}\beta_{aib}^{rrv}(S_{ia}^{rr*} + U_{ia}^{rr}) V_{ba}^{vr} V_{ia}^{rr} \leq &\sum_{a\neq r}^{n_r}\sum_{v=1}^{M}\sum_{b=1}^{n_v} c_{ia}^{rr}\beta_{aib}^{rrv}(S_{ia}^{rr*} g_i^r(\delta_a^r) + g_i^r(\delta_a^r))(V_{ia}^{rr})^2 \\
&+ \sum_{a\neq r}^{n_r}\sum_{v=1}^{M}\sum_{b=1}^{n_v} c_{ia}^{rr}\beta_{aib}^{rrv}\left(\frac{S_{ia}^{rr*}}{g_i^r(\delta_a^r)} + \frac{\bar{B}^2}{g_i^r(\delta_a^r)}\right)(V_{bi}^{vr})^2,
\end{aligned}
$$

and

$$
\begin{aligned}
&2\sum_{u\neq r}^{M}\sum_{a=1}^{n_u}\sum_{v=1}^{M}\sum_{b=1}^{n_v} c_{ia}^{ru}\beta_{aib}^{urv}(S_{ia}^{ru*}+U_{ia}^{ru})V_{ba}^{vu}V_{ia}^{ru}\\
&\quad\leq \sum_{u\neq r}^{M}\sum_{a=1}^{n_u}\sum_{v=1}^{M}\sum_{b=1}^{n_v} c_{ia}^{ru}\beta_{aib}^{urv}(S_{ia}^{ru*}g_i^r(\delta_a^u)+g_i^r(\delta_a^u))(V_{ia}^{ru})^2\\
&\qquad+\sum_{u\neq r}^{M}\sum_{a=1}^{n_u}\sum_{v=1}^{M}\sum_{b=1}^{n_v} c_{ia}^{ru}\beta_{aib}^{urv}\left(\frac{S_{ia}^{ru*}}{g_i^r(\delta_a^u)}+\frac{\bar{B}^2}{g_i^r(\delta_a^u)}\right)(V_{ba}^{vu})^2.
\end{aligned}\tag{7.67}
$$

By (7.63)–(7.67), (7.62), and repeated usage of (7.46) and inequality (7.66) coupled with some algebraic manipulations and simplifications, we have the inequality

$$
\begin{aligned}
LV_1(\tilde{x}_{00}^{00}) \leq \sum_{r=1}^{M}\sum_{i=1}^{n_r}\Bigg\{&\Bigg[2\sum_{u=1}^{M}\sum_{a=1}^{n_u}\mu_{ia}^{ru}+2\sum_{a\neq i}^{n_r}\frac{(\sigma_{ia}^{rr})^2}{\mu_{ii}^{rr}}+2\sum_{u\neq r}^{M}\sum_{a=1}^{n_r}\frac{(\gamma_{ia}^{ru})^2}{\mu_{ii}^{rr}}+3\mu_{ii}^{rr}+\mu_{ii}^{rr}e^{-\delta_i^r T_i^r}\\
&-2(\gamma_i^r+\sigma_i^r+\delta_i^r)\Bigg](U_{ii}^{rr})^2\\
&+\Bigg[(2+c_{ii}^{rr})\sum_{u=1}^{M}\sum_{a=1}^{n_u}\mu_{ia}^{ru}+\sum_{a\neq r}^{n_r}(2+c_{ia}^{rr})\frac{(\sigma_{ia}^{rr})^2}{\mu_{ii}^{rr}}+\sum_{u=1}^{M}\sum_{a=1}^{n_u}(2+c_{ia}^{ru})\frac{(\gamma_{ia}^{ru})^2}{\mu_{ii}^{rr}}+\mu_{ii}^{rr}e^{-\delta_i^r T_i^r}\\
&+\frac{(\varrho_i^r+d_i^r)^2}{\mu_{ii}^{rr}}+4\frac{(\gamma_i^r+\sigma_i^r+\delta_i^r)^2}{\mu_{ii}^{rr}}+\frac{(\varrho_i^r)^2}{\mu_{ii}^{rr}}+c_{ii}^{rr}\sum_{v=1}^{M}\sum_{b=1}^{n_r}\beta_{iib}^{rrv}(S_{ii}^{rr*}\mu_{ii}^{rr}+\mu_{ii}^{rr})\\
&-2(c_{ii}^{rr}+1)(\varrho_i^r+\gamma_i^r+\sigma_i^r+\delta_i^r+d_i^r)\Bigg](V_{ii}^{rr})^2\\
&+\Bigg[\sum_{u=1}^{M}\sum_{a=1}^{n_u}\mu_{ia}^{ru}+\sum_{a\neq i}^{n_r}\frac{(\sigma_{ia}^{rr})^2}{\mu_{ii}^{rr}}+\sum_{u\neq r}^{M}\sum_{a=1i}^{n_r}\frac{(\gamma_{ia}^{ru})^2}{\mu_{ii}^{rr}}+\mu_{ii}^{rr}e^{-\delta_i^r T_i^r}+\mu_{ii}^{rr}\\
&-2(\gamma_i^r+\sigma_i^r+\delta_i^r)\Bigg](W_{ii}^{rr})^2\Bigg\}\\
&+\sum_{r=1}^{M}\sum_{i=1}^{n_r}\sum_{a\neq i}^{n_r}\Bigg\{\Bigg[2\frac{(\rho_{ia}^{rr})^2}{\mu_{ia}^{rr}}+2\mu_{ii}^{rr}+2\mu_{ia}^{rr}+\mu_{ia}^{rr}e^{-\delta_a^r T_i^r}-2(\rho_{ia}^{rr}+\delta_a^r)\Bigg](U_{ia}^{rr})^2\\
&+\Bigg[(2+c_{ii}^{rr})\frac{(\rho_{ia}^{rr})^2}{\mu_{ia}^{rr}}+(2+c_{ia}^{rr})\mu_{ii}^{rr}+\frac{(\varrho_a^r+d_a^r)^2}{\mu_{ia}^{rr}}+4\frac{(\rho_{ia}^{rr}+\delta_a^r)^2}{\mu_{ia}^{rr}}\\
&+c_{ia}^{rr}\sum_{v=1}^{M}\sum_{b=1}^{n_r}\beta_{aib}^{rrv}(S_{ia}^{rr*}\mu_{ia}^{rr}+\mu_{ia}^{rr})-2(c_{ia}^{rr}+1)(\varrho_a^r+\rho_{ia}^{rr}+\delta_a^r+d_a^r)\Bigg](V_{ia}^{rr})^2\\
&+\Bigg[\frac{(\rho_{ia}^{rr})^2}{\mu_{ia}^{rr}}+\mu_{ii}^{rr}+\mu_{ia}^{rr}+\mu_{ia}^{rr}e^{-\delta_a^r T_i^r}-2(\rho_{ia}^{rr}+\delta_a^r)\Bigg](W_{ia}^{rr})^2\Bigg\}\\
&+\sum_{r=1}^{M}\sum_{i=1}^{n_r}\sum_{u\neq r}^{M}\sum_{a=1}^{n_u}\Bigg\{\Bigg[2\frac{(\rho_{ia}^{ru})^2}{\mu_{ia}^{rr}}+2\mu_{ii}^{rr}+2\mu_{ia}^{ru}+\mu_{ia}^{ru}e^{-\delta_a^u T_i^r}-2(\rho_{ia}^{ru}+\delta_a^u)\Bigg](U_{ia}^{ru})^2
\end{aligned}
$$

$$+\left[(2+c_{ii}^{rr})\frac{(\rho_{ia}^{ru})^2}{\mu_{ia}^{ru}}+(2+c_{ia}^{ru})\mu_{ii}^{rr}+\frac{(\varrho_a^u+d_a^u)^2}{\mu_{ia}^{ru}}+4\frac{(\rho_{ia}^{ru}+\delta_a^u)^2}{\mu_{ia}^{ru}}\right.$$
$$\left.+c_{ia}^{ru}\sum_{v=1}^{M}\sum_{b=1}^{n_r}\beta_{aib}^{urv}(S_{ia}^{ru*}\mu_{ia}^{ru}+\mu_{ia}^{ru})-2(c_{ia}^{ru}+1)(\varrho_a^u+\rho_{ia}^{ru}+\delta_a^u+d_a^u)\right](V_{ia}^{ru})^2$$
$$\left.+\left[\frac{(\rho_{ia}^{ru})^2}{\mu_{ia}^{ru}}+\mu_{ii}^{rr}+\mu_{ia}^{ru}+\mu_{ia}^{ru}e^{-\delta_a^u T_i^r}-2(\rho_{ia}^{ru}+\delta_a^u)\right](W_{ia}^{ru})^2\right\}$$
$$+3\sum_{r=1}^{M}\sum_{i=1}^{n_r}\sum_{u=1}^{M}\sum_{a=1}^{n_u}\left[\frac{(\varrho_a^u)^2}{\mu_{ia}^{ru}}e^{-\delta_a^u T_i^r}\right](V_{ia}^{ru}(t-T_i^r))^2$$
$$+\sum_{r=1}^{M}\sum_{i=1}^{n_r}c_{ii}^{rr}\sum_{v=1}^{M}\sum_{b=1}^{n_r}\left[\beta_{iib}^{rrv}\left(\frac{S_{ii}^{rr*}}{\mu_{ii}^{rr}}+\frac{\bar{B}^2}{\mu_{ii}^{rr}}\right)+(\nu_{iib}^{rrv})^2(S_{ii}^{rr*}+\bar{B})^2\right](V_{bi}^{vr})^2$$
$$+\sum_{r=1}^{M}\sum_{i=1}^{n_r}\sum_{a\neq i}^{n_r}c_{ia}^{rr}\left[\sum_{v=1}^{M}\sum_{b=1}^{n_r}\beta_{aib}^{rrv}\left(\frac{S_{ia}^{rr*}}{\mu_{ia}^{rr}}+\frac{\bar{B}^2}{\mu_{ia}^{rr}}\right)+(\nu_{aib}^{rrv})^2(S_{ia}^{rr*}+\bar{B})^2\right](V_{ba}^{vr})^2$$
$$+\sum_{r=1}^{M}\sum_{i=1}^{n_r}\sum_{u\neq r}^{M}\sum_{a=1}^{n_r}c_{ia}^{ru}\left[\sum_{v=1}^{M}\sum_{b=1}^{n_r}\beta_{aib}^{urv}\left(\frac{S_{ia}^{ru*}}{\mu_{ia}^{ru}}+\frac{\bar{B}^2}{\mu_{ia}^{ru}}\right)+(\nu_{aib}^{urv})^2(S_{ia}^{ru*}+\bar{B})^2\right](V_{ba}^{vu})^2, \quad (7.68)$$

where $\mu_{ia}^{ru}=g_i^r(\delta_a^u)$, and g_i^r is appropriately defined by (7.66). For all $r,u\in I(1,M)$, $i\in I(1,n_r)$, and $a\in I(1,n_u)$, we define the constants d_{ai}^{ur}, ϕ_{ia}^{ru}, ψ_{ia}^{ru}, and φ_{ia}^{ru} as follows:

$$d_{ai}^{ur}=\sum_{v=1}^{M}\sum_{b=1}^{n_v}c_{ba}^{vu}\beta_{abi}^{uvr}\left(\frac{S_{ba}^{vu*}+\bar{B}^2}{\mu_{ba}^{vu}}\right)+\sum_{v=1}^{M}\sum_{b=1}^{n_v}c_{ba}^{vu}(\nu_{abi}^{uvr})^2(S_{ba}^{vu*}+\bar{B})^2 \quad (7.69)$$

for some positive numbers c_{ia}^{ru}, $r,u\in I^r(1,M)$, $i\in I(1,n)$, $a\in I_i^r(1,n_r)$,

$$\phi_{ia}^{ru}=\begin{cases}2(\gamma_i^r+\sigma_i^r+\delta_i^r)(1-\mathfrak{U}_{ia}^{ru}), & u=r,a=i,\\ 2(\rho_{ia}^{rr}+\delta_a^r)(1-\mathfrak{U}_{ia}^{ru}), & u=r,\ a\neq i,\\ 2(\rho_{ia}^{ru}+\delta_a^u)(1-\mathfrak{U}_{ia}^{ru}), & u\neq r,\end{cases} \quad (7.70)$$

$$\psi_{ia}^{ru}=\begin{cases}2(\varrho_i^r+\gamma_i^r+\sigma_i^r+\delta_i^r+d_i^r)[c_{ii}^{rr}(1-\mathfrak{V}_{ii}^{rr})+(1-\frac{1}{2}\mathfrak{C}_{ii}^{rr})], & u=r,\ a=i,\\ 2(\varrho_a^r+\rho_{ia}^{rr}+\delta_a^r+d_a^r)[c_{ia}^{rr}(1-\mathfrak{V}_{ia}^{rr})+(1-\frac{1}{2}\mathfrak{C}_{ia}^{rr})], & u=r,\ a\neq i,\\ 2(\varrho_a^u+\rho_{ia}^{ru}+\delta_a^u+d_a^u)[c_{ia}^{ru}(1-\mathfrak{V}_{ia}^{ru})+(1-\frac{1}{2}\mathfrak{C}_{ia}^{ru})], & u\neq r,\end{cases} \quad (7.71)$$

and

$$\varphi_{ia}^{ru}=\begin{cases}2(\gamma_i^r+\sigma_i^r+\delta_i^r)(1-\mathfrak{W}_{ia}^{ru}), & u=r,\ a=i,\\ 2(\rho_{ia}^{rr}+\delta_a^r)(1-\mathfrak{W}_{ia}^{ru}), & u=r,\ a\neq i,\\ 2(\rho_{ia}^{ru}+\delta_a^u)(1-\mathfrak{W}_{ia}^{ru}), & u\neq r,\end{cases} \quad (7.72)$$

where $\mathfrak{V}_{ia}^{ru}$ is given in (7.58),

$$\mathfrak{U}_{ia}^{ru} = \begin{cases} \dfrac{[\sum_{u=1}^{M}\sum_{a=1}^{n_u}\mu_{ia}^{ru} + \sum_{a\neq i}^{n_r}\frac{(\sigma_{ia}^{rr})^2}{\mu_{ii}^{rr}} + \sum_{a\neq r}^{M}\sum_{a=1}^{n_r}\frac{(\gamma_{ia}^{ru})^2}{\mu_{ii}^{rr}} + \frac{3}{2}\mu_{ii}^{rr} + \frac{1}{2}\mu_{ii}^{rr}e^{-\delta_i^r T_i^r}]}{(\gamma_i^r + \sigma_i^r + \delta_i^r)}, & u = r,\ i = a, \\ \dfrac{[\frac{(\rho_{ia}^{rr})^2}{\mu_{ia}^{rr}} + \mu_{ii}^{rr} + \mu_{ia}^{rr} + \frac{1}{2}\mu_{ia}^{rr}e^{-\delta_a^r T_i^r}]}{(\rho_{ia}^{rr} + \delta_a^r)}, & u = r,\ a \neq i, \\ \dfrac{[\frac{(\rho_{ia}^{ru})^2}{\mu_{ia}^{ru}} + \mu_{ii}^{rr} + \mu_{ia}^{ru} + \frac{1}{2}\mu_{ia}^{ru}e^{-\delta_a^u T_i^r}]}{(\rho_{ia}^{ru} + \delta_a^u)}, & u \neq r, \end{cases} \tag{7.73}$$

$$\mathfrak{W}_{ia}^{ru} = \begin{cases} \dfrac{[\frac{1}{2}\sum_{u=1}^{M}\sum_{a=1}^{n_u}\mu_{ia}^{ru} + \frac{1}{2}\sum_{u\neq r}^{M}\sum_{a=1}^{n_r}\frac{(\gamma_{ia}^{ru})^2}{\mu_{ii}^{rr}} + \frac{1}{2}\sum_{a\neq i}^{n_r}\frac{(\sigma_{ia}^{rr})^2}{\mu_{ii}^{rr}} + \frac{1}{2}\mu_{ii}^{rr} + \frac{1}{2}\mu_{ii}^{rr}e^{-\delta_i^r T_i^r}]}{(\gamma_i^r + \sigma_i^r + \delta_i^r)}, & u = r,\ a = i, \\ \dfrac{[\frac{1}{2}\frac{(\rho_{ia}^{rr})^2}{\mu_{ia}^{rr}} + \frac{1}{2}\mu_{ii}^{rr} + \frac{1}{2}\mu_{ia}^{rr} + \frac{1}{2}\mu_{ia}^{rr}e^{-\delta_a^r T_i^r}]}{(\rho_{ia}^{rr} + \delta_a^r)}, & u = r,\ a \neq i, \\ \dfrac{[\frac{1}{2}\frac{(\rho_{ia}^{ru})^2}{\mu_{ia}^{ru}} + \frac{1}{2}\mu_{ii}^{rr} + \frac{1}{2}\mu_{ia}^{ru} + \frac{1}{2}\mu_{ia}^{ru}e^{-\delta_a^u T_i^r}]}{(\rho_{ia}^{ru} + \delta_a^u)}, & u \neq r, \end{cases} \tag{7.74}$$

and

$$\mathfrak{E}_{ia}^{ru} = \begin{cases} \Bigg[\dfrac{2\sum_{u=1}^{M}\sum_{a=1}^{n_u}\mu_{ia}^{ru} + \sum_{a\neq r}^{n_r}(2+c_{ia}^{rr})\frac{(\sigma_{ia}^{rr})^2}{\mu_{ii}^{rr}} + \sum_{u=1}^{M}\sum_{a=1}^{n_u}(2+c_{ia}^{ru})\frac{(\gamma_{ia}^{ru})^2}{\mu_{ii}^{rr}} + \mu_{ii}^{rr}e^{-\delta_i^r T_i^r}}{(\varrho_i^r + \gamma_i^r + \sigma_i^r + \delta_i^r + d_i^r)} \\ \quad + \dfrac{\frac{(\varrho_i^r + d_i^r)^2}{\mu_{ii}^{rr}} + 4\frac{(\gamma_i^r + \sigma_i^r + \delta_i^r)^2}{\mu_{ii}^{rr}} + \frac{(\varrho_i^r)^2}{\mu_{ii}^{rr}} + 3\frac{(\varrho_i^r)^2}{\mu_{ii}^{rr}}e^{-\delta_i^r T_i^r}}{(\varrho_i^r + \gamma_i^r + \sigma_i^r + \delta_i^r + d_i^r)}\Bigg], & u = r, a = i, \\ \Bigg[\dfrac{(2+c_{ii}^{rr})\frac{(\rho_{ia}^{rr})^2}{\mu_{ia}^{rr}} + 2\mu_{ii}^{rr} + \frac{(\varrho_a^r + d_a^r)^2}{\mu_{ia}^{rr}} + 4\frac{(\rho_{ia}^{rr} + \delta_a^r)^2}{\mu_{ia}^{rr}} + 3\frac{(\varrho_a^r)^2}{\mu_{ia}^{rr}}e^{-\delta_a^r T_i^r}}{(\varrho_a^r + \rho_{ia}^{rr} + \delta_a^r + d_a^r)}\Bigg], & u = r, a \neq i, \\ \Bigg[\dfrac{(2+c_{ii}^{rr})\frac{(\rho_{ia}^{ru})^2}{\mu_{ia}^{ru}} + 2\mu_{ii}^{rr} + \frac{(\varrho_a^u + d_a^u)^2}{\mu_{ia}^{ru}} + 4\frac{(\rho_{ia}^{ru} + \delta_a^u)^2}{\mu_{ia}^{ru}} + 3\frac{(\varrho_a^u)^2}{\mu_{ia}^{ru}}e^{-\delta_a^u T_i^r}}{(\varrho_a^u + \rho_{ia}^{ru} + \delta_a^u + d_a^u)}\Bigg], & u \neq r. \end{cases}$$

By (7.56), (7.57), (7.68), and (7.69), the differential operator LV [53, 24] applied to the Lyapunov functional (7.56), and some further algebraic manipulations we have the inequality

$$\begin{aligned} LV(\tilde{x}_{00}^{00}) \leq \sum_{r=1}^{M}\sum_{i=1}^{n_r} -\Big\{ &[\phi_{ii}^{rr}(U_{ii}^{rr})^2 + \psi_{ii}^{rr}(V_{ii}^{rr})^2 + \varphi_{ii}^{rr}(W_{ii}^{rr})^2] \\ &+ \sum_{a\neq r}^{n_r}[\phi_{ia}^{rr}(U_{ia}^{rr})^2 + \psi_{ia}^{rr}(V_{ia}^{rr})^2 + \varphi_{ia}^{rr}(W_{ia}^{rr})^2] \\ &+ \sum_{u\neq r}^{M}\sum_{a=1}^{n_u}[\phi_{ia}^{ru}(U_{ia}^{ru})^2 + \psi_{ia}^{ru}(V_{ia}^{ru})^2 + \varphi_{ia}^{ru}(W_{ia}^{ru})^2]\Big\}. \end{aligned} \tag{7.75}$$

Under the assumption on T_i^r, it follows that $\mathfrak{U}_{ia}^{ru} \leq 1$ and $\mathfrak{W}_{ia}^{ru} \leq 1$ for all $u \in I(1, M)$, $a \in I(1, n_r)$. Moreover, under the assumption on $\mathfrak{V}_{ia}^{ru}$, it is clear that ϕ_{ia}^{ru}, ψ_{ia}^{ru}, and φ_{ia}^{ru} are positive for suitable choices of the constants $c_{ia}^{ru} > 0$. Thus this proves inequality (7.59). Now (7.60) follows from (7.59) and (7.54), that is,

$$LV(\tilde{x}_{00}^{00}) \leq -cV_1(\tilde{x}_{00}^{00}),$$

where $c = \frac{\min_{1\leq r,u\leq M, 1\leq i\leq n_r, 1\leq a\leq n_u}\{\phi_{ia}^{ru}, \psi_{ia}^{ru}, \varphi_{ia}^{ru}\}}{\max_{1\leq r,u\leq M, 1\leq i\leq n_r, 1\leq a\leq n_u}\{C_{ia}^{ru}+2\}}$. This completes the proof. □

We now formally state the stochastic stability theorems for the disease-free equilibria.

Theorem 7.4.3. *Let $r, u \in I(1, M)$, $i \in I(1, n_r)$, and $a \in I(1, n_u)$. Let us assume that the hypotheses of Lemma 7.4.2 are satisfied. Then the disease-free solutions E_{ia}^{ru} are asymptotically stable in the large. Moreover, the solutions E_{ia}^{ru} are exponentially mean square stable.*

Proof. By an application of comparison result [53, 24] the proof of stochastic asymptotic stability follows immediately. Moreover, the disease-free equilibrium state is exponentially mean square stable. □

We now consider the following corollary to Theorem 7.4.3.

Corollary 7.4.4. *Let $r \in I(1, M)$ and $i \in I(1, n_r)$. Assume that $\sigma_i^r = \gamma_i^r = 0$ for all $r \in I(1, M)$ and $i \in I(1, n_r)$. Suppose that*

$$\begin{cases} \delta_i^r > \max((\sum_{u=1}^{M} \sum_{a=1}^{n_u} \mu_{ia}^{ru} + \frac{3}{2}\mu_{ii}^{rr}), (\frac{1}{2} \sum_{u=1}^{M} \sum_{a=1}^{n_u} \mu_{ia}^{ru} + \frac{1}{2}\mu_{ii}^{rr})), & u = r,\ a = i, \\ (\rho_{ia}^{rr} + \delta_a^r) > \max((\frac{(\rho_{ia}^{rr})^2}{\mu_{ia}^{rr}} + \mu_{ii}^{rr} + \mu_{ia}^{rr}), (\frac{1}{2} \frac{(\rho_{ia}^{rr})^2}{\mu_{ia}^{rr}} + \frac{1}{2}\mu_{ii}^{rr} + \frac{1}{2}\mu_{ia}^{rr})), & u = r, a \neq i, \\ (\rho_{ia}^{ru} + \delta_a^u) > \max((\frac{(\rho_{ia}^{ru})^2}{\mu_{ia}^{ru}} + \mu_{ii}^{rr} + \mu_{ia}^{ru}), (\frac{1}{2} \frac{(\rho_{ia}^{ru})^2}{\mu_{ia}^{ru}} + \frac{1}{2}\mu_{ii}^{rr} + \frac{1}{2}\mu_{ia}^{ru})), & u \neq r. \end{cases}$$

Furthermore, let

$$\mathfrak{M}_{ia}^{ru} = \begin{cases} \max\Big(\frac{1}{\delta_i^r} \log\Big(\frac{\mu_{ii}^{rr}}{2(\delta_i^r - (\sum_{u=1}^{M} \sum_{a=1}^{n_u} \mu_{ia}^{ru} + \frac{3}{2}\mu_{ii}^{rr}))}\Big), \frac{1}{\delta_i^r} \log\Big(\frac{\mu_{ii}^{rr}}{2(\delta_i^r - (\frac{1}{2} \sum_{u=1}^{M} \sum_{a=1}^{n_u} \mu_{ia}^{ru} + \frac{1}{2}\mu_{ii}^{rr}))}\Big)\Big), \\ \quad u = r,\ i = a, \\ \max\Big(\frac{1}{\delta_a^r} \log\Big(\frac{\mu_{ia}^{rr}}{2((\rho_{ia}^{rr} + \delta_a^r) - [\frac{(\rho_{ia}^{rr})^2}{\mu_{ia}^{rr}} + \mu_{ii}^{rr} + \mu_{ia}^{rr}])}\Big), \frac{1}{\delta_a^r} \log\Big(\frac{\mu_{ia}^{rr}}{2((\rho_{ia}^{rr} + \delta_a^r) - [\frac{1}{2} \frac{(\rho_{ia}^{rr})^2}{\mu_{ia}^{rr}} + \frac{1}{2}\mu_{ii}^{rr} + \frac{1}{2}\mu_{ia}^{rr}])}\Big)\Big), \\ \quad u = r,\ a \neq i, \\ \max\Big(\frac{1}{\delta_a^u} \log\Big(\frac{\mu_{ia}^{ru}}{2((\rho_{ia}^{ru} + \delta_a^u) - [\frac{(\rho_{ia}^{ru})^2}{\mu_{ia}^{ru}} + \mu_{ii}^{rr} + \mu_{ia}^{ru}])}\Big), \frac{1}{\delta_a^u} \log\Big(\frac{\mu_{ia}^{ru}}{2((\rho_{ia}^{ru} + \delta_a^u) - [\frac{1}{2} \frac{(\rho_{ia}^{ru})^2}{\mu_{ia}^{ru}} + \frac{1}{2}\mu_{ii}^{rr} + \frac{1}{2}\mu_{ia}^{ru}])}\Big)\Big), \\ \quad u \neq r, \end{cases} \tag{7.76}$$

and

$$\mathfrak{V}_{ia}^{ru} = \begin{cases} \frac{\frac{1}{2} \sum_{u=1}^{M} \sum_{a=1}^{n_u} \mu_{ia}^{ru} + \frac{1}{2} \sum_{v=1}^{M} \sum_{b=1}^{n_v} \beta_{iib}^{rrv} (S_{ii}^{rr*} \mu_{ii}^{rr} + \mu_{ii}^{rr})}{\varrho_i^r + \gamma_i^r + \sigma_i^r + \delta_i^r + d_i^r}, & a = i,\ u = r, \\ \frac{\frac{1}{2}\mu_{ii}^{rr} + \frac{1}{2} \sum_{v=1}^{M} \sum_{b=1}^{n_v} \beta_{aib}^{rrv} (S_{ia}^{rr*} \mu_{ia}^{rr} + \mu_{ia}^{rr})}{\varrho_a^r + \rho_{ia}^{rr} + \delta_a^r + d_a^r}, & a \neq i,\ u = r, \\ \frac{\frac{1}{2}\mu_{ii}^{rr} + \frac{1}{2} \sum_{v=1}^{M} \sum_{b=1}^{n_v} \beta_{aib}^{urv} (S_{ii}^{ru*} \mu_{ia}^{ru} + \mu_{ia}^{ru})}{\varrho_a^u + \rho_{ia}^{ru} + \delta_a^u + d_a^u}, & u \neq r, \end{cases} \tag{7.77}$$

for some suitably defined positive number μ_{ia}^{ru}, depending on δ_a^u, for all $r, u \in I^r(1, M)$, $i \in I(1, n)$, and $a \in I_i^r(1, n_r)$. The equilibrium state E_{ii}^{rr} is stochastically asymptotically stable, provided that $\mathfrak{V}_{ia}^{ru} < 1$ and $T_i^r \geq \max_{1 \leq r, u \leq M, 1 \leq i \leq n_r, 1 \leq a \leq n_u} (\mathfrak{M}_{ia}^{ru})$.

Proof. Follows immediately from the hypotheses of Lemma 7.4.2 (letting $\sigma_i^r = \gamma_i^r = 0$), Theorem 7.4.3, and some algebraic manipulations. □

7.4.3 Discussion of the results for the disease eradication in the delay two-scale network SIRS epidemic model

The presented results for the two-scale network delay SIRS SDE model depend on the underlying system parameters. In particular, the sufficient conditions are algebraically simple, computationally attractive, and explicit in terms of the rate parameters. As a result, several scenarios can be discussed and exhibit practical course of action to control the disease. For simplicity, we present an illustration as follows: the conditions $\sigma_i^r = \gamma_i^r = 0$ for all r, i in Corollary 7.4.4 signify that the arbitrary site s_i^r is a "sink" in the context of compartmental systems (see [21, 20]) for all other sites in the inter- and intraregional accessible domains. This scenario is displayed in Figure 7.1. The condition $T_i^r \geq \max_{1\leq r,u\leq M, 1\leq i\leq n_r, 1\leq a\leq n_u}(\mathfrak{M}_{ia}^{ru})$ is a threshold condition for the immunity delay period of residents of site s_i^r in the region C_r, controlling the stochastic asymptotic stability of the disease-free equilibrium. Furthermore, the condition $\mathfrak{V}_{ia}^{ru} < 1$ signifies that the magnitude of disease inhibitory processes, for example, the magnitude of the recovery process, is greater than the disease transmission process. Indeed, observe from (7.58) that $\mathfrak{V}_{ia}^{ru}$ is expressed in terms of the *basic reproduction number* (*BRN*) $\mathfrak{R}_{ia_0}^{ru}$ for the disease as follows:

$$\mathfrak{V}_{ia}^{ru} = \begin{cases} \dfrac{\frac{1}{2}\sum_{u=1}^{M}\sum_{a=1}^{n_u}\mu_{ia}^{ru} + \frac{1}{2}\sum_{v=1}^{M}\sum_{b=1}^{n_v}\beta_{iib}^{rrv}(S_{ii}^{rr*}\mu_{ii}^{rr} + \mu_{ii}^{rr}) + \frac{1}{2}d_{ii}^{rr}}{\varrho_i^r + \gamma_i^r + \sigma_i^r + \delta_i^r + d_i^r}, & a = i,\ u = r, \\ \dfrac{\frac{1}{2}\mu_{ii}^{rr} + \frac{1}{2}\sum_{v=1}^{M}\sum_{b=1}^{n_v}\beta_{aib}^{rrv}(S_{ia}^{rr*}\mu_{ia}^{rr} + \mu_{ia}^{rr}) + \frac{1}{2}d_{ai}^{rr}}{\varrho_a^r + \rho_{ia}^{rr} + \delta_a^r + d_a^r}, & a \neq i,\ u = r, \\ \dfrac{\frac{1}{2}\mu_{ii}^{rr} + \frac{1}{2}\sum_{v=1}^{M}\sum_{b=1}^{n_v}\beta_{aib}^{urv}(S_{ii}^{ru*}\mu_{ia}^{ru} + \mu_{ia}^{ru}) + \frac{1}{2}d_{ai}^{ur}}{\varrho_a^u + \rho_{ia}^{ru} + \delta_a^u + d_a^u}, & u \neq r, \end{cases} \tag{7.78}$$

expressed further as

$$\mathfrak{V}_{ia}^{ru} = \begin{cases} \dfrac{\frac{1}{2}\sum_{u=1}^{M}\sum_{a=1}^{n_u}\mu_{ia}^{ru} + \frac{1}{2}\sum_{v=1}^{M}\sum_{b=1}^{n_v}\beta_{iib}^{rrv}(\mu_{ii}^{rr}) + \frac{1}{2}d_{ii}^{rr}}{\varrho_i^r + \gamma_i^r + \sigma_i^r + \delta_i^r + d_i^r} + \frac{1}{2}\mathfrak{R}_{ii_0}^{rr}, & a = i,\ u = r, \\ \dfrac{\frac{1}{2}\mu_{ii}^{rr} + \frac{1}{2}\sum_{v=1}^{M}\sum_{b=1}^{n_v}\beta_{aib}^{rrv}(\mu_{ia}^{rr}) + \frac{1}{2}d_{ai}^{rr}}{\varrho_a^r + \rho_{ia}^{rr} + \delta_a^r + d_a^r} + \frac{1}{2}\mathfrak{R}_{ia_0}^{rr}, & a \neq i,\ u = r, \\ \dfrac{\frac{1}{2}\mu_{ii}^{rr} + \frac{1}{2}\sum_{v=1}^{M}\sum_{b=1}^{n_v}\beta_{aib}^{urv}(\mu_{ia}^{ru}) + \frac{1}{2}d_{ai}^{ur}}{\varrho_a^u + \rho_{ia}^{ru} + \delta_a^u + d_a^u} + \frac{1}{2}\mathfrak{R}_{ia_0}^{ru}, & u \neq r, \end{cases} \tag{7.79}$$

where

$$\mathfrak{R}_{ia_0}^{ru} = \begin{cases} \dfrac{\sum_{v=1}^{M}\sum_{b=1}^{n_v}\beta_{iib}^{rrv}(S_{ii}^{rr*}\mu_{ii}^{rr})}{\varrho_i^r + \gamma_i^r + \sigma_i^r + \delta_i^r + d_i^r}, & a = i,\ u = r, \\ \dfrac{\sum_{v=1}^{M}\sum_{b=1}^{n_v}\beta_{aib}^{rrv}(S_{ia}^{rr*}\mu_{ia}^{rr})}{\varrho_a^r + \rho_{ia}^{rr} + \delta_a^r + d_a^r}, & a \neq i,\ u = r, \\ \dfrac{\sum_{v=1}^{M}\sum_{b=1}^{n_v}\beta_{aib}^{urv}(S_{ii}^{ru*}\mu_{ia}^{ru})}{\varrho_a^u + \rho_{ia}^{ru} + \delta_a^u + d_a^u}, & u \neq r, \end{cases} \tag{7.80}$$

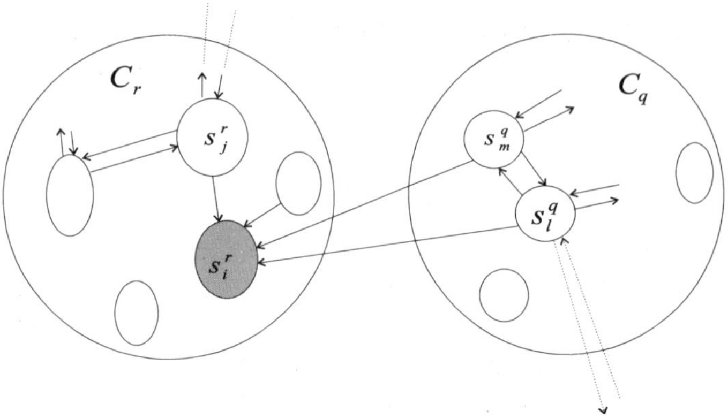

Figure 7.1: Shows that residents of site s_i^r are present only at their home site s_i^r. Hence they isolate every site from their inter- and intraregional accessible domain $C(s_i^r)$. Site s_i^r is a "sink" in the context of the compartmental system [21, 20]. The arrows represent a transport network between any two sites and regions, and the dotted lines and arrows indicate connection with other sites and regions.

that is,

$$\mathfrak{R}_{ia_0}^{ru} \propto \begin{cases} \frac{\sum_{v=1}^{M}\sum_{b=1}^{n_v}\beta_{iib}^{rrv}(S_{ii}^{rr*})}{\varrho_i^r+\gamma_i^r+\sigma_i^r+\delta_i^r+d_i^r}, & a=i,\ u=r, \\ \frac{\sum_{v=1}^{M}\sum_{b=1}^{n_v}\beta_{aib}^{rrv}(S_{ia}^{rr*})}{\varrho_a^r+\rho_{ia}^{rr}+\delta_a^r+d_a^r}, & a\neq i,\ u=r, \\ \frac{\sum_{v=1}^{M}\sum_{b=1}^{n_v}\beta_{aib}^{urv}(S_{ii}^{ru*})}{\varrho_a^u+\rho_{ia}^{ru}+\delta_a^u+d_a^u}, & u\neq r. \end{cases} \tag{7.81}$$

Note that the terms on the right-hand-side of (7.81) are the average number of new infections in the DFE $\beta_{aib}^{urv}(S_{ii}^{ru*})$ over the average lifespan $\frac{1}{\varrho_a^u+\rho_{ia}^{ru}+\delta_a^u+d_a^u}$ at the site s_a^u, $u, r \in I(1, M)$, $i \in I(1, n_r)$, $a \in I(1, n_u)$. Thus the condition for disease eradication $\mathfrak{V}_{ia}^{ru} < 1$ suggests that the BRNs $\mathfrak{R}_{ia_0}^{ru} < 1$ at all sites s_a^u interconnected with the site s_i^r, $u, r \in I(1, M)$, $i \in I(1, n_r)$, $a \in I(1, n_u)$.

In other words, the disease is eradicated in the population if by some biomedical means the acquired immunity span is enhanced to last longer than $T_i^r \geq \max_{1\leq r,u\leq M, 1\leq i\leq n_r, 1\leq a\leq n_u}(\mathfrak{M}_{ia}^{ru})$ for all residents of site s_i^r in the region C_r and if in addition, there are additional controls suggested by the condition for the BRNS, $\mathfrak{R}_{ia_0}^{ru} < 1$ at all sites: at home s_i^r, at intraregional sites $s_j^r, j \neq i$, and at interregional sites s_a^u, $u \neq r$, to fight the disease.

7.5 Simulation results for a two-scale network delay SIRS influenza epidemic dynamic model

In this section, we extend the stochastic influenza epidemic model introduced in Section 6.8 by incorporating a delay period representing the temporary natural immunity of the recovered population. Influenza is characterized by short-lived, strain-dependent immunity [27]. The hierarchical population structure, human mobility process, influenza transmission dynamics, and the birth and death processes from Section 6.8 are retained in this extension. Additionally, the parameter specifications defined in Section 6.8 remain applicable. Below we describe the recovery process and the acquisition of temporary immunity. For a detailed discussion of the influenza scenario, we refer the reader to Section 6.8.

We assume that residents of site s_a^u in the region C_u recover from the disease and subsequently acquire temporary natural immunity to the specific influenza strain. In the absence of empirical data for the recovery and disease-related mortality processes, we set the recovery rate and disease mortality rate as $\varrho_a^u = 0.05067$ and $d_a^u = 0.01838$, respectively, for $u, a, i = 1, 2, 3$.

Furthermore, we assume that the average natural immunity period is identical for recovered individuals across all sites in the two-scale population structure. Specifically, for all residents of site s_i^r in the region C_r, irrespective of their location within the intra- or interregional accessible domain, the natural immunity period is set as $T_i^r = 1$ for all $r, i = 1, 2, 3$.

7.5.1 Mathematical algorithm and simulation results

We apply the standard Euler–Maruyama stochastic approximation scheme [19] to generate the trajectories for the residents of sites s_1^1, s_1^2, and s_1^3 in the regions C_1, C_2, and C_3, respectively, for the different population disease classifications (S, I, R) and current locations at some sites in the intra- and interregional accessible domain of the sites. We summarize the Euler–Maruyama method steps to obtain strong solution approximations to a system of stochastic delay differential equations as follows. Given a scalar autonomous stochastic delay differential equation

$$\begin{aligned} dX(t) &= f(X(t), X(t-\tau))dt + g(X(t), X(t-\tau))dW(t), \quad T_0 \le t \le T, \\ X(t) &= \varphi_0(t), \quad t \in [T_0 - \tau, T_0], \end{aligned} \tag{7.82}$$

where $\varphi_0(t)$ is a measurable random variable on $C([T_0 - \tau, T_0], \mathbb{R})$ and $T_0, T \ge 0$. Let $T_0 = t_0 \le t_1 \le t_2 \le \cdots \le t_n = T$ be a regular partition of $[T_0, T]$, where $h = t_j - t_{j-1} = \frac{T-T_0}{N}$, $t_j = T_0 + jh, j = 0, \ldots, N$, and N is a positive integer. Furthermore, suppose there exists a positive integer N_τ such that the delay parameter $\tau = N_\tau h$. The Euler–Maruyama method takes the form

$$\begin{aligned} &X(t_{j-N_\tau}) = \varphi_0(t_j - \tau), \quad j - N_\tau \le 0, \\ &X(t_{j+1}) = X(t_j) + hf(X(t_j), X(t_{j-N_\tau})) + g(X(t_j), X(t_{j-N_\tau}))\Delta W(t_{j+1}), \quad j = 0, 1, \ldots, N-1, \end{aligned} \tag{7.83}$$

where $\Delta W(t_{j+1}) = (W(t_{j+1}) - W(t_j)), j = 0, \ldots, N-1$, are an independent Gaussian $N(0, h)$ random variables. Using (7.82) as a general form of the equations in system (7.16)–(7.18) in the context of this example, the algorithm to execute the Euler–Maruyama method to find the solution process of (7.82) consists of the following steps.

Step 1: **Parameter specification.** The system rate parameters for the epidemic model (7.16)–(7.18) represented in this example are specified in Section 6.8.

Step 2: **Initial conditions:** The initial solutions are approximated using $X(t_{j-N_\tau}) = \varphi_0(t_j - \tau), j = 0, \dots N_\tau, X \in \{S_{ia}^{ru}, I_{ia}^{ru}, R_{ia}^{ru}\}$. In this example, we set $\tau = T_i^r = 1$ for all $r, i = 1, 2, 3$, $T_0 = 0$, and $T = 1$. This implies from the definition of τ and N_τ in (7.83) that $N_\tau = N$. Furthermore, from (7.19) and (7.82) we use the following convenient initial conditions for the simulation process: for $r, u, i, a \in I(1,3)$, $\varphi_{iak}^{ru} \in C([-1,0], \mathbb{R}_+)$, $k = 1, 2, 3$,

$$S_{ia}^{ru}(t) = \varphi_{ia1}^{ru}(t) = \begin{cases} 9, & r = u,\ i = a, \\ 8, & r = u,\ i \neq a, \\ 7, & r \neq u, \end{cases} \quad t \in [-1, 0],$$

$$I_{ia}^{ru}(t) = \varphi_{ia2}^{ru}(t) = \begin{cases} 6, & r = u,\ i = a, \\ 4, & r = u,\ i \neq a, \\ 3, & r \neq u, \end{cases} \quad t \in [-1, 0],$$

and $R_{ia}^{ru}(t) = \varphi_{ia3}^{ru}(t) = 2$, $t \in [-1, 0]$, $r, u, i, a \in I(1, 3)$.

Step 3: **Generate Brownian path:** The standard Brownian motion or normalized Wiener process $W(t)$ is generated over the time interval $[T_0, T]$. Let $T_0 = \tau_0 \leq \tau_1 \leq \tau_2 \leq \cdots \leq \tau_n = T$ be a regular partition of $[T_0, T]$, where $\delta\tau = \tau_{j+1} - \tau_j = \frac{T - T_0}{N}$, $\tau_j = T_0 + j\delta\tau, j = 0, \dots, N$, and N is a positive integer. The Brownian motion is generated as the solution to the difference equation

$$\begin{cases} W(0) = 0, \\ W(\tau_{j+1}) = W(\tau_j) + dW(\tau_{j+1}), \quad j = 0, \dots, N-1, \end{cases} \tag{7.84}$$

where $dW(\tau_{j+1}), j = 0, \dots, N-1$, are the independent $\sqrt{\delta t} N(0,1)$ Gaussian random variables. Furthermore, for this example with $T_0 = 0$ and $T = 1$, we simulated 500 sample points for the Brownian motion over the interval $[0, 1]$.

Step 4: **Generate solution Path for the susceptible, infectious, and removal populations (states).** Using (7.82) as a general representation of each equation in system (7.16)–(7.18), we use discretization (7.83) to find solution paths for each equation in the system. For convenience, we choose $h = R\delta\tau$ [18], where $R \geq 1$, and $\delta\tau$ is defined in Step 3. Moreover, from (7.83) it follows that

$$\Delta W(t_{j+1}) = W(t_{j+1}) - W(t_j) = W(T_0 + (j+1)R\delta\tau) - W(T_0 + jR\delta\tau) = \sum_{k=jR}^{jR+R-1} dW_{k+1}, \tag{7.85}$$

where dW_{k+1} is given by the Brownian path (7.84). We choose $R = 1$ for this example. Moreover, from (7.83), (7.84), and (7.85), we obtain trajectories for susceptible, infectious, and removal populations of residents of sites s_1^1, s_1^2, and s_1^3 in the regions C_1, C_2, and C_3 over the time interval $[0, 1]$. The trajectories for the residents of sites s_1^1, s_1^2, and s_1^3 in the regions C_1, C_2, and C_3 are exhibited in Figures 7.2, 7.3, and 7.4, respectively.

Remark 7.5.1. From Figure 7.2, Figures (a), (b), and (c) represent the trajectories of the different disease classes of residents of site s_1^1 at home. Figures (d), (e), and (f) represent the trajectories of the different disease classes of residents of site s_1^1 visiting site s_2^1 in the home region C_1. These two groups of figures are representative of the disease dynamics of influenza affecting the residents of site s_1^1 at the intraregional level. Figures (g), (h),

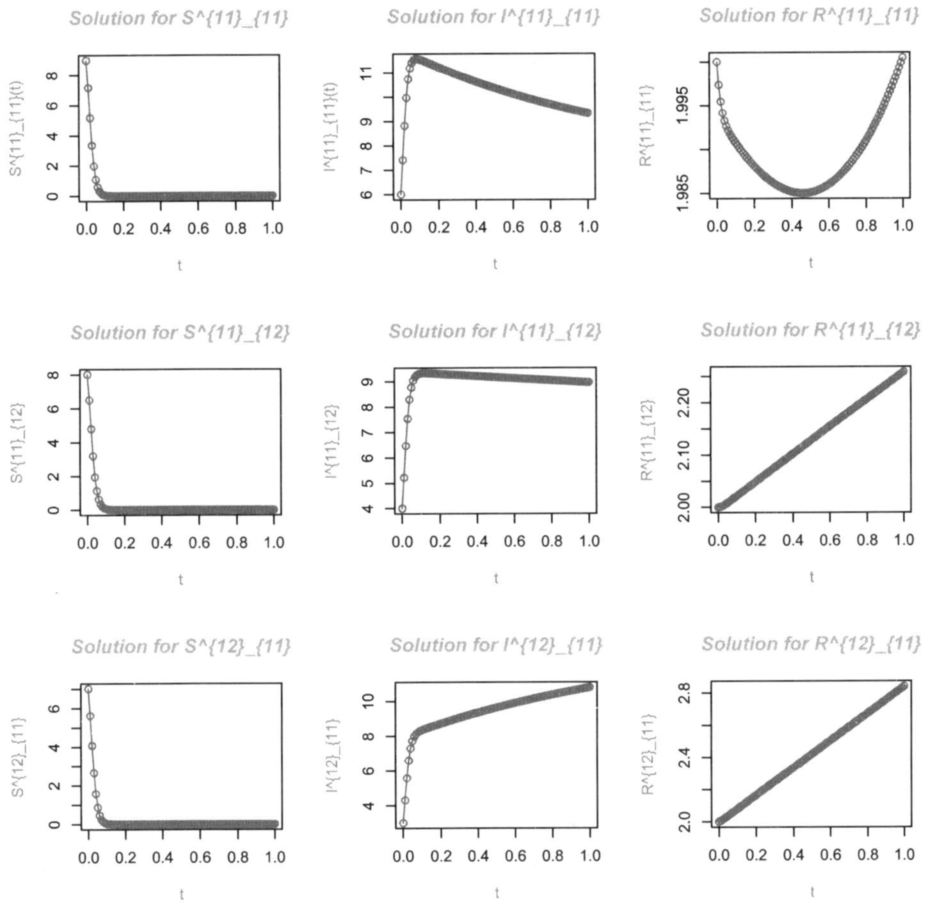

Figure 7.2: Trajectories of the disease classification (S, I, R) for residents of site s_1^1 in the region C_1 at their current location in the two-scale spatial patch dynamic structure. See Remark 7.5.1 for more comments on this figure.

and (i) represent the trajectories of the different disease classes of residents of site s_1^1 visiting site s_1^2 in the region C_2. These figures reflect the behavior of the disease affecting the residents of site s_1^1 at the interregional level.

Furthermore, we observe that the trajectories of the susceptible (S) and infectious (I) populations saturate to their equilibrium states. We further note that the trajectory paths are random in character, but because of the scale of the pictures presented in this figure, they seem to be smooth.

Remark 7.5.2. From Figure 7.3, we note that Figures (a), (b), and (c) represent the trajectories of the different disease classes of residents of site s_1^2 at home. Figures (d), (e),

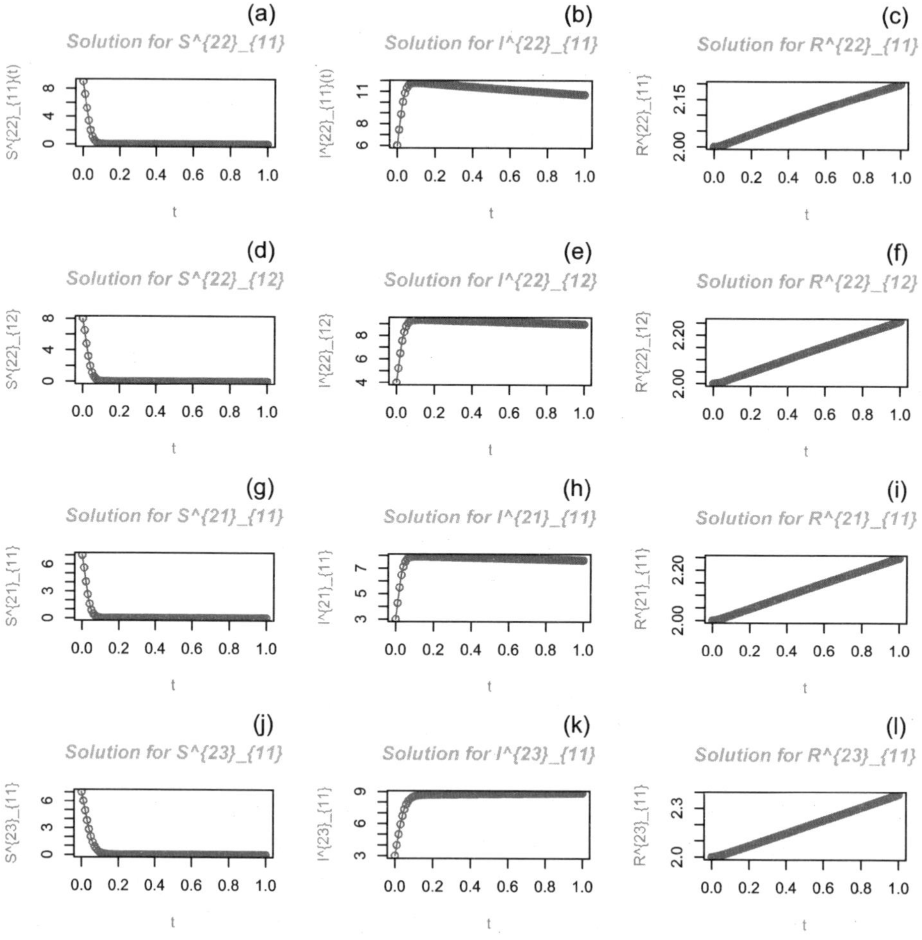

Figure 7.3: Trajectories of the disease classification (S, I, R) for residents of site s_1^2 in the region C_2 at their current location in the two-scale spatial patch dynamic structure. See Remark 7.5.1 for more comments on this figure.

and (f) represent the trajectories of the different disease classes of residents of site s_1^2 visiting site s_2^2 in the home region C_2. These two groups of figures are representative of the disease dynamics of influenza affecting the residents of site s_1^2 at the intraregional level. Figures (g), (h), and (i) represent the trajectories of the different disease classes of residents of site s_1^2 visiting site s_1^1 in the region C_1. Figures (j), (k), and (l) represent the trajectories of the different disease classes of residents of site s_1^2 visiting site s_1^3 in the region C_3. These last two groups of figures reflect the behavior of the disease affecting the residence of site s_1^2 at the interregional level. Furthermore, we observe that the trajectories of the susceptible (S) and infectious (I) populations saturate to their equilibrium states. We further note that the trajectory paths are random in character, but because of the scale of the pictures presented in this figure, they seem to be smooth.

Remark 7.5.3. From Figure 7.4, we note that Figures (a), (b), and (c) represent the trajectories of the different disease classes of residents of site s_1^3 at home. Figures (d), (e), and (f) represent the trajectories of the different disease classes of residents of site s_1^3 visiting site s_2^3 in the home region C_3. These two groups of figures are representatives of the disease dynamics of influenza affecting the residents of site s_1^3 at the intraregional level. Figures (g), (h), and (i) represent the trajectories of the different disease classes of residents of site s_1^3 visiting site s_1^1 in the region C_1. Figures (j), (k), and (l) represent the trajectories of the different disease classes of residents of site s_1^3 visiting site s_1^2 in the region C_2. The last two groups of figures reflect the behavior of the disease affecting the residence of site s_1^3 at the interregional level. Furthermore, we observe that the trajectories of the susceptible (S) and infectious (I) populations saturate to their equilibrium states. We further note that the trajectory paths are random in character, but because of the scale of the pictures presented in this figure, they seem to be smooth.

We make the following comparative remark on the trends in the trajectories of the naturally immune populations represented in Figures 7.2, 7.3, and 7.4, and on the trends of the naturally immune populations represented in Figures 6.4, 6.5, and 6.6.

Remark 7.5.4. We observe significant differences between the trajectories of the naturally immune populations R_{11}^{11} in Figure 7.2, R_{11}^{22} in Figure 7.3, and R_{11}^{33} in Figure 7.4 and the corresponding trajectories of R_{11}^{11} in [47, Figure 6.4], R_{11}^{22} in [47, Figure 6.5], and R_{11}^{33} in [47, Figure 6.6]. The trajectories of the naturally immune populations R_{11}^{11} in Figure 7.2, R_{11}^{22} in Figure 7.3, and R_{11}^{33} in Figure 7.4 exhibit a growth trend in the naturally immune population that initially decreases from the initial state and finally increases over time. This further exhibits the fact that natural immunity and the fluctuating environment influence the growth trends of the endemic population. Moreover, the trajectory of the naturally immune populations R_{11}^{11} in Figure 7.2 indicates a periodic solution over time with period equal to the length of the immunity period.

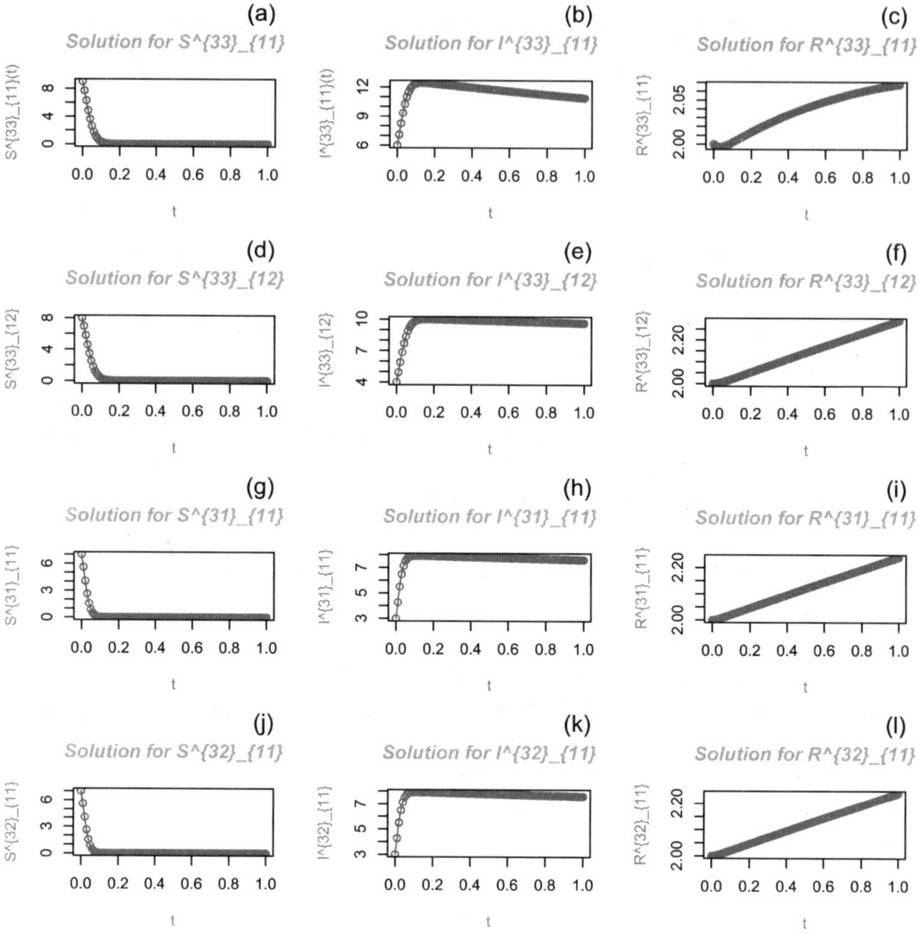

Figure 7.4: Trajectories of the disease classification (S, I, R) for residents of site s_1^3 in the region C_3 at their current location in the two-scale spatial patch dynamic structure. See Remark 7.5.2 for more comments on this figure.

7.6 Discussion and other improvements

This chapter has provided a detailed framework for modeling disease epidemic dynamics that include a fixed temporary delay for acquired immunity in finite-scale populations. It also accounts for environmental fluctuations modeled as white noise, which captures the variability in disease transmission rates due to external environmental factors. By integrating these features the chapter highlights how delays and environmental variability influence the overall disease dynamics.

The findings in Theorem 7.4.3 and Lemma 7.4.2 underscore the existence of a critical threshold delay time, which, together with other system parameters, determines the stability of the disease-free equilibrium and the possibility of eradicating the disease from the population. Although these results are mathematically significant, their practical relevance is limited by the unrealistic assumption of a constant acquired immunity period. In real-world scenarios, the duration of immunity varies across individuals due to factors such as immune system differences, disease progression, and external conditions. This variability undermines the applicability of the results to real populations.

Although introducing a fixed delay for acquired immunity represents an improvement over instantaneous transitions from the recovered to the susceptible state, it remains a simplified representation. A more realistic model would treat the acquired immunity period as a random variable. This approach allows for a more nuanced representation, capturing a wide range of immunity durations—from permanent immunity to short-lived immunity in individuals with weaker immune systems. Such variability is critical for accurately reflecting real-world epidemic dynamics.

To address this limitation, the next chapter explores a more general framework where the acquired immunity delay is modeled as a distributed delay with a continuous probability distribution. This refinement provides a more comprehensive representation of the variability in immunity durations within the population. By incorporating this distributed delay the model gains greater flexibility and realism, making it more relevant for practical applications, such as designing targeted intervention strategies and evaluating the long-term impact of vaccination programs.

8 Modeling the stochastic dynamics of infectious disease epidemics with distributed delay SDEs in finite-scale network populations

8.1 Introduction

Building on the refinements suggested in Section 7.6, this chapter advances the epidemic dynamic model with temporary immunity delay from Chapter 7 by incorporating a more nuanced representation of varying acquired immunity periods. In reality, the duration of natural immunity after recovery from an infectious disease is neither uniform across individuals nor consistent across diseases. This variability arises from differences in individual immune system responses and the unique antigenic profiles of diseases. For instance, measles typically confers lifelong immunity upon recovery [28], whereas influenza provides only transient, strain-specific immunity.

To capture this variability, distributed time delays are introduced into the mean-field differential equations governing epidemic dynamics. Distributed delays enable the representation of immunity periods as random variables with a specified probability distribution, reflecting the heterogeneity among individuals. The use of distributed delays in both deterministic and stochastic models is well established in the literature (see [37, 33, 36]), and these techniques have proven essential in improving the accuracy and applicability of epidemic models.

Stochastic models with distributed delays provide a more realistic framework for understanding the complexities of infectious disease dynamics. For example, Wanduku [36, 35, 37] explored the influence of random factors, such as fluctuations in foreign aid supply and poverty rates, on the prevalence of HIV/AIDS. In these studies, delays were modeled for critical events, including the time until infected individuals begin antiretroviral therapy (ART) and the time until AIDS develops in untreated cases. Notable conclusions from these studies include the finding that shorter expected delays in ART initiation significantly improve disease control outcomes. Additionally, low levels of environmental noise in aid supply and poverty rates, coupled with sufficient support, are critical to preventing the disease from becoming unmanageable. These results highlight the value of incorporating random delays and environmental variability into epidemic modeling for devising effective control strategies.

Revisiting the SIRS epidemic model in Example 7.2.1, this chapter seeks to integrate the randomness associated with the acquired immunity period into the model. By treating the immunity duration as a random variable this enhanced model reflects the inherent diversity in immunity loss among individuals. Such a framework not only improves the realism of the epidemic model but also facilitates a deeper understanding of how immunity dynamics interact with environmental stochasticity and intervention strategies in a two-scale network population. This stochastic extension serves as a foun-

https://doi.org/10.1515/9783111613215-008

dation for further analytical exploration and practical application in disease modeling.

Example 8.1.1. Recall Example 7.2.1 and the dimensionless SIRS model with temporary delay T given by (7.14):

$$\begin{aligned} ds(t) &= [\mu s(t) - \beta_{IS} s(t)i(t) + \alpha_{IR} i(t-T)e^{-\mu T} - \mu s(t)]dt + \sigma_S s(t)dw_S(t),\\ di(t) &= [\beta_{IS} s(t)i(t) - \alpha_{IR} i(t) - \mu i(t)]dt + \sigma_I i(t)dw_I(t),\\ dr(t) &= [\alpha_{IR} i(t) - \alpha_{IR} I(t-T)e^{-\mu T} - \mu r(t)]dt + \sigma_R r(t)dw_R(t), \end{aligned} \tag{8.1}$$

where $\phi_k \in C([t_0 - T, t_0])$, $k = 1, 2, 3$, are $\mathfrak{F}_0$-measurable continuous bounded positive real-valued functions, and $C([t_0-T, t_0], \mathbb{R}_+)$ is the Banach space equipped with the supremum norm

$$\|\phi_k\|_\infty = \sup_{t_0-T\le t\le t_0} |\phi_k(t)|. \tag{8.2}$$

Now (a) assume that the parameter T is a continuous random variable representing time to event (in this case the random time until a recovered individual loses acquired immunity and return to the susceptible state) with distribution $\nu(d\tau)$ and probability density function $f_T(\tau)$ with support $\tau \in [t_0, h]$, where $h > t_0 \ge 0$, and $\nu(d\tau) = f_T(\tau)d\tau$; (b) note that the random variable T depends on the stochastic process $x(t) = (s(t), i(t), r(t))$, $t \ge t_0$, as T is time until the next transition into state s from r.

Under statements (a) and (b), applying the conditional expectation operator $\mathbb{E}_{T|x(t)}[\cdot]$ with respect to the random variable T given $x(t)$ and the Wiener process $w(t) = (w_S, w_I, w_R)(t)$ to system (8.1) leads to the new system

$$\begin{aligned} ds(t) &= \left[\mu s(t) - \beta_{IS} s(t)i(t) + \alpha_{IR}\int_{t_0}^{h} i(t-\tau)e^{-\mu\tau} f_T(\tau)d\tau - \mu s(t)\right]dt + \sigma_S s(t)dw_S(t),\\ di(t) &= [\beta_{IS} s(t)i(t) - \alpha_{IR} i(t) - \mu i(t)]dt + \sigma_I i(t)dw_I(t),\\ dr(t) &= \left[\alpha_{IR} i(t) - \alpha_{IR}\int_{t_0}^{h} i(t-\tau)e^{-\mu\tau} f_T(\tau)d\tau - \mu r(t)\right]dt + \sigma_R r(t)dw_R(t), \end{aligned} \tag{8.3}$$

where $\phi_k \in C([t_0 - h, t_0])$, $k = 1, 2, 3$, are $\mathfrak{F}_0$-measurable continuous bounded positive real-valued functions, and $C([t_0-h, t_0], \mathbb{R}_+)$ is the Banach space equipped with the supremum norm

$$\|\phi_k\|_\infty = \sup_{t_0-h\le t\le t_0} |\phi_k(t)|. \tag{8.4}$$

The primary objective of this chapter is to develop a comprehensive modeling framework for stochastic differential equation (SDE) epidemic models that incorporate distributed delays within finite-scale network populations. This framework addresses the variability in acquired immunity periods and their influence on disease dynamics. Additionally, we conduct a comparative analysis between the fixed delay and distributed delay cases, focusing on their impact on the SIRS epidemic dynamics in finite-scale network populations.

Although certain analytical techniques and results presented in this chapter may overlap with those in Chapter 7, this repetition is both intentional and methodical. It ensures that each chapter remains self-contained and accessible to the readers with diverse interests and varying levels of familiarity with the material. By revisiting foundational concepts while introducing advanced ideas, we aim to provide a cohesive and structured presentation of the progression from fixed delay models to the more versatile distributed delay framework.

8.2 Derivation of the finite-scale network SIRS disease epidemic SDE with distributed delay

In this section, we model the epidemic dynamics in a two-scale network population, incorporating distributed delays to account for varying immunity periods after recovery from a disease. We assume that for all $r \in I(1, M)$ and $i \in I(1, n_r)$, infectious (I_{ia}^{ru}) residents of site s_i^r in the region C_r visiting site s_a^u in the region C_u recover from the disease and acquire immunity against the disease immediately after recovery. However, over time, these recovered individuals lose their immunity and become susceptible to the disease again. The time until immunity wanes is represented as a random variable $T_i^r \in [0, \infty)$, where the immunity period T_i^r is an infinite random variable. From (7.1) we incorporate the varying acquired immunity time delay into the stochastic epidemic dynamic model (7.16)–(7.18) using the average of (7.1) given by

$$\mathbb{E}[\varrho_a^u I_{ia}^{ru}(t - T_i^r) e^{-\delta_a^u T_i^r}] = \varrho_a^u \int_0^\infty I_{ia}^{ru}(t - s) f_{ia}^{ru}(s) e^{-\delta_a^u s} ds, \tag{8.5}$$

where all terms $e^{-\delta_a^u s}$ are earlier defined, and $f_{ia}^{ru}(s)$ is the integral kernel representing the probability density function (pdf) of the random variable T_i^r of the time until the loss of acquired immunity by residents of site s_i^r in the region C_r. Note that the notation of the pdf $f_{ia}^{ru}(s)$ signifies that before recovery and loss of immunity, the resident of s_i^r was infected while visiting site s_a^u in the region C_u. Moreover, the properties of pdfs hold, i. e., $\int_0^\infty f_{ia}^{ru}(s) ds = 1$, and $f_{ia}^{ru}(s) \geq 0$, $s \in [0, \infty)$. The two-scale network stochastic SIRS disease epidemic model with distributed delays obtained from (8.5) and (7.16)–(7.18) is given as follows:

$$dS_{ia}^{ru} = \begin{cases} [B_i^r + \sum_{k=1}^{n_r} \rho_{ik}^{rr} S_{ik}^{rr} + \sum_{q\neq r}^{M} \sum_{a=1}^{n_q} \rho_{ia}^{rq} S_{ia}^{rq} + \varrho_i^r \int_0^\infty I_{ii}^{rr}(t-s) f_{ii}^{rr}(s) e^{-\delta_i^r s} ds \\ \quad -(\gamma_i^r + \sigma_i^r + \delta_i^r) S_{ii}^{rr} - \sum_{u=1}^{M} \sum_{a=1}^{n_u} \beta_{iia}^{rru} S_{ii}^{rr} I_{ai}^{ur}]dt \\ -[\sum_{u=1}^{M} \sum_{a=1}^{n_u} \nu_{iia}^{rru} S_{ii}^{rr} I_{ai}^{ur} dw_{iia}^{rru}(t)], \quad u = r, a = i \\ [\sigma_{ij}^{rr} S_{ii}^{rr} + \varrho_j^r \int_0^\infty I_{ij}^{rr}(t-s) f_{ij}^{rr}(s) e^{-\delta_j^r s} ds - (\rho_{ij}^{rr} + \delta_j^r) S_{ij}^{rr} - \sum_{u=1}^{M} \sum_{a=1}^{n_u} \beta_{jia}^{rru} S_{ij}^{rr} I_{aj}^{ur}]dt \\ \quad - [\sum_{u=1}^{M} \sum_{a=1}^{n_u} \nu_{jia}^{rru} S_{ij}^{rr} I_{aj}^{ur} dw_{jia}^{rru}(t)], \quad u = r,\ a = j,\ j \neq i, \\ [\gamma_{il}^{rq} S_{ii}^{rr} + \varrho_l^q \int_0^\infty I_{il}^{rq}(t-s) f_{il}^{rq}(s) e^{-\delta_l^q s} ds - (\rho_{il}^{rq} + \delta_l^q) S_{il}^{rq} \\ \quad - \sum_{u=1}^{M} \sum_{a=1}^{n_u} \beta_{lia}^{qru} S_{il}^{rq} I_{al}^{uq}]dt - [\sum_{u=1}^{M} \sum_{a=1}^{n_u} \nu_{lia}^{qru} S_{il}^{rq} I_{al}^{uq} dw_{lia}^{qru}(t)], \\ \quad u = q,\ a = l,\ q \neq r, \end{cases} \tag{8.6}$$

$$dI_{ia}^{ru} = \begin{cases} [\sum_{k=1}^{n_r} \rho_{ik}^{rr} I_{ik}^{rr} + \sum_{q\neq r}^{M} \sum_{a=1}^{n_q} \rho_{ia}^{rq} I_{ia}^{rq} - \varrho_i^r I_{ii}^{rr} \\ \quad - (\gamma_i^r + \sigma_i^r + \delta_i^r + d_i^r) I_{ii}^{rr} + \sum_{u=1}^{M} \sum_{a=1}^{n_u} \beta_{iia}^{rru} S_{ii}^{rr} I_{ai}^{ur}]dt \\ \quad + [\sum_{u=1}^{M} \sum_{a=1}^{n_u} \nu_{iia}^{rru} S_{ii}^{rr} I_{ai}^{ur} dw_{iia}^{rru}(t)], \quad u = r,\ a = i, \\ [\sigma_{ij}^{rr} I_{ii}^{rr} - \varrho_j^r I_{ij}^{rr} - (\rho_{ij}^{rr} + \delta_j^r + d_j^r) I_{ij}^{rr} + \sum_{u=1}^{M} \sum_{a=1}^{n_u} \beta_{jia}^{rru} S_{ij}^{rr} I_{aj}^{ur}]dt \\ \quad + [\sum_{u=1}^{M} \sum_{a=1}^{n_u} \nu_{jia}^{rru} S_{ij}^{rr} I_{aj}^{ur} dw_{jia}^{rru}(t)], \quad u = r,\ a = j,\ j \neq i, \\ [\gamma_{il}^{rq} I_{ii}^{rr} - \varrho_l^q I_{il}^{rq} - (\rho_{il}^{rq} + \delta_l^q + d_l^q) I_{il}^{rq} + \sum_{u=1}^{M} \sum_{a=1}^{n_u} \beta_{lia}^{qru} S_{il}^{rq} I_{al}^{uq}]dt \\ \quad + [\sum_{u=1}^{M} \sum_{a=1}^{n_u} \nu_{lia}^{qru} S_{il}^{rq} I_{al}^{uq} dw_{lia}^{qru}(t)], \quad u = q,\ a = l,\ q \neq r, \end{cases} \tag{8.7}$$

$$R_{ia}^{ru} = \begin{cases} [\sum_{k=1}^{n_r} \rho_{ik}^{rr} R_{ik}^{rr} + \sum_{q\neq r}^{M} \sum_{l=1}^{n_q} \rho_{il}^{rq} R_{il}^{rq} + \varrho_i^r I_{ii}^{rr} - \varrho_i^r \int_0^\infty I_{ii}^{rr}(t-s) f_{ii}^{rr}(s) e^{-\delta_i^r s} ds \\ \quad - (\gamma_i^r + \sigma_i^r + \delta_i^r) R_{ii}^{rr}]dt, \quad u = r,\ a = i, \\ [\sigma_{ij}^{rr} R_{ii}^{rr} + \varrho_j^r I_{ij}^{rr} - \varrho_j^r \int_0^\infty I_{ij}^{rr}(t-s) f_{ij}^{rr}(s) e^{-\delta_j^r s} ds \\ \quad - (\rho_{ij}^{rr} + \delta_j^r) R_{ij}^{rr}]dt, \quad u = r,\ a = j,\ j \neq i, \\ [\gamma_{il}^{rq} R_{ii}^{rr} + \varrho_l^q I_{il}^{rq} - \varrho_l^q \int_0^\infty I_{il}^{rq}(t-s) f_{il}^{rq}(s) e^{-\delta_l^q s} ds \\ \quad - (\rho_{il}^{rq} + \delta_l^q) R_{il}^{rq}]dt, \quad u = q,\ a = l,\ q \neq r, \end{cases} \tag{8.8}$$

where all parameters are previously defined. Furthermore, for all $r \in I(1, M)$ and $i \in I(1, n_r)$, we have the following initial conditions:

$$\begin{gathered} (S_{ia}^{ru}(t), I_{ia}^{ru}(t), R_{ia}^{ru}(t)) = (\varphi_{ia1}^{ru}(t), \varphi_{ia2}^{ru}(t), \varphi_{ia3}^{ru}(t)), \quad t \in [-\infty, t_0], \\ \varphi_{iak}^{ru} \in C([-\infty, t_0], \mathbb{R}_+), \quad k = 1,2,3,\ r, q \in I(1, M),\ a \in I(1, n_u),\ i \in I(1, n_r), \\ \varphi_{iak}^{ru}(t_0) > 0, \quad k = 1,2,3, \\ \varphi_k \in \mathcal{U}_C^g \subset C((-\infty, t_0], \mathbb{R}_+), \quad k = 1,2,3,4, \quad \varphi_k(t_0) > 0, \quad k = 1,2,3,4, \end{gathered} \tag{8.9}$$

where $\mathcal{U}_C^g$ is some fading memory sub-Banach space of the Banach space $C((-\infty, t_0], \mathbb{R}_+)$ endowed with the norm

$$\|\varphi\|_g = \sup_{t \le t_0} \frac{|\varphi(t)|}{g(t)}, \tag{8.10}$$

and g is a continuous function with the following properties:

(P1) $g((-\infty, t_0]) \subseteq [1, \infty)$, nonincreasing, and $g(t_0) = 1$.
(P2) $\lim_{u \to 0^-} \frac{g(t+u)}{g(t)} = 1$, uniformly on $[t_0, \infty)$; $\lim_{t \to -\infty} g(t) = \infty$.

An example of such a function is $g(t) = e^{-at}$, $a > 0$. Note that for any g satisfying (P1)–(P2), the Banach space $C((-\infty, t_0], \mathbb{R}_+)$ is continuously embedded in $\mathcal{U}_C^g$, which allows for structural properties for $C((-\infty, t_0], \mathbb{R}_+)$ with the uniform norm to hold in $\mathcal{U}_C^g$ with the $\|\cdot\|_g$ norm. Moreover, $\varphi \in \mathcal{U}_C^g$ if and only if $\|\varphi\|_g < \infty$ and $|\varphi(t)|g(t)$ is uniformly continuous on $(-\infty, t_0]$.

Furthermore, the random continuous functions φ_{iak}^{ru}, $k = 1, 2, 3$, are $\mathcal{F}_0$-measurable or independent of $w(t)$ for all $t \geq t_0$.

It follows from (6.45) and system (8.6)–(8.8) that for $i \in I(1, n_r)$, $l \in I_i^r(1, n_q)$, $r \in I(1, M)$, and $q \in I^r(1, M)$,

$$dy_{il}^{rq} = \begin{cases} [B_i^r + \sum_{k \neq i}^{n_r} \rho_{ik}^{rr} y_{ik}^{rr} + \sum_{q \neq r}^{M} \sum_{a=1}^{n_q} \rho_{ia}^{rq} y_{ia}^{rq} - (\gamma_i^r + \sigma_i^r + \delta_i^r) y_{ii}^{rr} - d_i^r I_{ii}^{rr}]dt, & q = r,\ l = i, \\ [\sigma_{ij}^{rr} y_{ii}^{rr} - (\rho_{ij}^{rr} + \delta_j^r) y_{ij}^{rr} - d_j^r I_{ij}^{rr}]dt, & q = r,\ a = j,\ i \neq j, \\ [\gamma_{il}^{rq} y_{ii}^{rr} - (\rho_{il}^{rq} + \delta_l^q) y_{il}^{rq} - d_l^q I_{il}^{rq}]dt, & q \neq r,\ y_{il}^{rq}(t_0) \geq 0. \end{cases} \tag{8.11}$$

8.3 Model validation results for the SIRS stochastic dynamic model with distributed delay

In the following, we state and prove the existence of a positive solution process for the delayed system (8.6)–(8.8). We utilize the Lyapunov energy function method [7] to establish the results. We observe from (8.6)–(8.8) that (8.8) decouples from the first two equations in the system. Therefore it suffices to prove the existence of a positive solution process for $(S_{ia}^{ru}, I_{ia}^{ru})$. We utilize notations (6.45) and keep in mind that $X_{ia}^{ru} = (S_{ia}^{ru}, I_{ia}^{ru})^T$.

Theorem 8.3.1. *Let $r, u \in I(1, M)$, $i \in I(1, n_r)$, and $a \in I(1, n_u)$. Given any initial conditions (8.9) and (8.10), there exists a unique solution process $X_{ia}^{ru}(t, w) = (S_{ia}^{ru}(t, w), I_{ia}^{ru}(t, w))^T$ satisfying (8.6) and (8.7) for all $t \geq t_0$. Moreover, the solution process is positive for all $t \geq t_0$ a. s., that is, $S_{ia}^{ru}(t, w) > 0$ and $I_{ia(t,w)}^{ru} > 0$ for all $t \geq t_0$ a. s.*

Proof. It is easy to see that the coefficients of (8.6) and (8.7) satisfy the local Lipschitz condition for the given initial data (8.9). Therefore there exists a unique maximal local solution $X_{ia}^{ru}(t, w)$ on $t \in [-\infty, \tau_e(w)]$, where $\tau_e(w)$ is the first hitting time or the explosion time [53]. We show subsequently that $S_{ia}^{ru}(t, w), I_{ia}^{ru}(t, w) > 0$ for all $t \in [-\infty, \tau_e(w)]$ almost surely. We define the stopping time

$$\begin{cases} \tau_+ = \sup\{t \in (t_0, \tau_e(w)) : S_{ia}^{ru}|_{[t_0,t]} > 0 \text{ and } I_{ia}^{ru}|_{[t_0,t]} > 0\}, \\ \tau_+(t) = \min(t, \tau_+) \quad \text{for } t \geq t_0, \end{cases} \tag{8.12}$$

and we show that $\tau_+(t) = \tau_e(w)$ a. s. Suppose on the contrary that $P(\tau_+(t) < \tau_e(w)) > 0$. Let $w \in \{\tau_+(t) < \tau_e(w)\}$ and $t \in [t_0, \tau_+(t))$. Define

$$\begin{cases} V(X_{00}^{00}) = \sum_{r=1}^{M}\sum_{i=1}^{n_r}\sum_{u=1}^{M}\sum_{a=1}^{n_u} V(X_{ia}^{ru}), \\ V(X_{ia}^{ru}) = \ln(S_{ia}^{ru}) + \ln(I_{ia}^{ru}), \quad t \le \tau_+(t). \end{cases} \tag{8.13}$$

We rewrite (8.13) as follows:

$$V(X_{00}^{00}) = \sum_{r=1}^{M}\sum_{i=1}^{n_r}\left[V(X_{ii}^{rr}) + \sum_{j\neq i}^{n_r} V(X_{ij}^{rr}) + \sum_{q\neq r}^{M}\sum_{l=1}^{n_q} V(X_{il}^{rq})\right], \tag{8.14}$$

which further implies that

$$dV(X_{00}^{00}) = \sum_{r=1}^{M}\sum_{i=1}^{n_r}\left[dV(X_{ii}^{rr}) + \sum_{j\neq i}^{n_r} dV(X_{ij}^{rr}) + \sum_{q\neq r}^{M}\sum_{l=1}^{n_q} dV(X_{il}^{rq})\right], \tag{8.15}$$

where dV is the Itô–Doob differential operator with respect to system (8.6)–(8.8). We express the terms on the right-hand side of (8.15) as follows:

Site level: From (8.13) the terms on the right-hand side of (8.15) for the case of $u = r$, $a = i$ are

$$\begin{aligned} dV(X_{ii}^{rr}) = &\left[\frac{B_i^r}{S_{ii}^{rr}} + \sum_{k\neq i}^{n_r}\rho_{ik}^{rr}\frac{S_{ik}^{rr}}{S_{ii}^{rr}} + \sum_{q\neq r}^{M}\sum_{l=1}^{n_q}\rho_{ia}^{rq}\frac{S_{ia}^{rq}}{S_{ii}^{rr}} + \frac{\varrho_i^r}{S_{ii}^{rr}}\int_0^\infty I_{ii}^{rr}(t-s)f_{ii}^{rr}(s)e^{-\delta_i^r s}ds\right. \\ &\left. - (\gamma_i^r + \sigma_i^r + \delta_i^r) - \sum_{u=1}^{M}\sum_{a=1}^{n_u}\beta_{iia}^{rru}I_{ai}^{ur} - \frac{1}{2}\sum_{u=1}^{M}\sum_{a=1}^{n_u}(v_{iia}^{rru})^2(I_{ai}^{ur})^2\right]dt \\ &\times\left[\sum_{k\neq i}^{n_r}\rho_{ik}^{rr}\frac{I_{ik}^{rr}}{S_{ii}^{rr}} + \sum_{q\neq r}^{M}\sum_{l=1}^{n_q}\rho_{ia}^{rq}\frac{I_{ia}^{rq}}{S_{ii}^{rr}} - \varrho_i^r - (\gamma_i^r + \sigma_i^r + \delta_i^r + d_i^r)\right. \\ &\left. - \sum_{u=1}^{M}\sum_{a=1}^{n_u}\beta_{iia}^{rru}\frac{S_{ii}^{rr}}{I_{ii}^{rr}}I_{ai}^{ur} - \frac{1}{2}\sum_{u=1}^{M}\sum_{a=1}^{n_u}(v_{iia}^{rru})^2\frac{(S_{ii}^{rr})^2}{(I_{ii}^{rr})^2}(I_{ai}^{ur})^2\right]dt \\ &- \sum_{u=1}^{M}\sum_{a=1}^{n_u}v_{iia}^{rru}I_{ai}^{ur}dw_{iia}^{rru}(t) + \sum_{u=1}^{M}\sum_{a=1}^{n_u}v_{iia}^{rru}\frac{S_{ii}^{rr}}{I_{ii}^{rr}}I_{ai}^{ur}dw_{iia}^{rru}(t). \end{aligned} \tag{8.16}$$

Intraregional level: From (8.13) the terms on the right-hand side of (8.15) for the case of $u = r$, $a = j, j \neq i$ are

$$\begin{aligned} dV(X_{ij}^{rr}) = &\left[\sigma_{ij}^{rr}\frac{S_{ii}^{rr}}{S_{ij}^{rr}} + \frac{\varrho_j^r}{S_{ij}^{rr}}\int_0^\infty I_{ij}^{rr}(t-s)f_{ij}^{rr}(s)e^{-\delta_j^r s}ds\right. \\ &\left. - (\rho_{ij}^{rr} + \delta_j^r) - \sum_{u=1}^{M}\sum_{a=1}^{n_u}\beta_{jia}^{rru}I_{aj}^{ur} - \frac{1}{2}\sum_{u=1}^{M}\sum_{a=1}^{n_u}(v_{jia}^{rru})^2(I_{aj}^{ur})^2\right]dt \end{aligned}$$

$$
\begin{aligned}
&+\left[\sigma_{ij}^{rr}\frac{I_{ii}^{rr}}{I_{ij}^{rr}}-\varrho_j^r-(\rho_{ij}^{rr}+\delta_j^r+d_j^r)\right.\\
&\left.+\sum_{u=1}^{M}\sum_{a=1}^{n_u}\beta_{jia}^{rru}\frac{S_{ij}^{rr}}{I_{ij}^{rr}}I_{aj}^{ur}-\frac{1}{2}\sum_{u=1}^{M}\sum_{a=1}^{n_u}(v_{jia}^{rru})^2\frac{(S_{ij}^{rr})^2}{(I_{ij}^{rr})^2}(I_{aj}^{ur})^2\right]dt\\
&-\sum_{u=1}^{M}\sum_{a=1}^{n_u}v_{jia}^{rru}I_{aj}^{ur}dw_{jia}^{rru}(t)+\sum_{u=1}^{M}\sum_{a=1}^{n_u}v_{jia}^{rru}\frac{S_{ij}^{rr}}{I_{ij}^{rr}}I_{aj}^{ur}dw_{jia}^{rru}(t).
\end{aligned}
\tag{8.17}
$$

Regional level: From (8.13) the terms on the right-hand side of (8.15) for the case of $u = q, q \neq r, a = l$ are

$$
\begin{aligned}
dV(X_{il}^{rq})=&\left[\gamma_{il}^{rq}\frac{S_{ii}^{rr}}{S_{iq}^{rq}}+\frac{\varrho_l^q}{S_{il}^{rq}}\int_0^\infty I_{il}^{rq}(t-s)f_{il}^{rq}(s)e^{-\delta_l^q s}ds\right.\\
&\left.-(\rho_{il}^{rq}+\delta_l^q)-\sum_{u=1}^{M}\sum_{a=1}^{n_u}\beta_{lia}^{qru}I_{al}^{uq}-\frac{1}{2}\sum_{u=1}^{M}\sum_{a=1}^{n_u}(v_{lia}^{qru})^2(I_{al}^{uq})^2\right]dt\\
&+\left[\gamma_{il}^{rq}\frac{I_{ii}^{rr}}{I_{il}^{rq}}-\varrho_l^q-(\rho_{il}^{rq}+\delta_l^q+d_l^q)\right.\\
&\left.+\sum_{u=1}^{M}\sum_{a=1}^{n_u}\beta_{lia}^{qru}\frac{S_{il}^{rq}}{I_{il}^{rq}}I_{al}^{uq}-\frac{1}{2}\sum_{u=1}^{M}\sum_{a=1}^{n_u}(v_{lia}^{qru})^2\frac{(S_{il}^{rq})^2}{(I_{il}^{rq})^2}(I_{al}^{uq})^2\right]dt\\
&-\sum_{u=1}^{M}\sum_{a=1}^{n_u}v_{lia}^{qru}I_{al}^{uq}dw_{lia}^{qru}(t)+\sum_{u=1}^{M}\sum_{a=1}^{n_u}v_{lia}^{qru}\frac{S_{il}^{rq}}{I_{il}^{rq}}I_{al}^{uq}dw_{lia}^{qru}(t).
\end{aligned}
\tag{8.18}
$$

It follows from (8.16)–(8.18), (8.15), and (8.12) that for $t < \tau_+(t)$,

$$
\begin{aligned}
&V(X_{00}^{00}(t))-V(X_{00}^{00}(t_0))\\
&\geq\sum_{r=1}^{M}\sum_{i=1}^{n_r}\int_{t_0}^{t}\left[\frac{\varrho_i^r}{S_{ii}^{rr}}\int_0^\infty I_{ii}^{rr}(t-s)f_{ii}^{rr}(s)e^{-\delta_i^r s}ds-(\gamma_i^r+\sigma_i^r+\delta_i^r)\right.\\
&\left.-\sum_{u=1}^{M}\sum_{a=1}^{n_u}\beta_{iia}^{rru}I_{ai}^{ur}-\frac{1}{2}\sum_{u=1}^{M}\sum_{a=1}^{n_u}(v_{iia}^{rru})^2(I_{ai}^{ur})^2\right]ds\\
&+\sum_{r=1}^{M}\sum_{i=1}^{n_r}\int_{t_0}^{t}\left[-\varrho_i^r-(\gamma_i^r+\sigma_i^r+\delta_i^r+d_i^r)\right.\\
&\left.-\sum_{u=1}^{M}\sum_{a=1}^{n_u}\beta_{iia}^{rru}\frac{S_{ii}^{rr}}{I_{ii}^{rr}}I_{ai}^{ur}-\frac{1}{2}\sum_{u=1}^{M}\sum_{a=1}^{n_u}(v_{iia}^{rru})^2\frac{(S_{ii}^{rr})^2}{(I_{ii}^{rr})^2}(I_{ai}^{ur})^2\right]ds-\sum_{r=1}^{M}\sum_{i=1}^{n_r}\sum_{u=1}^{M}\sum_{a=1}^{n_u}\int_{t_0}^{t}v_{iia}^{rru}I_{ai}^{ur}dw_{iia}^{rru}(s)\\
&+\sum_{r=1}^{M}\sum_{i=1}^{n_r}\sum_{u=1}^{M}\sum_{a=1}^{n_u}\int_{t_0}^{t}v_{iia}^{rru}\frac{S_{ii}^{rr}}{I_{ii}^{rr}}I_{ai}^{ur}dw_{iia}^{rru}(s)\sum_{r=1}^{M}\sum_{i=1}^{n_r}\sum_{j\neq i}^{n_r}\int_{t_0}^{t}\left[\frac{\varrho_j^r}{S_{ij}^{rr}}\int_0^\infty I_{ij}^{rr}(t-s)f_{ij}^{rr}(s)e^{-\delta_j^r s}ds\right.
\end{aligned}
$$

$$
\begin{aligned}
&- (\rho_{ij}^{rr} + \delta_j^r) - \sum_{u=1}^{M}\sum_{a=1}^{n_u} \beta_{jia}^{rru} I_{aj}^{ur} - \frac{1}{2}\sum_{u=1}^{M}\sum_{a=1}^{n_u} (v_{jia}^{rru})^2 (I_{aj}^{ur})^2 \Bigg] ds \\
&+ \sum_{r=1}^{M}\sum_{i=1}^{n_r}\sum_{j\neq i}^{n_r} \int_{t_0}^{t} \Bigg[-\varrho_j^r - (\rho_{ij}^{rr} + \delta_j^r + d_j^r) - \frac{1}{2}\sum_{u=1}^{M}\sum_{a=1}^{n_u} (v_{jia}^{rru})^2 \frac{(S_{ij}^{rr})^2}{(I_{ij}^{rr})^2} (I_{aj}^{ur})^2 \Bigg] ds \\
&- \sum_{r=1}^{M}\sum_{i=1}^{n_r}\sum_{j\neq i}^{n_r}\sum_{u=1}^{M}\sum_{a=1}^{n_u} \int_{t_0}^{t} v_{jia}^{rru} I_{aj}^{ur} dw_{jia}^{rru}(s) + \sum_{r=1}^{M}\sum_{i=1}^{n_r}\sum_{j\neq i}^{n_r}\sum_{u=1}^{M}\sum_{a=1}^{n_u} \int_{t_0}^{t} v_{jia}^{rru} \frac{S_{ij}^{rr}}{I_{ij}^{rr}} I_{aj}^{ur} dw_{jia}^{rru}(s) \\
&+ \sum_{r=1}^{M}\sum_{i=1}^{n_r}\sum_{q\neq r}^{M}\sum_{l=1}^{n_q} \int_{t_0}^{t} \Bigg[\frac{\varrho_l^q}{S_{il}^{rq}} \int_0^{\infty} I_{il}^{rq}(t-s) f_{il}^{rq}(s) e^{-\delta_l^q s} ds - (\rho_{il}^{rq} + \delta_l^q) \\
&- \sum_{u=1}^{M}\sum_{a=1}^{n_u} \beta_{lia}^{qru} I_{al}^{uq} - \frac{1}{2}\sum_{u=1}^{M}\sum_{a=1}^{n_u} (v_{lia}^{qru})^2 (I_{al}^{uq})^2 \Bigg] ds \\
&+ \sum_{r=1}^{M}\sum_{i=1}^{n_r}\sum_{q\neq r}^{M}\sum_{l=1}^{n_q} \int_{t_0}^{t} \Bigg[-\varrho_l^q - (\rho_{il}^{rq} + \delta_l^q + d_l^q) - \frac{1}{2}\sum_{u=1}^{M}\sum_{a=1}^{n_u} (v_{lia}^{qru})^2 \frac{(S_{il}^{rq})^2}{(I_{il}^{rq})^2} (I_{al}^{uq})^2 \Bigg] ds \\
&- \sum_{r=1}^{M}\sum_{i=1}^{n_r}\sum_{q\neq r}^{M}\sum_{l=1}^{n_q}\sum_{u=1}^{M}\sum_{a=1}^{n_u} \int_{t_0}^{t} v_{lia}^{qru} I_{al}^{uq} dw_{lia}^{qru}(s) + \sum_{r=1}^{M}\sum_{i=1}^{n_r}\sum_{q\neq r}^{M}\sum_{l=1}^{n_q}\sum_{u=1}^{M}\sum_{a=1}^{n_u} \int_{t_0}^{t} v_{lia}^{qru} \frac{S_{il}^{rq}}{I_{il}^{rq}} I_{al}^{uq} dw_{lia}^{qru}(s).
\end{aligned}
\tag{8.19}
$$

Taking the limit of (8.19) as $t \to \tau_+(t)$, it follows from (8.13) and (8.12) that the left-hand side $V(X_{00}^{00}(t)) - V(X_{00}^{00}(t_0)) \le -\infty$ (since by (8.13) and (8.12), $V(X_{ia}^{ru}(\tau_+(t))) = \ln S_{ia}^{ru}(\tau_+(t)) + \ln I_{ia}^{ru}(\tau_+(t)) = -\infty$). This contradicts the finiteness of the right-hand side of inequality (8.19). Hence $\tau_+(t) = \tau_e(w)$ a. s. We show subsequently that $\tau_e(w) = \infty$.

Let $k > 0$ be a positive integer such that $\|\varphi_{00}^{00}\|_1 \le k$, where the vector of initial values $\varphi_{00}^{00} = (\varphi_{ia}^{ru})_{1\le r,u\le M, 1\le i\le n_r, 1\le a\le n_u} \in \mathbb{R}^{2n^2}$ is defined in (8.9). Furthermore, $\|\cdot\|_1$ is the p-sum norm (6.46) for the case $p = 1$. We define the stopping time

$$
\begin{cases}
\tau_k = \sup\{t \in [t_0, \tau_e) : \|X_{00}^{00}(s)\|_1 \le k, s \in [0,t]\}, \\
\tau_k(t) = \min(t, \tau_k),
\end{cases}
\tag{8.20}
$$

where, by (6.46),

$$
\|X_{00}^{00}(s)\|_1 = \sum_{r=1}^{M}\sum_{u=1}^{M}\sum_{i=1}^{n_r}\sum_{a=1}^{n_u} (S_{ia}^{ru}(s) + I_{ia}^{ru}(s)). \tag{8.21}
$$

It is easy to see that as $k \to \infty$, τ_k increases. Set $\lim_{k\to\infty} \tau_k(t) = \tau_\infty$. Then $\tau_\infty \le \tau_e$ a. s. We show in the following that (1) $\tau_e = \tau_\infty$ a. s. $\Leftrightarrow P(\tau_e \neq \tau_\infty) = 0$ and (2) $\tau_\infty = \infty$ a. s. $\Leftrightarrow P(\tau_\infty = \infty) = 1$.

Suppose on the contrary that $P(\tau_\infty < \tau_e) > 0$. Let $w \in \{\tau_\infty < \tau_e\}$ and $t \le \tau_\infty$. In the same way as (8.13) and (8.15), define

$$\begin{cases} V_1(X_{00}^{00}) = \sum_{r=1}^{M}\sum_{i=1}^{n_r}\sum_{u=1}^{M}\sum_{a=1}^{n_u} V(X_{ia}^{ru}), \\ V_1(X_{ia}^{ru}) = e^{\delta_a^u t}(S_{ia}^{ru} + I_{ia}^{ru}), \quad t \le \tau_k(t). \end{cases} \tag{8.22}$$

From (8.22), using expression (8.15), the Itô–Doob differential dV_1 with respect to system (8.6)–(8.8) is given as follows.

Site level: From (8.22) the terms of the right-hand side of (8.15) for the case of $u = r$, $a = i$ are

$$\begin{aligned} &dV_1(X_{ii}^{rr}) \\ &= e^{\delta_i^r t}\left[B_i^r + \sum_{k\neq i}^{n_r}\rho_{ik}^{rr}S_{ik}^{rr} + \sum_{q\neq r}^{M}\sum_{l=1}^{n_q}\rho_{ia}^{rq}S_{ia}^{rq} + \varrho_i^r\int_0^\infty I_{ii}^{rr}(t-s)f_{ii}^{rr}(s)e^{-\delta_i^r s}ds - (\gamma_i^r + \sigma_i^r)S_{ii}^{rr}\right]dt \\ &+ e^{\delta_i^r t}\left[\sum_{k\neq i}^{n_r}\rho_{ik}^{rr}I_{ik}^{rr} + \sum_{q\neq r}^{M}\sum_{l=1}^{n_q}\rho_{ia}^{rq}I_{ia}^{rq} - \varrho_i^r I_{ii}^{rr} - (\gamma_i^r + \sigma_i^r + d_i^r)I_{ii}^{rr}\right]dt. \end{aligned} \tag{8.23}$$

Intraregional level: From (8.22) the terms of the right-hand side of (8.15) for the case of $u = r$, $a = j, j \neq i$ are

$$\begin{aligned} dV_1(X_{ij}^{rr}) &= e^{\delta_i^r t}\left[\sigma_{ij}^{rr}S_{ii}^{rr} + \varrho_j^r\int_0^\infty I_{ij}^{rr}(t-s)f_{ij}^{rr}(s)e^{-\delta_j^r s}ds - \rho_{ij}^{rr}S_{ij}^{rr}\right]dt \\ &+ e^{\delta_j^r t}[\sigma_{ij}^{rr}I_{ii}^{rr} + \varrho_j^r I_{ij}^{rr} - (\rho_{ij}^{rr} + d_j^r)I_{ij}^{rr}]dt. \end{aligned} \tag{8.24}$$

Regional level: From (8.22) the terms of the right-hand side of (8.15) for the case of $u = q, q \neq r, a = l$ are

$$\begin{aligned} dV_1(X_{il}^{rq}) &= e^{\delta_l^q t}\left[\gamma_{il}^{rq}S_{ii}^{rr} + \varrho_l^q\int_0^\infty I_{il}^{rq}(t-s)f_{il}^{rq}(s)e^{-\delta_l^q s}ds - \rho_{il}^{rq}S_{il}^{rq}\right]dt \\ &+ e^{\delta_l^q t}[\gamma_{il}^{rq}I_{ii}^{rr} + \varrho_l^q I_{il}^{rq} - (\rho_{il}^{rq} + d_l^q)I_{il}^{rq}]dt. \end{aligned} \tag{8.25}$$

From (8.23)–(8.25), integrating (8.15) over $[t_0, \tau]$ leads to

$$\begin{aligned} &V_1(X_{00}^{00}(\tau)) \\ &= V_1(X_{00}^{00}(t_0)) + \sum_{r=1}^{M}\sum_{i=1}^{n_r}\int_{t_0}^{\tau} e^{\delta_i^r s}\left[B_i^r + \sum_{k\neq i}^{n_r}\rho_{ik}^{rr}S_{ik}^{rr} + \sum_{q\neq r}^{M}\sum_{l=1}^{n_q}\rho_{ia}^{rq}S_{ia}^{rq} + \varrho_i^r\int_0^\infty I_{ii}^{rr}(t-s)f_{ii}^{rr}(s)e^{-\delta_i^r s}ds\right. \\ &\left.- (\gamma_i^r + \sigma_i^r)S_{ii}^{rr}\right]ds + \sum_{r=1}^{M}\sum_{i=1}^{n_r}\int_{t_0}^{\tau} e^{\delta_i^r s}\left[\sum_{k\neq i}^{n_r}\rho_{ik}^{rr}I_{ik}^{rr} + \sum_{q\neq r}^{M}\sum_{l=1}^{n_q}\rho_{ia}^{rq}I_{ia}^{rq} - \varrho_i^r I_{ii}^{rr} - (\gamma_i^r + \sigma_i^r + d_i^r)I_{ii}^{rr}\right]ds \\ &+ \sum_{r=1}^{M}\sum_{i=1}^{n_r}\sum_{j\neq i}^{n_r}\int_{t_0}^{\tau} e^{\delta_i^r s}\left[\sigma_{ij}^{rr}S_{ii}^{rr} + \varrho_j^r\int_0^\infty I_{ij}^{rr}(t-s)f_{ij}^{rr}(s)e^{-\delta_j^r s}ds - \rho_{ij}^{rr}S_{ij}^{rr}\right]ds \end{aligned}$$

$$+\sum_{r=1}^{M}\sum_{i=1}^{n_r}\sum_{j\neq i}^{n_r}\int_{t_0}^{\tau} e^{\delta_j^r s}[\sigma_{ij}^{rr}I_{ii}^{rr}-\varrho_j^r I_{ij}^{rr}-(\rho_{ij}^{rr}+d_j^r)I_{ij}^{rr}]ds$$

$$+\sum_{r=1}^{M}\sum_{i=1}^{n_r}\sum_{q\neq r}^{n_r}\sum_{l=1}^{n_q}\int_{t_0}^{\tau} e^{\delta_l^q s}\left[\gamma_{il}^{rq}S_{ii}^{rr}+\varrho_l^q\int_{0}^{\infty} I_{il}^{rq}(t-s)f_{il}^{rq}(s)e^{-\delta_l^q s}ds-\rho_{il}^{rq}S_{il}^{rq}\right]ds$$

$$+\sum_{r=1}^{M}\sum_{i=1}^{n_r}\sum_{q\neq r}^{n_r}\sum_{l=1}^{n_q}\int_{t_0}^{\tau} e^{\delta_l^q s}[\gamma_{il}^{rq}I_{ii}^{rr}-\varrho_l^q I_{il}^{rq}-(\rho_{il}^{rq}+d_l^q)I_{il}^{rq}]ds. \tag{8.26}$$

From (8.26) we let $\tau=\tau_k(t)$, where $\tau_k(t)$ is defined in (8.20). It is easy to see from (8.26), (8.20), (8.21), and (8.22) that

$$k=\|X_{00}^{00}(\tau_k(t))\|_1\leq V_1(X_{00}^{00}(\tau_k(t))). \tag{8.27}$$

Taking the limit of (8.27) as $k\to\infty$ leads to a contradiction because the left-hand side of inequality (8.27) is infinite and the right-hand side is finite. Hence $\tau_e=\tau_\infty$ a. s. In the following, we show that $\tau_e=\tau_\infty=\infty$ a. s.

Let $w\in\{\tau_e<\infty\}$. Applying some algebraic manipulations and simplifications to (8.26), we have

$$\begin{aligned}
&I_{\{\tau_e<\infty\}}V_1(X_{00}^{00}(\tau))\\
&=I_{\{\tau_e<\infty\}}V_1(X_{00}^{00}(t_0))+I_{\{\tau_e<\infty\}}\sum_{r=1}^{M}\sum_{i=1}^{n_r}\frac{B_i^r}{\delta_i^r}(e^{\delta_i^r\tau}-1)\\
&+I_{\{\tau_e<\infty\}}\sum_{r=1}^{M}\sum_{i=1}^{n_r}\sum_{q=1}^{M}\sum_{l=1}^{n_q}\int_{0}^{\infty} f_{il}^{rq}(t)\left[\varrho_l^q\int_{-t}^{t_0} I_{il}^{rq}(s)e^{\delta_l^q s}ds-\varrho_l^q\int_{\tau-t}^{\tau} I_{il}^{rq}(s)e^{\delta_l^q s}ds\right]dt\\
&-I_{\{\tau_e<\infty\}}\sum_{r=1}^{M}\sum_{i=1}^{n_r}\int_{t_0}^{\tau}\left[\sigma_i^r e^{\delta_i^r s}-\sum_{j\neq i}^{n_r}\sigma_{ij}^{rr}e^{\delta_j^r s}\right](S_{ii}^{rr}+I_{ii}^{rr})ds\\
&-I_{\{\tau_e<\infty\}}\sum_{r=1}^{M}\sum_{i=1}^{n_r}\int_{t_0}^{\tau}\left[\gamma_i^r e^{\delta_i^r s}-\sum_{q=1}^{M}\sum_{l=1}^{n_q}\gamma_{il}^{rq}e^{\delta_l^q s}\right](S_{ii}^{rr}+I_{ii}^{rr})ds\\
&-I_{\{\tau_e<\infty\}}\sum_{r=1}^{M}\sum_{i=1}^{n_r}d_i^r\int_{t_0}^{\tau} I_{ii}^{rr}e^{\delta_i^r s}ds\\
&-I_{\{\tau_e<\infty\}}\sum_{r=1}^{M}\sum_{i=1}^{n_r}\sum_{j\neq i}^{n_r}d_j^r\int_{t_0}^{\tau} I_{ij}^{rr}e^{\delta_j^r s}ds\\
&-I_{\{\tau_e<\infty\}}\sum_{r=1}^{M}\sum_{i=1}^{n_r}\sum_{q=1}^{M}\sum_{l=1}^{n_q}d_l^q\int_{t_0}^{\tau} I_{il}^{rq}e^{\delta_l^q s}ds,
\end{aligned} \tag{8.28}$$

where I_A is the indicator function of a set A.

Recalling [45], we have $\sigma_i^r = \sum_{j\neq i}^{n_r} \sigma_{ij}^{rr}$ and $\gamma_i^r = \sum_{q\neq r}^{M} \sum_{l=1}^{n_q} \gamma_{il}^{rq}$. Hence the fourth and fifth terms on the right-hand side of (8.28) are such that $[\sigma_i^r e^{\delta_i^r s} - \sum_{j\neq i}^{n_r} \sigma_{ij}^{rr} e^{\delta_j^r s}] \geq 0, \delta_i^r \geq \delta_j^r$, $j \neq i$, and $[\gamma_i^r e^{\delta_i^r s} - \sum_{q=1}^{M} \sum_{l=1}^{n_q} \gamma_{il}^{rq} e^{\delta_l^q s}] \geq 0, \delta_i^r \geq \delta_l^q, q \neq r, l \in I(1, n_q)$. We now let $\tau = \tau_k(t) \wedge T$ in (8.28), $T > 0$, where $\tau_k(t)$ is defined in (8.20). The expected value of (8.28) is estimated as follows:

$$
\begin{aligned}
&E[I_{\{\tau_e<\infty\}} V_1(X_{00}^{00}(\tau_k(t) \wedge T))] \\
&\quad \leq V_1(X_{00}^{00}(t_0)) + \sum_{i=1}^{n_r} \frac{B_i^r}{\delta_i^r} e^{\delta_i^r \tau_k(t) \wedge T} + \sum_{r=1}^{M} \sum_{i=1}^{n_r} \sum_{q=1}^{M} \sum_{l=1}^{n_q} \int_0^{\infty} f_{il}^{rq}(t) \left[\varrho_l^q \int_{-t}^{t_0} \varphi_{il2}^{rq}(s) e^{\delta_l^q s} ds \right] dt. \qquad (8.29)
\end{aligned}
$$

Furthermore, from (8.21), (8.22), and the definition of the indicator function I_A it follows that

$$
I_{\{\tau_e<\infty, \tau_k(t) \leq T\}} \|X_{00}^{00}(\tau_k(t))\|_1 \leq I_{\{\tau_e<\infty\}} V_1(X_{00}^{00}(\tau_k(t) \wedge T)). \qquad (8.30)
$$

It follows from (8.29), (8.30), and (8.20) that

$$
\begin{aligned}
&P(\{\tau_e < \infty, \tau_k(t) \leq T\})k \\
&\quad = E[I_{\{\tau_e<\infty, \tau_k(t) \leq T\}} \|X_{00}^{00}(\tau_k(t))\|_1] \\
&\quad \leq E[I_{\{\tau_e<\infty\}} V(X_{00}^{00}(\tau_k(t) \wedge T))] \\
&\quad \leq V_1(X_{00}^{00}(t_0)) + \sum_{i=1}^{n_r} \frac{B_i^r}{\delta_i^r} e^{\delta_i^r T} + \sum_{r=1}^{M} \sum_{i=1}^{n_r} \sum_{q=1}^{M} \sum_{l=1}^{n_q} \int_0^{\infty} f_{il}^{rq}(t) \left[\varrho_l^q \int_{-t}^{t_0} \varphi_{il2}^{rq}(s) e^{\delta_l^q s} ds \right] dt. \qquad (8.31)
\end{aligned}
$$

It follows immediately from (8.31) that $P(\{\tau_e < \infty, \tau_\infty \leq T\}) \to 0$ as $k \to \infty$. Furthermore, since $T < \infty$ is arbitrary, we conclude that $P(\{\tau_e < \infty, \tau_\infty < \infty\}) = 0$.

Finally, by the total probability principle,

$$
\begin{aligned}
P(\{\tau_e < \infty\}) &= P(\{\tau_e < \infty, \tau_\infty = \infty\}) + P(\{\tau_e < \infty, \tau_\infty < \infty\}) \\
&\leq P(\{\tau_e \neq \tau_\infty\}) + P(\{\tau_e < \infty, \tau_\infty < \infty\}) \\
&= 0. \qquad (8.32)
\end{aligned}
$$

Thus $\tau_e = \tau_\infty = \infty$ a. s., as required. □

Remark 8.3.1. For all $r \in I(1, M)$ and $i \in I(1, n_r)$, Theorem 8.3.1 signifies that the number of residents of site s_i^r of all categories present at home site s_i^r or visiting intra- and interregional sites s_j^r and s_l^q, respectively, are nonnegative. This implies that the total number of residents of site s_i^r present at the home site and also visiting sites in regions in their intra- and interregional accessible domains [47], given by the sum $N_{i0}^{rr}(t) = \sum_{u=1}^{M} \sum_{a=1}^{n_u} y_{ia}^{ru}$, is nonnegative. Moreover, the total effective population [47], defined by $eff(N_{i0}^{rr})(t) = \sum_{u=1}^{M} \sum_{a=1}^{n_u} y_{ai}^{ur}$, at any site s_i^r in the region C_r is also nonnegative at all times $t \geq t_0$.

The following result defines an upper bound for the solution process of system (8.6)–(8.8). We utilize Theorem 8.3.1 to establish this result.

Theorem 8.3.2. *Suppose the hypotheses of Theorem* 8.3.1 *are satisfied. Let* $\mu = \min_{1\le u\le M, 1\le a\le n_u}(\delta_a^u)$. *If*

$$\sum_{r=1}^{M}\sum_{u=1}^{M}\sum_{i=1}^{n_r}\sum_{a=1}^{n_u} y_{ia}^{ru}(t_0) \le \frac{1}{\mu}\sum_{r=1}^{M}\sum_{i=1}^{n_r} B_i^r, \tag{8.33}$$

then

$$\sum_{r=1}^{M}\sum_{u=1}^{M}\sum_{i=1}^{n_r}\sum_{a=1}^{n_u} y_{ia}^{ru}(t) \le \frac{1}{\mu}\sum_{r=1}^{M}\sum_{i=1}^{n_r} B_i^r \quad \textit{for } t \ge t_0 \quad \textit{a. s.} \tag{8.34}$$

Proof. See [47, Lemma 3.2]. □

Remark 8.3.2. From Theorems 8.3.1 and 8.3.2 we conclude that a closed ball $\bar{\mathfrak{B}}_{R^{3n^2}}(\vec{0}; r)$ in R^{3n^2} under the sum norm $\|\cdot\|_1$ centered at the origin $\vec{0} \in R^{3n^2}$ with radius $r = \frac{1}{\mu}\sum_{r=1}^{M}\sum_{i=1}^{n_r} B_i^r$ is self-invariant with regard to a two-scale network dynamics of human epidemic process (8.6)–(8.8) that is under the influence of human mobility process [45], that is,

$$\bar{\mathfrak{B}}_{R^{3n^2}}(\vec{0}; r) = \left\{(S_{ia}^{ru}, I_{ia}^{ru}, R_{ia}^{ru}) : y_{ia}^{ru}(t) \ge 0 \text{ and } \|x_{00}^{00}\|_1 = \sum_{r=1}^{M}\sum_{u=1}^{M}\sum_{i=1}^{n_r}\sum_{a=1}^{n_u} y_{ia}^{ru}(t) \le \frac{1}{\mu}\sum_{r=1}^{M}\sum_{i=1}^{n_r} B_i^r\right\} \tag{8.35}$$

is a positive self-invariant set for system (8.6)–(8.8). We denote

$$\bar{B} \equiv \frac{1}{\mu}\sum_{r=1}^{M}\sum_{i=1}^{n_r} B_i^r. \tag{8.36}$$

8.4 Stability of the DFE and stochastic stability of the delay SIRS epidemic dynamic model

The disease-free equilibrium state (DFE) is obtained similarly as in the earlier chapters. The DFE for system (8.6)–(8.8) is given as follows. For $r, u \in I(1, M)$, $i \in I(1, n_r)$, and $a \in I(1, n_u)$, let

$$D_i^r = \gamma_i^r + \sigma_i^r + \delta_i^r - \sum_{a=1}^{n_r} \frac{\rho_{ia}^{rr}\sigma_{ia}^{rr}}{\rho_{ia}^{rr} + \delta_a^r} - \sum_{u\ne r}^{M}\sum_{a=1}^{n_u} \frac{\rho_{ia}^{rr}\gamma_{ia}^{ru}}{\rho_{ia}^{ru} + \delta_a^u} > 0. \tag{8.37}$$

Furthermore, let $(S_{ia}^{ru*}, I_{ia}^{ru*}, R_{ia}^{ru*})$ be the equilibrium state of the delayed system (8.6)–(8.8). It is easy to see that the disease-free equilibrium state is $E_{ia}^{ru} = (S_{ia}^{ru*}, 0, 0)$, where

$$S_{ia}^{ru*} = \begin{cases} \frac{B_i^r}{D_i^r}, & u = r,\ a = i, \\ \frac{B_i^r}{D_i^r} \frac{\sigma_{ij}^{rr}}{\rho_{ij}^{rr}+\delta_j^r}, & u = r,\ a \neq i, \\ \frac{B_i^r}{D_i^r} \frac{\gamma_{ia}^{ru}}{\rho_{ia}^{ru}+\delta_a^u}, & u \neq r. \end{cases} \tag{8.38}$$

The asymptotic stability property of E_{ia}^{ru} will be established by verifying the conditions of the stochastic version of the Lyapunov second method given in Section 1.4. Transform system (8.6)–(8.8) with respect to the equilibrium state $(S_{ia}^{ru*}, 0, 0)$ using

$$\begin{cases} U_{ia}^{ru} = S_{ia}^{ru} - S_{ia}^{ru*}, \\ V_{ia}^{ru} = I_{ia}^{ru}, \\ W_{ia}^{ru} = R_{ia}^{ru}. \end{cases} \tag{8.39}$$

By this transformation system (8.6)–(8.8) is transformed into

$$dU_{il}^{rq} = \begin{cases} [\sum_{q=1}^{M}\sum_{a=1}^{n_q}\rho_{ia}^{rq}U_{ia}^{rq} + \varrho_i^r\int_0^\infty V_{ii}^{rr}(t-s)f_{ii}^{rr}(s)e^{-\delta_i^r s}ds \\ \quad - (\gamma_i^r + \sigma_i^r + \delta_i^r)U_{ii}^{rr} - \sum_{u=1}^{M}\sum_{a=1}^{n_u}\beta_{iia}^{rru}(S_{ii}^{rr*} + U_{ii}^{rr})V_{ai}^{ur}]dt \\ \quad - [\sum_{u=1}^{M}\sum_{a=1}^{n_u}\nu_{iia}^{rru}(S_{ii}^{rr*} + U_{ii}^{rr})V_{ai}^{ur}dw_{iia}^{rru}(t)], \quad q = r,\ l = i, \\ [\sigma_{ij}^{rr}U_{ii}^{rr} + \varrho_j^r\int_0^\infty V_{ij}^{rr}(t-s)f_{ij}^{rr}(s)e^{-\delta_j^r s}ds - (\rho_{ij}^{rr} + \delta_j^r)U_{ij}^{rr} \\ \quad - \sum_{u=1}^{M}\sum_{a=1}^{n_u}\beta_{jia}^{rru}(S_{ij}^{rr*} + U_{ij}^{rr})V_{aj}^{ur}]dt \\ \quad - [\sum_{u=1}^{M}\sum_{a=1}^{n_u}\nu_{jia}^{rru}(S_{ij}^{rr*} + U_{ij}^{rr})V_{aj}^{ur}dw_{jia}^{rru}(t)], \quad q = r,\ l = j,\ j \neq i, \\ [\gamma_{il}^{rq}U_{ii}^{rr} + \varrho_l^q\int_0^\infty V_{il}^{rq}(t-s)f_{il}^{rq}(s)e^{-\delta_l^q s}ds - (\rho_{il}^{rq} + \delta_l^q)U_{il}^{rq} \\ \quad - \sum_{u=1}^{M}\sum_{a=1}^{n_u}\beta_{lia}^{qru}S_{il}^{rq}I_{al}^{uq}]dt - [\sum_{u=1}^{M}\sum_{a=1}^{n_u}\nu_{lia}^{qru}(S_{il}^{rq*} + U_{il}^{rq})V_{al}^{uq}dw_{lia}^{qru}(t)], \quad q \neq r, \end{cases} \tag{8.40}$$

$$dV_{il}^{rq} = \begin{cases} [\sum_{q=1}^{M}\sum_{a=1}^{n_q}\rho_{ia}^{rq}V_{ia}^{rq} - (\varrho_i^r + \gamma_i^r + \sigma_i^r + \delta_i^r + d_i^r)W_{ii}^{rr} \\ \quad + \sum_{u=1}^{M}\sum_{a=1}^{n_u}\beta_{iia}^{rru}(S_{ii}^{rr*} + U_{ii}^{rr})V_{ai}^{ur}]dt \\ \quad + [\sum_{u=1}^{M}\sum_{a=1}^{n_u}\nu_{iia}^{rru}(S_{ii}^{rr*} + U_{ii}^{rr})V_{ai}^{ur}dw_{iia}^{rru}(t)], \quad q = r,\ l = i, \\ [\sigma_{ij}^{rr}V_{ii}^{rr} - (\varrho_j^r + \rho_{ij}^{rr} + \delta_j^r + d_j^r)V_{ij}^{rr} + \sum_{u=1}^{M}\sum_{a=1}^{n_u}\beta_{jia}^{rru}(S_{ij}^{rr*} + U_{ij}^{rr})V_{aj}^{ur}]dt \\ \quad + [\sum_{u=1}^{M}\sum_{a=1}^{n_u}\nu_{jia}^{rru}(S_{ij}^{rr*} + U_{ij}^{rr})V_{aj}^{ur}dw_{jia}^{rru}(t)], \quad q = r,\ l = j,\ j \neq i, \\ [\gamma_{il}^{rq}V_{ii}^{rr} - (\varrho_l^q + \rho_{il}^{rq} + \delta_l^q + d_l^q)V_{il}^{rq} + \sum_{u=1}^{M}\sum_{a=1}^{n_u}\beta_{lia}^{qru}(S_{il}^{rq*} + U_{il}^{rq})V_{al}^{uq}]dt \\ \quad + [\sum_{u=1}^{M}\sum_{a=1}^{n_u}\nu_{lia}^{qru}(S_{il}^{rq*} + U_{il}^{rq})V_{al}^{uq}dw_{lia}^{qru}(t)], \quad q \neq r, \end{cases} \tag{8.41}$$

and

$$dW_{il}^{rq} = \begin{cases} [\sum_{q=1}^{M}\sum_{l=1}^{n_q}\rho_{il}^{rq}W_{il}^{rq} + \varrho_i^r V_{ii}^{rr} - \varrho_i^r\int_0^\infty V_{ii}^{rr}(t-s)f_{ii}^{rr}(s)e^{-\delta_i^r s}ds - (\gamma_i^r + \sigma_i^r + \delta_i^r)W_{ii}^{rr}]dt, \\ \quad q = r,\ l = i, \\ [\sigma_{ij}^{rr}W_{ii}^{rr} + \varrho_j^r V_{ij}^{rr} - \varrho_j^r\int_0^\infty V_{ij}^{rr}(t-s)f_{ij}^{rr}(s)e^{-\delta_j^r s}ds - (\rho_{ij}^{rr} + \delta_j^r)W_{ij}^{rr}]dt, \\ \quad q = r,\ l = j,\ j \neq i, \\ [\gamma_{il}^{rq}W_{ii}^{rr} + \varrho_l^q V_{il}^{rq} - \varrho_l^q\int_0^\infty V_{il}^{rq}(t-s)f_{il}^{rq}(s)e^{-\delta_l^q s}ds - (\rho_{il}^{rq} + \delta_l^q)W_{il}^{rq}]dt, \quad q \neq r. \end{cases} \tag{8.42}$$

We state and prove the following lemmas that would be useful in the proofs of the stability results.

Lemma 8.4.1. *Let $V_1 : \mathbb{R}^{3n^2} \times \mathbb{R}_+ \to \mathbb{R}_+$ be the function defined by*

$$\begin{cases} V_1(\tilde{x}_{00}^{00}) = \sum_{r=1}^{M}\sum_{u=1}^{M}\sum_{i=1}^{n_r}\sum_{a=1}^{n_u} V(\tilde{x}_{ia}^{ru}), \\ V_1(\tilde{x}_{ia}^{ru}) = (S_{ia}^{ru} - S_{ia}^{ru*} + I_{ia}^{ru})^2 + c_{ia}^{ru}(I_{ia}^{ru})^2 + (R_{ia}^{ru})^2 \\ \tilde{x}_{00}^{00} = (U_{ia}^{ru}, V_{ia}^{ru}, W_{ia}^{ru})^T, \quad \text{and} \quad c_{ia}^{ru} \geq 0. \end{cases} \tag{8.43}$$

Then $V_1 \in C^{2,1}(\mathbb{R}^{3n^2} \times \mathbb{R}_+, \mathbb{R}_+)$, and it satisfies

$$b(\|\tilde{x}_{00}^{00}\|) \leq V_1(\tilde{x}_{00}^{00}(t)) \leq a(\|\tilde{x}_{00}^{00}\|), \tag{8.44}$$

where

$$b(\|\tilde{x}_{00}^{00}\|) = \min_{1\leq r,u\leq M, 1\leq i\leq n_r, 1\leq a\leq n_u}\left\{\frac{c_{ia}^{ru}}{2 + c_{ia}^{ru}}\right\}\sum_{r=1}^{M}\sum_{u=1}^{M}\sum_{i=1}^{n_r}\sum_{a=1}^{n_u}[(U_{ia}^{ru})^2 + (V_{ia}^{ru})^2 + (W_{ia}^{ru})^2],$$

and

$$a(\|\tilde{x}_{00}^{00}\|) = \max_{1\leq r,u\leq M, 1\leq i\leq n_r, 1\leq a\leq n_u}\{c_{ia}^{ru} + 2\}\sum_{r=1}^{M}\sum_{u=1}^{M}\sum_{i=1}^{n_r}\sum_{a=1}^{n_u}[(U_{ia}^{ru})^2 + (V_{ia}^{ru})^2 + (W_{ia}^{ru})^2]. \tag{8.45}$$

Proof. See Chapter 6, Lemma 8.4.1, or [47, Lemma 4.1]. □

Remark 8.4.1. Lemma 8.4.1 shows that the Lyapunov function V defined in (8.43) is a positive definite, decrescent, and radially unbounded (see (8.44)) function [53, 24].

We now state the following lemma.

Lemma 8.4.2. *Assume that the hypothesis of Lemma* 8.4.1 *is satisfied. Define the Lyapunov functional*

$$V = V_1 + V_2, \tag{8.46}$$

where V_1 is defined by (8.43)*, and*

$$V_2 = 3\sum_{r=1}^{M}\sum_{i=1}^{n_r}\sum_{u=1}^{M}\sum_{a=1}^{n_u}\left[\frac{(\varrho_a^u)^2}{\mu_{ia}^{ru}}\int_0^\infty\left(f_{ia}^{rr}(s)e^{-2\delta_a^r s}\int_{t-s}^{t}(V_{ia}^{ru}(\theta))^2 d\theta\right)ds\right]. \tag{8.47}$$

Furthermore, let

$$\mathfrak{U}_{ia}^{ru} = \begin{cases} \dfrac{[\sum_{u=1}^{M}\sum_{a=1}^{n_u}\mu_{ia}^{ru}+\sum_{a\neq i}^{n_r}\frac{(\sigma_{ia}^{rr})^2}{\mu_{ii}^{rr}}+\sum_{a\neq r}^{M}\sum_{a=1}^{n_r}\frac{(\gamma_{ia}^{ru})^2}{\mu_{ii}^{rr}}+2\mu_{ii}^{rr}]}{(\gamma_i^r+\sigma_i^r+\delta_i^r)}, & u=r,\ i=a, \\ \dfrac{[\frac{(\rho_{ia}^{rr})^2}{\mu_{ia}^{rr}}+\mu_{ii}^{rr}+\frac{3}{2}\mu_{ia}^{rr}]}{(\rho_{ia}^{rr}+\delta_a^r)}, & u=r,\ a\neq i, \\ \dfrac{[\frac{(\rho_{ia}^{ru})^2}{\mu_{ia}^{ru}}+\mu_{ii}^{rr}+\frac{3}{2}\mu_{ia}^{ru}]}{(\rho_{ia}^{ru}+\delta_a^u)}, & u\neq r, \end{cases} \tag{8.48}$$

$$\mathfrak{V}_{ia}^{ru} = \begin{cases} \dfrac{\frac{1}{2}\sum_{u=1}^{M}\sum_{a=1}^{n_u}\mu_{ia}^{ru}+\frac{1}{2}\sum_{v=1}^{M}\sum_{b=1}^{n_v}\beta_{iib}^{rrv}(S_{ii}^{rr*}\mu_{ii}^{rr}+\mu_{ii}^{rr})+\frac{1}{2}d_{ii}^{rr}}{\varrho_i^r+\gamma_i^r+\sigma_i^r+\delta_i^r+d_i^r}, & a=i,\ u=r, \\ \dfrac{\frac{1}{2}\mu_{ii}^{rr}+\frac{1}{2}\sum_{v=1}^{M}\sum_{b=1}^{n_v}\beta_{aib}^{rrv}(S_{ia}^{rr*}\mu_{ia}^{rr}+\mu_{ia}^{rr})+\frac{1}{2}d_{ai}^{rr}}{\varrho_a^r+\rho_{ia}^{rr}+\delta_a^r+d_a^r}, & a\neq i,\ u=r, \\ \dfrac{\frac{1}{2}\mu_{ii}^{rr}+\frac{1}{2}\sum_{v=1}^{M}\sum_{b=1}^{n_v}\beta_{aib}^{urv}(S_{ii}^{ru*}\mu_{ia}^{ru}+\mu_{ia}^{ru})+\frac{1}{2}d_{ai}^{ur}}{\varrho_a^u+\rho_{ia}^{ru}+\delta_a^u+d_a^u}, & u\neq r, \end{cases} \tag{8.49}$$

and

$$\mathfrak{W}_{ia}^{ru} = \begin{cases} \dfrac{[\frac{1}{2}\sum_{u=1}^{M}\sum_{a=1}^{n_u}\mu_{ia}^{ru}+\frac{1}{2}\sum_{u\neq r}^{M}\sum_{a=1}^{n_r}\frac{(\gamma_{ia}^{ru})^2}{\mu_{ii}^{rr}}+\frac{1}{2}\sum_{a\neq i}^{n_r}\frac{(\sigma_{ia}^{rr})^2}{\mu_{ii}^{rr}}+\mu_{ii}^{rr}]}{(\gamma_i^r+\sigma_i^r+\delta_i^r)}, & u=r,\ a=i, \\ \dfrac{[\frac{1}{2}\frac{(\rho_{ia}^{rr})^2}{\mu_{ia}^{rr}}+\frac{1}{2}\mu_{ii}^{rr}+\mu_{ia}^{rr}]}{(\rho_{ia}^{rr}+\delta_a^r)}, & u=r,\ a\neq i, \\ \dfrac{[\frac{1}{2}\frac{(\rho_{ia}^{ru})^2}{\mu_{ia}^{ru}}+\frac{1}{2}\mu_{ii}^{rr}+\mu_{ia}^{ru}]}{(\rho_{ia}^{ru}+\delta_a^u)}, & u\neq r, \end{cases} \tag{8.50}$$

for some suitably defined positive number μ_{ia}^{ru}, depending on δ_a^u, for all $r, u \in I^r(1, M)$, $i \in I(1, n)$, and $a \in I_i^r(1, n_r)$. Assume that $\mathfrak{U}_{ia}^{ru} \le 1$, $\mathfrak{V}_{ia}^{ru} < 1$, and $\mathfrak{W}_{ia}^{ru} \le 1$. Then there exist positive numbers ϕ_{ia}^{ru}, ψ_{ia}^{ru}, and φ_{ia}^{ru} such that the differential operator LV associated with Itô–Doob-type stochastic system (8.6)–(8.8) satisfies the inequality

$$\begin{aligned} LV(\tilde{x}_{00}^{00}) \le \sum_{r=1}^{M}\sum_{i=1}^{n_r}\Big[&-[\phi_{ii}^{rr}(U_{ii}^{rr})^2+\psi_{ii}^{rr}(V_{ii}^{rr})^2+\varphi_{ii}^{rr}(W_{ii}^{rr})^2] \\ &-\sum_{a\neq i}^{n_r}[\phi_{ia}^{rr}(U_{ia}^{rr})^2+\psi_{ia}^{rr}(V_{ia}^{rr})^2+\varphi_{ia}^{rr}(W_{ia}^{rr})^2] \\ &-\sum_{u\neq r}^{M}\sum_{a=1}^{n_u}[\phi_{ia}^{ru}(U_{ia}^{rr})^2+\psi_{ia}^{ru}(V_{ia}^{ru})^2+\varphi_{ia}^{ru}(W_{ia}^{ru})^2]\Big]. \end{aligned} \tag{8.51}$$

Moreover,

$$LV(\tilde{x}_{00}^{00}) \le -cV_1(\tilde{x}_{00}^{00}), \tag{8.52}$$

where a positive constant c is defined by

$$c = \frac{\min_{1\le r,u\le M, 1\le i\le n_r, 1\le a\le n_u}\{\phi_{ia}^{ru}, \psi_{ia}^{ru}, \varphi_{ia}^{ru}\}}{\max_{1\le r,u\le M, 1\le i\le n_r, 1\le a\le n_u}\{C_{ia}^{ru}+2\}}. \tag{8.53}$$

Proof. The computation of differential operator [53, 24] applied to the Lyapunov function V_1 in (8.43) with respect to the large-scale system of Itô–Doob-type stochastic differential equation (8.6)–(8.8) is as follows:

$$LV_1(\tilde{x}_{00}^{00}) = \sum_{r=1}^{M}\sum_{i=1}^{n_r}\left[LV_1(\tilde{x}_{ii}^{rr}) + \sum_{j\neq i}^{n_r} LV_1(\tilde{x}_{ij}^{rr}) + \sum_{u\neq r}^{M}\sum_{a=1}^{n_u} LV_1(\tilde{x}_{ia}^{ru})\right], \tag{8.54}$$

where

$$\begin{aligned}
LV_1(\tilde{x}_{ii}^{rr}) &= 2\sum_{u=1}^{M}\sum_{a=1}^{n_u}\Big[(1+C_{ii}^{rr})\rho_{ia}^{ru}V_{ia}^{ru}V_{ii}^{rr} + \rho_{ia}^{ru}U_{ia}^{ru}U_{ii}^{rr} + \rho_{ia}^{ru}V_{ia}^{ru}U_{ii}^{rr} + \rho_{ia}^{ru}U_{ia}^{ru}V_{ii}^{rr}\\
&\quad + \rho_{ia}^{ru}W_{ia}^{ru}W_{ii}^{rr}\Big]\\
&\quad + 2\varrho_i^r U_{ii}^{rr}\int_0^\infty V_{ii}^{rr}(t-s)f_{ii}^{rr}(s)e^{-\delta_i^r s}ds + 2\varrho_i^r V_{ii}^{rr}\int_0^\infty V_{ii}^{rr}(t-s)f_{ii}^{rr}(s)e^{-\delta_i^r s}ds\\
&\quad - 2\varrho_i^r W_{ii}^{rr}\int_0^\infty V_{ii}^{rr}(t-s)f_{ii}^{rr}(s)e^{-\delta_i^r s}ds - 2\varrho_i^r V_{ii}^{rr}W_{ii}^{rr}\\
&\quad - 2[(\varrho_i^r + d_i^r) + 2(\gamma_i^r + \sigma_i^r + \delta_i^r)]V_{ii}^{rr}U_{ii}^{rr} - 2(\gamma_i^r + \sigma_i^r + \delta_i^r)(U_{ii}^{rr})^2\\
&\quad - 2[(c_{ii}^{rr}+1)\varrho_i^r + 2(c_{ii}^{rr}+1)(\gamma_i^r + \sigma_i^r + \delta_i^r + d_i^r)](V_{ii}^{rr})^2\\
&\quad - 2(\gamma_i^r + \sigma_i^r + \alpha_i^r + \delta_i^r)(W_{ii}^{rr})^2 + 2c_{ii}^{rr}\sum_{u=1}^{M}\sum_{a=1}^{n_u}\beta_{iia}^{rru}(S_{ii}^{rr*} + U_{ii}^{rr})V_{ai}^{ur}V_{ii}^{rr}\\
&\quad + c_{ii}^{rr}\sum_{u=1}^{M}\sum_{a=1}^{n_u}(v_{iia}^{rru})^2(S_{ii}^{rr*} + U_{ii}^{rr})^2(V_{ai}^{ur})^2, \quad u=r,\ a=i,
\end{aligned}\tag{8.55}$$

$$\begin{aligned}
\sum_{a\neq i}^{n_r} LV_1(\tilde{x}_{ia}^{rr}) &= \sum_{a\neq r}^{n_r}\Big\{2(1+c_{ia}^{rr})\sigma_{ia}^{rr}V_{ia}^{rr}V_{ii}^{rr} + 2\sigma_{ia}^{rr}U_{ia}^{rr}U_{ii}^{rr} + 2\sigma_{ia}^{rr}V_{ia}^{rr}U_{ii}^{rr} + 2\sigma_{ia}^{rr}U_{ia}^{rr}V_{ii}^{rr}\\
&\quad + 2\sigma_{ia}^{rr}W_{ia}^{rr}W_{ii}^{rr} + 2\varrho_a^r U_{ia}^{rr}\int_0^\infty V_{ia}^{rr}(t-s)f_{ia}^{rr}(s)e^{-\delta_a^r s}ds\\
&\quad + 2\varrho_a^r V_{ia}^{rr}\int_0^\infty V_{ia}^{rr}(t-s)f_{ia}^{rr}(s)e^{-\delta_a^r s}ds - 2\varrho_a^r W_{ia}^{rr}\int_0^\infty V_{ia}^{rr}(t-s)f_{ia}^{rr}(s)e^{-\delta_a^r s}ds\\
&\quad - 2[(c_{ia}^{rr}+1)\varrho_a^r + 2(c_{ia}^{rr}+1)(\rho_{ia}^{rr} + \delta_a^r)](V_{ia}^{rr})^2\\
&\quad - 2(\rho_{ia}^{rr} + \delta_a^r)(U_{ia}^{rr})^2 - 2(\rho_{ia}^{rr} + \delta_a^r)(W_{ia}^{rr})^2 + 2\varrho_a^r V_{ia}^{rr}W_{ia}^{rr}\\
&\quad - 2[(\varrho_a^r + d_a^r) + 2(\rho_{ia}^{rr} + \delta_a^r)]V_{ia}^{rr}U_{ia}^{rr}\Big\}\\
&\quad + 2\sum_{a\neq i}^{n_r} c_{ia}^{rr}\sum_{v=1}^{M}\sum_{b=1}^{n_v}\beta_{aib}^{rrv}(S_{ia}^{rr*} + U_{ia}^{rr})V_{ba}^{vr}V_{ia}^{rr}
\end{aligned}$$

$$+\sum_{a\neq i}^{n_r} c_{ia}^{rr} \sum_{v=1}^{M}\sum_{b=1}^{n_v} (v_{aib}^{rrv})^2 (S_{ia}^{rr*} + U_{ia}^{rr})^2 (V_{ba}^{vr})^2, \quad u = r,\ a \neq i, \tag{8.56}$$

$$\begin{aligned}
\sum_{u\neq r}^{M}\sum_{a=1}^{n_r} LV_1(\tilde{x}_{ia}^{ru}) = \sum_{u\neq r}^{M}\sum_{a=1}^{n_u} \Bigg\{ & 2(1 + c_{ia}^{ru})\gamma_{ia}^{ru} V_{ia}^{ru} V_{ii}^{rr} + 2\gamma_{ia}^{ru} U_{ia}^{ru} U_{ii}^{rr} + 2\gamma_{ia}^{ru} V_{ia}^{ru} U_{ii}^{rr} + 2\gamma_{ia}^{ru} U_{ia}^{ru} V_{ii}^{rr} \\
& + 2\gamma_{ia}^{ru} W_{ia}^{ru} W_{ii}^{rr} + 2\varrho_a^u U_{ia}^{ru} \int_0^\infty V_{ia}^{ru}(t-s) f_{ia}^{ru}(s) e^{-\delta_a^u s} ds \\
& + 2\varrho_a^u V_{ia}^{ru} \int_0^\infty V_{ia}^{ru}(t-s) f_{ia}^{ru}(s) e^{-\delta_a^u s} ds - 2\varrho_a^u W_{ia}^{ru} \int_0^\infty V_{ia}^{ru}(t-s) f_{ia}^{ru}(s) e^{-\delta_a^u s} ds \\
& - 2[(c_{ia}^{ru} + 1)\varrho_a^u + 2(c_{ia}^{ru} + 1)(\rho_{ia}^{ru} + \delta_a^u + d_a^u)](V_{ia}^{ru})^2 - 2(\rho_{ia}^{ru} + \delta_a^u)(U_{ia}^{ru})^2 \\
& - 2(\rho_{ia}^{ru} + \alpha_a^u + \delta_a^u)(W_{ia}^{rr})^2 + 2\varrho_a^u V_{ia}^{ru} W_{ia}^{ru} \\
& - 2[(\varrho_a^u + d_a^u) + 2(\rho_{ia}^{ru} + \delta_a^u)] V_{ia}^{ru} U_{ia}^{ru} \Bigg\} \\
& + 2\sum_{u\neq r}^{M}\sum_{a=1}^{n_u} c_{ia}^{ru} \sum_{v=1}^{M}\sum_{b=1}^{n_v} \beta_{aib}^{urv} (S_{ia}^{ru*} + U_{ia}^{ru}) V_{ba}^{vu} V_{ia}^{ru} \\
& + \sum_{u\neq r}^{M}\sum_{a=1}^{n_r} c_{ia}^{ru} \sum_{v=1}^{M}\sum_{b=1}^{n_v} (v_{aib}^{urv})^2 (S_{ia}^{ru*} + U_{ia}^{ru})^2 (V_{ba}^{vu})^2, \quad u \neq r.
\end{aligned} \tag{8.57}$$

Using (8.36) and the algebraic inequality

$$2ab \leq \frac{a^2}{g(c)} + b^2 g(c), \tag{8.58}$$

where $a, b, c \in \mathbb{R}$, and the function g is such that $g(c) > 0$, the fourteenth term in (8.55)–(8.57) is estimated as follows:

$$\begin{aligned}
2\sum_{v=1}^{M}\sum_{b=1}^{n_v} c_{ii}^{rr} \beta_{iib}^{rrv} (S_{ii}^{rr*} + U_{ii}^{rr}) V_{bi}^{vr} V_{ii}^{rr} \leq & \sum_{v=1}^{M}\sum_{b=1}^{n_v} c_{ii}^{rr} \beta_{iib}^{rrv} (S_{ii}^{rr*} g_i^r(\delta_i^r) + g_i^r(\delta_i^r))(V_{ii}^{rr})^2 \\
& + \sum_{v=1}^{M}\sum_{b=1}^{n_v} c_{ii}^{rr} \beta_{iib}^{rrv} \left(\frac{S_{ii}^{rr*}}{g_i^r(\delta_i^r)} + \frac{\bar{B}^2}{g_i^r(\delta_i^r)} \right) (V_{bi}^{vr})^2, \\
2\sum_{a\neq r}^{n_r}\sum_{v=1}^{M}\sum_{b=1}^{n_v} c_{ia}^{rr} \beta_{aib}^{rrv} (S_{ia}^{rr*} + U_{ia}^{rr}) V_{ba}^{vr} V_{ia}^{rr} \leq & \sum_{a\neq r}^{n_r}\sum_{v=1}^{M}\sum_{b=1}^{n_v} c_{ia}^{rr} \beta_{aib}^{rrv} (S_{ia}^{rr*} g_i^r(\delta_a^r) + g_i^r(\delta_a^r))(V_{ia}^{rr})^2 \\
& + \sum_{a\neq r}^{n_r}\sum_{v=1}^{M}\sum_{b=1}^{n_v} c_{ia}^{rr} \beta_{aib}^{rrv} \left(\frac{S_{ia}^{rr*}}{g_i^r(\delta_a^r)} + \frac{\bar{B}^2}{g_i^r(\delta_a^r)} \right) (V_{bi}^{vr})^2,
\end{aligned}$$

and

$$2\sum_{u\neq r}^{M}\sum_{a=1}^{n_u}\sum_{v=1}^{M}\sum_{b=1}^{n_v} c_{ia}^{ru} \beta_{aib}^{urv} (S_{ia}^{ru*} + U_{ia}^{ru}) V_{ba}^{vu} V_{ia}^{ru}$$

$$\leq \sum_{u\neq r}^{M}\sum_{a=1}^{n_u}\sum_{v=1}^{M}\sum_{b=1}^{n_v} c_{ia}^{ru}\beta_{aib}^{urv}(S_{ia}^{ru*}g_i^r(\delta_a^u)+g_i^r(\delta_a^u))(V_{ia}^{ru})^2$$
$$+\sum_{u\neq r}^{M}\sum_{a=1}^{n_u}\sum_{v=1}^{M}\sum_{b=1}^{n_v} c_{ia}^{ru}\beta_{aib}^{urv}\left(\frac{S_{ia}^{ru*}}{g_i^r(\delta_a^u)}+\frac{\bar{B}^2}{g_i^r(\delta_a^u)}\right)(V_{ba}^{vu})^2. \tag{8.59}$$

Furthermore, using the Cauchy–Schwarz and Hölder inequalities and (8.58), the sixth, seventh, and eighth terms in (8.55)–(8.57) are estimated as follows:

$$2\varrho_a^u A_{ia}^{ru}\int_0^\infty V_{ia}^{ru}(t-s)f_{ia}^{ru}(s)e^{-\delta_a^u s}ds \leq \frac{(\varrho_a^u)^2}{\mu_{ia}^{ru}}\int_0^\infty (V_{ia}^{ru}(t-s))^2 f_{ia}^{ru}(s)e^{-2\delta_a^u s}ds+\mu_{ia}^{ru}(A_{ia}^{ru})^2,$$
$$r,u\in I(1,M),\ i\in I(1,n_r),\ a\in I(1,n_u),\ A_{ia}^{ru}\in\{U_{ia}^{ru},V_{ia}^{ru},W_{ia}^{ru}\}. \tag{8.60}$$

By (8.55)–(8.59), (8.54), repeated usage of (8.36) and inequality (8.58), coupled with some algebraic manipulations and simplifications, we get the inequality

$$LV_1(\tilde{x}_{00}^{00}) \leq \sum_{r=1}^{M}\sum_{i=1}^{n_r}\Bigg\{\Bigg[2\sum_{u=1}^{M}\sum_{a=1}^{n_u}\mu_{ia}^{ru}+2\sum_{a\neq i}^{n_r}\frac{(\sigma_{ia}^{rr})^2}{\mu_{ii}^{rr}}+2\sum_{u\neq r}^{M}\sum_{a=1}^{n_r}\frac{(\gamma_{ia}^{ru})^2}{\mu_{ii}^{rr}}+4\mu_{ii}^{rr}$$
$$-2(\gamma_i^r+\sigma_i^r+\delta_i^r)\Bigg](U_{ii}^{rr})^2$$
$$+\Bigg[(2+c_{ii}^{rr})\sum_{u=1}^{M}\sum_{a=1}^{n_u}\mu_{ia}^{ru}+\sum_{a\neq r}^{n_r}(2+c_{ia}^{rr})\frac{(\sigma_{ia}^{rr})^2}{\mu_{ii}^{rr}}+\sum_{u=1}^{M}\sum_{a=1}^{n_u}(2+c_{ia}^{ru})\frac{(\gamma_{ia}^{ru})^2}{\mu_{ii}^{rr}}+\mu_{ii}^{rr}$$
$$+\frac{(\varrho_i^r+d_i^r)^2}{\mu_{ii}^{rr}}+4\frac{(\gamma_i^r+\sigma_i^r+\delta_i^r)^2}{\mu_{ii}^{rr}}+\frac{(\varrho_i^r)^2}{\mu_{ii}^{rr}}+c_{ii}^{rr}\sum_{v=1}^{M}\sum_{b=1}^{n_r}\beta_{iib}^{rrv}(S_{ii}^{rr*}\mu_{ii}^{rr}+\mu_{ii}^{rr})$$
$$-2(c_{ii}^{rr}+1)(\varrho_i^r+\gamma_i^r+\sigma_i^r+\delta_i^r+d_i^r)\Bigg](V_{ii}^{rr})^2$$
$$+\Bigg[\sum_{u=1}^{M}\sum_{a=1}^{n_u}\mu_{ia}^{ru}+\sum_{a\neq i}^{n_r}\frac{(\sigma_{ia}^{rr})^2}{\mu_{ii}^{rr}}+\sum_{u\neq r}^{M}\sum_{a=1i}^{n_r}\frac{(\gamma_{ia}^{ru})^2}{\mu_{ii}^{rr}}+2\mu_{ii}^{rr}-2(\gamma_i^r+\sigma_i^r+\delta_i^r)\Bigg](W_{ii}^{rr})^2\Bigg\}$$
$$+\sum_{r=1}^{M}\sum_{i=1}^{n_r}\sum_{a\neq i}^{n_r}\Bigg\{\Bigg[2\frac{(\rho_{ia}^{rr})^2}{\mu_{ia}^{rr}}+2\mu_{ii}^{rr}+3\mu_{ia}^{rr}-2(\rho_{ia}^{rr}+\delta_a^r)\Bigg](U_{ia}^{rr})^2$$
$$+\Bigg[(2+c_{ii}^{rr})\frac{(\rho_{ia}^{rr})^2}{\mu_{ia}^{rr}}+(2+c_{ia}^{rr})\mu_{ii}^{rr}+\frac{(\varrho_a^r+d_a^r)^2}{\mu_{ia}^{rr}}+4\frac{(\rho_{ia}^{rr}+\delta_a^r)^2}{\mu_{ia}^{rr}}+\mu_{ia}^{rr}$$
$$+c_{ia}^{rr}\sum_{v=1}^{M}\sum_{b=1}^{n_r}\beta_{aib}^{rrv}(S_{ia}^{rr*}\mu_{ia}^{rr}+\mu_{ia}^{rr})-2(c_{ia}^{rr}+1)(\varrho_a^r+\rho_{ia}^{rr}+\delta_a^r+d_a^r)\Bigg](V_{ia}^{rr})^2$$
$$+\Bigg[\frac{(\rho_{ia}^{rr})^2}{\mu_{ia}^{rr}}+\mu_{ii}^{rr}+2\mu_{ia}^{rr}-2(\rho_{ia}^{rr}+\delta_a^r)\Bigg](W_{ia}^{rr})^2\Bigg\}$$
$$+\sum_{r=1}^{M}\sum_{i=1}^{n_r}\sum_{u\neq r}^{M}\sum_{a=1}^{n_u}\Bigg\{\Bigg[2\frac{(\rho_{ia}^{ru})^2}{\mu_{ia}^{rr}}+2\mu_{ii}^{rr}+3\mu_{ia}^{ru}-2(\rho_{ia}^{ru}+\delta_a^u)\Bigg](U_{ia}^{ru})^2$$

$$
\begin{aligned}
&+\Bigg[(2+c_{ii}^{rr})\frac{(\rho_{ia}^{ru})^2}{\mu_{ia}^{ru}}+(2+c_{ia}^{ru})\mu_{ii}^{rr}+\frac{(\varrho_a^u+d_a^u)^2}{\mu_{ia}^{ru}}+4\frac{(\rho_{ia}^{ru}+\delta_a^u)^2}{\mu_{ia}^{ru}}+\mu_{ia}^{ru}\\
&+c_{ia}^{ru}\sum_{v=1}^{M}\sum_{b=1}^{n_r}\beta_{aib}^{urv}(S_{ia}^{ru*}\mu_{ia}^{ru}+\mu_{ia}^{ru})-2(c_{ia}^{ru}+1)(\varrho_a^u+\rho_{ia}^{ru}+\delta_a^u+d_a^u)\Bigg](V_{ia}^{ru})^2\\
&+\left[\frac{(\rho_{ia}^{ru})^2}{\mu_{ia}^{ru}}+\mu_{ii}^{rr}+2\mu_{ia}^{ru}-2(\rho_{ia}^{ru}+\delta_a^u)\right](W_{ia}^{ru})^2\Bigg\}\\
&+3\sum_{r=1}^{M}\sum_{i=1}^{n_r}\sum_{u=1}^{M}\sum_{a=1}^{n_u}\frac{(\varrho_a^u)^2}{\mu_{ia}^{ru}}\int_0^\infty (V_{ia}^{ru}(t-s))^2 f_{ia}^{ru}(s)e^{-2\delta_a^u s}ds\\
&+\sum_{r=1}^{M}\sum_{i=1}^{n_r}c_{ii}^{rr}\sum_{v=1}^{M}\sum_{b=1}^{n_r}\left[\beta_{iib}^{rrv}\left(\frac{S_{ii}^{rr*}}{\mu_{ii}^{rr}}+\frac{\bar{B}^2}{\mu_{ii}^{rr}}\right)+(\nu_{iib}^{rrv})^2(S_{ii}^{rr*}+\bar{B})^2\right](V_{bi}^{vr})^2\\
&+\sum_{r=1}^{M}\sum_{i=1}^{n_r}\sum_{a\neq i}^{n_r}c_{ia}^{rr}\left[\sum_{v=1}^{M}\sum_{b=1}^{n_r}\beta_{aib}^{rrv}\left(\frac{S_{ia}^{rr*}}{\mu_{ia}^{rr}}+\frac{\bar{B}^2}{\mu_{ia}^{rr}}\right)+(\nu_{aib}^{rrv})^2(S_{ia}^{rr*}+\bar{B})^2\right](V_{ba}^{vr})^2\\
&+\sum_{r=1}^{M}\sum_{i=1}^{n_r}\sum_{u\neq r}^{M}\sum_{a=1}^{n_r}c_{ia}^{ru}\left[\sum_{v=1}^{M}\sum_{b=1}^{n_r}\beta_{aib}^{urv}\left(\frac{S_{ia}^{ru*}}{\mu_{ia}^{ru}}+\frac{\bar{B}^2}{\mu_{ia}^{ru}}\right)+(\nu_{aib}^{urv})^2(S_{ia}^{ru*}+\bar{B})^2\right](V_{ba}^{vu})^2,
\end{aligned}
\tag{8.61}
$$

where $\mu_{ia}^{ru} = g_i^r(\delta_a^u)$, and g_i^r is appropriately defined by (8.58). For all $r, u \in I(1, M)$, $i \in I(1, n_r)$, and $a \in I(1, n_u)$, using (8.48), (8.49), and (8.50), we define the constants d_{ai}^{ur}, ϕ_{ia}^{ru}, ψ_{ia}^{ru}, and φ_{ia}^{ru} as follows:

$$
d_{ai}^{ur} = \sum_{v=1}^{M}\sum_{b=1}^{n_v} c_{ba}^{vu}\beta_{abi}^{uvr}\left(\frac{S_{ba}^{vu*}+\bar{B}^2}{\mu_{ba}^{vu}}\right)+\sum_{v=1}^{M}\sum_{b=1}^{n_v} c_{ba}^{vu}(\nu_{abi}^{uvr})^2(S_{ba}^{vu*}+\bar{B})^2 \tag{8.62}
$$

for some positive numbers c_{ia}^{ru} and all $r, u \in I^r(1, M)$, $i \in I(1, n)$, and $a \in I_i^r(1, n_r)$,

$$
\phi_{ia}^{ru} = \begin{cases} 2(\gamma_i^r+\sigma_i^r+\delta_i^r)(1-\mathfrak{U}_{ia}^{ru}), & u = r,\ a = i,\\ 2(\rho_{ia}^{rr}+\delta_a^r)(1-\mathfrak{U}_{ia}^{ru}), & u = r,\ a \neq i,\\ 2(\rho_{ia}^{ru}+\delta_a^u)(1-\mathfrak{U}_{ia}^{ru}), & u \neq r, \end{cases} \tag{8.63}
$$

$$
\psi_{ia}^{ru} = \begin{cases} 2(\varrho_i^r+\gamma_i^r+\sigma_i^r+\delta_i^r+d_i^r)[c_{ii}^{rr}(1-\mathfrak{V}_{ii}^{rr})+(1-\frac{1}{2}\mathfrak{C}_{ii}^{rr})], & u = r,\ a = i,\\ 2(\varrho_a^r+\rho_{ia}^{rr}+\delta_a^r+d_a^r)[c_{ia}^{rr}(1-\mathfrak{V}_{ia}^{rr})+(1-\frac{1}{2}\mathfrak{C}_{ia}^{rr})], & u = r,\ a \neq i,\\ 2(\varrho_a^u+\rho_{ia}^{ru}+\delta_a^u+d_a^u)[c_{ia}^{ru}(1-\mathfrak{V}_{ia}^{ru})+(1-\frac{1}{2}\mathfrak{C}_{ia}^{ru})], & u \neq r, \end{cases} \tag{8.64}
$$

and

$$
\varphi_{ia}^{ru} = \begin{cases} 2(\gamma_i^r+\sigma_i^r+\delta_i^r)(1-\mathfrak{W}_{ia}^{ru}), & u = r,\ a = i,\\ 2(\rho_{ia}^{rr}+\delta_a^r)(1-\mathfrak{W}_{ia}^{ru}), & u = r,\ a \neq i,\\ 2(\rho_{ia}^{ru}+\delta_a^u)(1-\mathfrak{W}_{ia}^{ru}), & u \neq r, \end{cases} \tag{8.65}
$$

where $\mathfrak{U}_{ia}^{ru}$, $\mathfrak{V}_{ia}^{ru}$, and $\mathfrak{W}_{ia}^{ru}$ are given in (8.48), (8.49), and (8.50), and

$$\mathfrak{E}_{ia}^{ru} = \begin{cases} \Big[\dfrac{2\sum_{u=1}^{M}\sum_{a=1}^{n_u}\mu_{ia}^{ru}+\sum_{a\neq r}^{n_r}(2+c_{ia}^{rr})\frac{(\sigma_{ia}^{rr})^2}{\mu_{ii}^{rr}}+\sum_{u=1}^{M}\sum_{a=1}^{n_u}(2+c_{ia}^{ru})\frac{(\gamma_{ia}^{ru})^2}{\mu_{ii}^{rr}}+\mu_{ii}^{rr}}{(\varrho_i^r+\gamma_i^r+\sigma_i^r+\delta_i^r+d_i^r)} \\ \quad + \dfrac{\frac{(\varrho_i^r+d_i^r)^2}{\mu_{ii}^{rr}}+4\frac{(\gamma_i^r+\sigma_i^r+\delta_i^r)^2}{\mu_{ii}^{rr}}+\frac{(\varrho_i^r)^2}{\mu_{ii}^{rr}}+3\frac{(\varrho_i^r)^2}{\mu_{ii}^{rr}}\int_0^\infty f_{ii}^{rr}(s)e^{-2\delta_i^r s}ds}{(\varrho_i^r+\gamma_i^r+\sigma_i^r+\delta_i^r+d_i^r)}\Big], & u=r,\ a=i, \\ \Big[\dfrac{(2+c_{ii}^{rr})\frac{(\rho_{ia}^{rr})^2}{\mu_{ia}^{rr}}+2\mu_{ii}^{rr}+\frac{(\varrho_a^r+d_a^r)^2}{\mu_{ia}^{rr}}+4\frac{(\rho_{ia}^{rr}+\delta_a^r)^2}{\mu_{ia}^{rr}}+\mu_{ia}^{rr}+3\frac{(\varrho_a^r)^2}{\mu_{ia}^{rr}}\int_0^\infty f_{ia}^{rr}(s)e^{-2\delta_a^r s}ds}{(\varrho_a^r+\rho_{ia}^{rr}+\delta_a^r+d_a^r)}\Big], & u=r,\ a\neq i, \\ \Big[\dfrac{(2+c_{ii}^{rr})\frac{(\rho_{ia}^{ru})^2}{\mu_{ia}^{ru}}+2\mu_{ii}^{rr}+\frac{(\varrho_a^u+d_a^u)^2}{\mu_{ia}^{ru}}+4\frac{(\rho_{ia}^{ru}+\delta_a^u)^2}{\mu_{ia}^{ru}}+\mu_{ia}^{ru}+3\frac{(\varrho_a^u)^2}{\mu_{ia}^{ru}}\int_0^\infty f_{ia}^{ru}(s)e^{-2\delta_a^u s}ds}{(\varrho_a^u+\rho_{ia}^{ru}+\delta_a^u+d_a^u)}\Big], & u\neq r. \end{cases}$$

From (8.46), (8.47), (8.61), and (8.62), the differential operator LV [53, 24] applied to the Lyapunov functional (8.46), and some further algebraic manipulations we get the inequality

$$\begin{aligned} LV(\tilde{x}_{00}^{00}) \leq \sum_{r=1}^{M}\sum_{i=1}^{n_r} -\Big\{ & [\phi_{ii}^{rr}(U_{ii}^{rr})^2+\psi_{ii}^{rr}(V_{ii}^{rr})^2+\varphi_{ii}^{rr}(W_{ii}^{rr})^2] \\ & + \sum_{a\neq r}^{n_r}[\phi_{ia}^{rr}(U_{ia}^{rr})^2+\psi_{ia}^{rr}(V_{ia}^{rr})^2+\varphi_{ia}^{rr}(W_{ia}^{rr})^2] \\ & + \sum_{u\neq r}^{M}\sum_{a=1}^{n_u}[\phi_{ia}^{ru}(U_{ia}^{ru})^2+\psi_{ia}^{ru}(V_{ia}^{ru})^2+\varphi_{ia}^{ru}(W_{ia}^{ru})^2]\Big\}. \end{aligned} \tag{8.66}$$

Under the assumptions on $\mathfrak{U}_{ia}^{ru}$, $\mathfrak{V}_{ia}^{ru}$, and $\mathfrak{W}_{ia}^{ru}$, it is clear that ϕ_{ia}^{ru}, ψ_{ia}^{ru}, and φ_{ia}^{ru} are positive for suitable choices of the constants $c_{ia}^{ru} > 0$. This proves inequality (8.51). Now the validity of (8.52) follows from (8.51) and (8.44), that is,

$$LV(\tilde{x}_{00}^{00}) \leq -cV_1(\tilde{x}_{00}^{00}),$$

where $c = \frac{\min_{1\leq r,u\leq M,1\leq i\leq n_r,1\leq a\leq n_u}\{\phi_{ia}^{ru},\psi_{ia}^{ru},\varphi_{ia}^{ru}\}}{\max_{1\leq r,u\leq M,1\leq i\leq n_r,1\leq a\leq n_u}\{C_{ia}^{ru}+2\}}$. This completes the proof. □

We now formally state the stochastic stability theorems for the disease-free equilibrium.

Theorem 8.4.3. *Let $r, u \in I(1, M)$, $i \in I(1, n_r)$, and $a \in I(1, n_u)$. Let us assume that the hypotheses of Lemma* 8.4.2 *are satisfied. Then the disease-free solutions E_{ia}^{ru} are asymptotically stable in the large. Moreover, the solutions E_{ia}^{ru} are exponentially mean square stable.*

Proof. By the application of comparison result [53, 24] the proof of stochastic asymptotic stability follows immediately. Moreover, the disease-free equilibrium state is exponentially mean square stable. □

We now consider the following corollary of Theorem 8.4.3.

Corollary 8.4.4. *Let $r \in I(1,M)$ and $i \in I(1,n_r)$. Assume that $\sigma_i^r = \gamma_i^r = 0$ for all $r \in I(1,M)$ and $i \in I(1,n_r)$. Denote*

$$\mathfrak{U}_{ia}^{ru} = \begin{cases} \dfrac{\frac{1}{\delta_i^r}}{\frac{1}{[\sum_{u=1}^{M}\sum_{a=1}^{n_u}\mu_{ia}^{ru}+2\mu_{ii}^{rr}]}}, & u=r,\ i=a, \\ \dfrac{[\frac{(\rho_{ia}^{rr})^2}{\mu_{ia}^{rr}}+\mu_{ii}^{rr}+\frac{3}{2}\mu_{ia}^{rr}]}{(\rho_{ia}^{rr}+\delta_a^r)}, & u=r,\ a\neq i, \\ \dfrac{[\frac{(\rho_{ia}^{ru})^2}{\mu_{ia}^{ru}}+\mu_{ii}^{rr}+\frac{3}{2}\mu_{ia}^{ru}]}{(\rho_{ia}^{ru}+\delta_a^u)}, & u\neq r, \end{cases} \tag{8.67}$$

$$\mathfrak{V}_{ia}^{ru} = \begin{cases} \dfrac{\frac{1}{2}\sum_{u=1}^{M}\sum_{a=1}^{n_u}\mu_{ia}^{ru}+\frac{1}{2}\sum_{v=1}^{M}\sum_{b=1}^{n_v}\beta_{iib}^{rrv}(S_{ii}^{rr*}\mu_{ii}^{rr}+\mu_{ii}^{rr})+\frac{1}{2}d_{ii}^{rr}}{\varrho_i^r+\delta_i^r+d_i^r}, & a=i,\ u=r, \\ \dfrac{\frac{1}{2}\mu_{ii}^{rr}+\frac{1}{2}\sum_{v=1}^{M}\sum_{b=1}^{n_v}\beta_{aib}^{rrv}(S_{ia}^{rr*}\mu_{ia}^{rr}+\mu_{ia}^{rr})+\frac{1}{2}d_{ai}^{rr}}{\varrho_a^r+\rho_{ia}^{rr}+\delta_a^r+d_a^r}, & a\neq i,\ u=r, \\ \dfrac{\frac{1}{2}\mu_{ii}^{rr}+\frac{1}{2}\sum_{v=1}^{M}\sum_{b=1}^{n_v}\beta_{aib}^{urv}(S_{ii}^{ru*}\mu_{ia}^{ru}+\mu_{ia}^{ru})+\frac{1}{2}d_{ai}^{ur}}{\varrho_a^u+\rho_{ia}^{ru}+\delta_a^u+d_a^u}, & u\neq r, \end{cases} \tag{8.68}$$

and

$$\mathfrak{W}_{ia}^{ru} = \begin{cases} \dfrac{\frac{1}{\delta_i^r}}{\frac{1}{[\frac{1}{2}\sum_{u=1}^{M}\sum_{a=1}^{n_u}\mu_{ia}^{ru}+\mu_{ii}^{rr}]}}, & u=r,\ a=i, \\ \dfrac{[\frac{1}{2}\frac{(\rho_{ia}^{rr})^2}{\mu_{ia}^{rr}}+\frac{1}{2}\mu_{ii}^{rr}+\mu_{ia}^{rr}]}{(\rho_{ia}^{rr}+\delta_a^r)}, & u=r,\ a\neq i, \\ \dfrac{[\frac{1}{2}\frac{(\rho_{ia}^{ru})^2}{\mu_{ia}^{ru}}+\frac{1}{2}\mu_{ii}^{rr}+\mu_{ia}^{ru}]}{(\rho_{ia}^{ru}+\delta_a^u)}, & u\neq r. \end{cases} \tag{8.69}$$

The equilibrium state E_{ii}^{rr} is stochastically asymptotically stable if $\mathfrak{U}_{ia}^{ru}, \mathfrak{W}_{ia}^{ru} \leq 1$ and $\mathfrak{V}_{ia}^{ru} < 1$ for all $u \in I^r(1,M)$ and $a \in I_i^r(1,n_u)$.

Proof. Follows immediately from the hypotheses of Lemma 8.4.2 (letting $\sigma_i^r = \gamma_i^r = 0$), Theorem 8.4.3, and some algebraic manipulations. □

8.4.1 Discussion of the stability results of the DFE and disease eradication

The results in Theorem 8.4.3 and Corollary 8.4.4 for the two-level large-scale delayed SIRS disease dynamics model depend on the system parameters. Specifically, the sufficient conditions are algebraically straightforward, computationally efficient, and explicitly defined in terms of the rate parameters. This simplicity enables the examination of a variety of scenarios, offering practical approaches to disease control.

For example, the conditions of $\sigma_i^r = \gamma_i^r = 0$ for all r, i in Corollary 8.4.4 signify that the arbitrary site s_i^r is a "sink" in the context of compartmental systems (see [20, 21]) for all other sites in the inter- and intraregional accessible domains. This scenario is displayed in Figure 8.1. The conditions $\mathfrak{U}_{ii}^{rr} \leq 1$ and $\mathfrak{W}_{ii}^{rr} \leq 1$ exhibit that the basic reproduction number is less than 1 as already described in Section 7.4.3 whenever the following conditions hold: $\mathfrak{V}_{ia}^{ru} < 1$, $u \in I(1, M)$, $a \in I(1, n_r)$, and $\mathfrak{U}_{ia}^{ru} \leq 1$, $\mathfrak{W}_{ia}^{ru} \leq 1$ for all $u = r, a \neq i$ and $u \neq r, a \in I(1, n_r)$. The major difference from the results in Section 7.4.3 is that there are no explicit thresholds obtained for the time delay in the system, and this is because of averaging out the delays in (8.5). Wanduku [34, 35] has made several suggestions on obtaining disease control insights in distributed delay models.

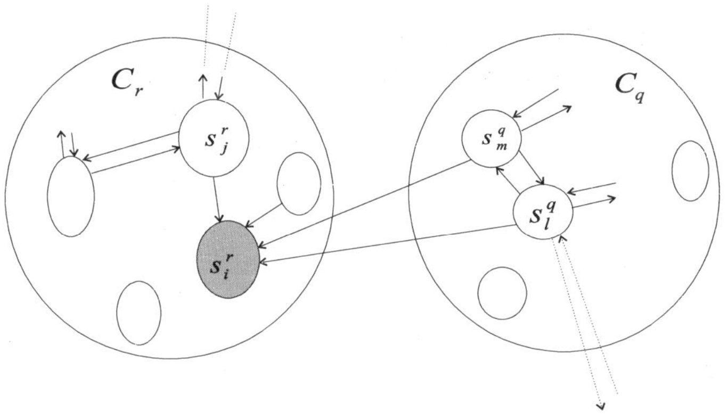

Figure 8.1: Shows that residents of site s_i^r are present only at their home site s_i^r. Hence they isolate every site from their inter- and intraregional accessible domain $C(s_i^r)$. Site s_i^r is a "sink" in the context of the compartmental system [20, 21]. The arrows represent a transport network between any two sites and regions, and the dotted lines and arrows indicate connection with other sites and regions.

Note that the stochastic delay SIRS disease epidemic model with temporary immunity period (7.16)–(7.18) studied in Chapter 7 is a particular case of (8.6)–(8.8) when we let the probability density function of the immunity period $f_{ia}^{ru}(s) = \delta(s - T_i^r)$, $r, u, i, a \in I(1, 3)$, where δ is the Dirac δ-function.

8.4.2 Other special two-scale mobility network patterns

Although Chapter 8 primarily focuses on modeling distributed infinite delays in the SIRS model—specifically the time lapse associated with acquired immunity before an individual, once removed, returns to the susceptible state—there are other important works on distributed delays that provide comprehensive insights into the impacts of memory effects in disease epidemics and the role of stochasticity in communicable disease dynamics.

For a deeper understanding, readers are referred to Wanduku [32], which models and incorporates finite distributed delays during the disease incubation phase in vector-borne epidemics within a two-scale network population. Additionally, this work explores several specialized human mobility networks and discusses spatial disease control mechanisms, shedding light on how these mobility networks influence disease eradication and persistence. These findings are particularly valuable for designing effective disease control policies.

Further related studies by Wanduku [36, 34] expand on how finite distributed delays are modeled in stochastic vector populations. In addition, the book by Ladde et al. [25], which is centered on delay differential equations, offers foundational principles and methods for analyzing delay-driven dynamics in epidemic modeling.

8.5 Future directions, challenges and open problems

This monograph has explored the intricate dynamics of multiscale network population modeling, integrating stochastic processes, hierarchical structures, and complex interactions within biological, social, and ecological systems. We have developed a comprehensive framework that accounts for heterogeneous connectivity, probabilistic transitions, and network-centric influences on population dynamics. Through rigorous analytical techniques and computational simulations, we have demonstrated how stochasticity and network topology shape system behavior, influencing epidemic spread, mobility patterns, and emergent social phenomena.

Key contributions of this work include:

- A unified framework for modeling hierarchical and multiscale population networks, capturing structural complexity and stochastic interactions.
- The integration of birth-and-death processes into compartmental models within hierarchical network populations, extending beyond mean-field dynamic Markovian approaches to better capture stochastic transitions and dependencies.
- Stability and control analysis of stochastic network systems, providing insights into long-term behavior and mitigation strategies.
- Establishing a strong foundation for statistical inference methods to estimate parameters and validate models against empirical data obtained from complex hierarchical population structures with mixed distributions.

These contributions provide a foundation for further advancements in the study of structured populations and their interactions in dynamic and uncertain environments.

8.5.1 Challenges and open problems

Despite the progress made in this work, several challenges remain open for future exploration:

1. **Higher-Order Interactions in Networks:** Traditional network models often assume pairwise interactions, but real-world systems exhibit group-level influences

(e. g., social clustering, co-infections). Extending the framework to incorporate higher-order interactions remains an important direction.
2. **Multiscale Data Integration:** The integration of real-world multiscale data—ranging from individual-level behavioral data to large-scale population mobility patterns—poses computational and methodological challenges. Advanced machine learning techniques and Bayesian inference approaches may provide solutions.
3. **Adaptive and Time-Varying Networks:** Many real-world networks are not static but evolve dynamically over time due to behavioral adaptations, policy interventions, and environmental changes. Developing models that incorporate adaptive network structures and time-dependent transition mechanisms is crucial.
4. **Robust Control and Intervention Strategies:** Designing effective intervention strategies in the presence of uncertainty, noise, and limited data is a significant challenge. Future work could focus on optimal control techniques, stochastic stabilization, and reinforcement learning approaches for decision-making in networked populations.
5. **Computational Efficiency and Scalability:** Large-scale network simulations require significant computational resources. Future research could explore parallel computing, reduced-order modeling, and efficient sampling methods to enhance computational feasibility.
6. **Extending Non-Markovian Models:** The assumption of exponential sojourn times in classical compartmental models is often unrealistic. Future research should develop numerical methods for solving non-Markovian compartmental models, extend theoretical stability and control results for systems with general lifetime distributions, and explore data-driven approaches for estimating non-exponential transition distributions from empirical observations. See Wanduku [44].

8.5.2 Future directions and concluding remarks

Building on the insights from this monograph, several promising avenues for future research emerge:
- **Bridging Theory and Empirical Validation:** While theoretical models provide valuable insights, their predictive power must be validated with real-world data. Collaborations with epidemiologists, sociologists, and ecologists will help refine models for practical applications.
- **Hybrid Modeling Approaches:** Combining stochastic models with agent-based simulations and deep learning frameworks could provide more flexible and data-driven representations of complex population dynamics.
- **Interdisciplinary Applications:** Beyond epidemiology and social sciences, multiscale network models can be applied to financial systems, supply chain networks, and ecological resilience studies, expanding the impact of these methodologies.

- **Development of Open-Source Tools:** Implementing computational frameworks and software tools that facilitate the simulation, analysis, and visualization of stochastic network models would benefit the broader research community.

The study of multiscale network population modeling is a rapidly evolving field with profound implications for understanding complex systems. This monograph has laid the groundwork for advancing both theoretical and applied aspects of the discipline. Future research will continue to refine these models, incorporating new mathematical techniques, computational advancements, and interdisciplinary insights. As our understanding deepens, these models will play an increasingly vital role in shaping policies, predicting outcomes, and mitigating risks in an interconnected world.

Bibliography

[1] Apostol, T. M.: Mathematical Analysis. Addison-Wesley Publishing Company Inc. (1974)

[2] Appiah, E. A., Ladde, G. S., Ladde, J. G.: Stochastic interconnected hybrid dynamic modeling for time-to-event processes. Stoch. Anal. Appl. 41(6), 1056–1098 (2022). https://doi.org/10.1080/07362994.2022.2121723

[3] Axler, S.: Linear Algebra Done Right, 2nd edn. Springer-Verlag, Berlin, New York (1997)

[4] Barabasi, A.: Network Science. Cambridge University Press, UK (2016)

[5] Casella, G., Berger, R. L.: Statistical Inference, 2nd edn. Brooks/Cole, Cengage Learning, Belmont, CA, USA (2002)

[6] Centers for Disease Control and Prevention: CDC Museum COVID-19 Timeline, Last Reviewed: March 15, 2023. https://www.cdc.gov/museum/timeline/covid19.html

[7] Chen, G., Li, T.: Stability of stochastic delayed SIR model. Stoch. Dyn. 9, 231–252 (2009)

[8] CIA World Factbook: http://www.indexmundi.com/g/g.aspx?c=dr&v=25

[9] CIA World Factbook: https://www.cia.gov/library/publications/the-world-factbook/rankorder/2102rank.html

[10] Dalal, N., Greenhalgh, D., Mao, X.: A stochastic model of AIDS and condom use. J. Math. Anal. Appl. 325(1), 36–53 (1 January 2007)

[11] Dalal, N., Greenhalgh, D., Mao, X.: A stochastic model for internal HIV dynamics. J. Math. Anal. Appl. 341(2), 1084–1101 (15 May 2008)

[12] Dudley, R. M.: Real Analysis and Probability. Cambridge University Press (2004)

[13] Dutta, H., Hattaf, K. (eds.): Advances in Epidemiological Modeling and Control of Viruses Academic Press (2023)

[14] Galanis, E., King, A. S., Varughese, P., Halperin, S. A.: Changing epidemiology and emerging risk groups for pertussis. CMAJ, Can. Med. Assoc. J. 174, 451–452 (2006)

[15] H1N1 Flu: International Situation Update (2009). http://www.cdc.gov/h1n1flu/updates/international/

[16] Haber, M. J., Shay, D. K., Davis, X. M., Patel, R., Jin, X., Weintraub, E., et al.: Effectiveness of interventions to reduce contact rates during a simulated influenza pandemic. Emerg. Infect. Dis. 13(4), 581–589 (2007)

[17] Hasminskii, R.: Stochastic Stability of Differential Equations. Sijthoff & Noordoff, Rockville, Maryland (1980)

[18] Higham, D. J.: An algorithmic introduction to numerical simulation of stochastic differential equations. SIAM Rev. 43(3), 525–546 (2001)

[19] Kloeden, P. E., Platen, E.: Numerical Solution of Stochastic Differential Equations. Stochastic Modelling and Applied Probability. Springer, Berlin Heidelberg (2011). https://books.google.com/books?id=BCvtssom1CMC

[20] Ladde, G. S.: Cellular systems – I. Stability of chemical systems. Math. Biosci. 29(3/4), 309–330 (1976)

[21] Ladde, G. S.: Cellular systems – II. Stability of compartmental systems. Math. Biosci. 30, 1–21 (1976)

[22] Ladde, G. S.: Competitive processes and comparism differential systems. Trans. Am. Math. Soc. 221, 391–402 (1976)

[23] Ladde, A., Ladde, G.: An Introduction to Differential: Equations Stochastic Modeling, Methods and Analysis, vol. 2. World Scientific (2013). https://doi.org/10.1142/8384

[24] Ladde, G. S., Lakshmikantham, V.: Random Differential Inequalities. Academic Press, New York (1980)

[25] Ladde, G. S., Lakshmikantham, V., Zhang, B. G.: Oscillation Theory of Differential Equations with Deviating Arguments. Monographs and Textbooks in Pure and Applied Mathematics. M. Dekker, (1987). https://books.google.com/books?id=7RerAAAAIAAJ

[26] Mallapaty, S.: COVID pandemic started in Wuhan market animals after all, suggests latest study. Nature 634(8032), 14–15 (2024 Oct). https://doi.org/10.1038/d41586-024-03026-9

https://doi.org/10.1515/9783111613215-009

[27] Mathews, J. D., McCaw, C. T., McVernon, J., McBryde, E. S., McCaw, J. M.: A biological model for influenza transmission: pandemic planning implications of asymptomatic infection and immunity. PLoS ONE 2, e1220 (2007)

[28] Mossonga, J., Muller, C. P.: Modelling measles re-emergence as a result of waning of immunity in vaccinated populations. Vaccine 21, 4597–4603 (2003)

[29] Murray, J. D.: Mathematical Biology. Springer (1989)

[30] Ratti, J. S., McWaters, M. S.: Precalculus: A Unit Circle Approach. Pearson Education Canada (2009)

[31] Sattenspiel, L., Herring, D. A.: Simulating the effect of quarantine on the spread of the 1918–1919 flu in central Canada. Bull. Math. Biol. 65, 1–26 (2003)

[32] Wanduku, D.: Complete global analysis of a two-scale network SIRS epidemic dynamic model with distributed delay and random perturbations. Appl. Math. Comput. 294, 49–76 (2017 Feb 1)

[33] Wanduku, D.: Threshold conditions for a family of epidemic dynamic models for malaria with distributed delays in a non-random environment. Int. J. Biomath. 11(6), 1850085 (2018)

[34] Wanduku, D.: The stochastic extinction and stability conditions for nonlinear malaria epidemics. Math. Biosci. Eng. 16(5), 3771–3806 (2019). https://doi.org/10.3934/mbe.2019187

[35] Wanduku, D.: A nonlinear multi-population behavioral model to assess the roles of education campaigns, random supply of aids, and delayed ART treatment in HIV/AIDS epidemics. Math. Biosci. Eng. 17(6), 6791–6837 (2020). https://doi.org/10.3934/mbe.2020354

[36] Wanduku, D.: The stationary distribution and stochastic persistence for a class of disease models: case study of malaria. Int. J. Biomath. 04(13), 2050024 (2020). https://www.worldscientific.com/doi/abs/10.1142/S1793524520500242

[37] Wanduku, D.: On the almost sure convergence of a stochastic process in a class of nonlinear multi-population behavioral models for HIV/AIDS with delayed ART treatment. Stoch. Anal. Appl. 39(5), 861–897 (2020). https://doi.org/10.1080/07362994.2020.1848593

[38] Wanduku, D.: The multilevel hierarchical data EM-algorithm. Applications to discrete-time Markov chain epidemic models. Heliyon 8(12), e12622 (2022). https://doi.org/10.1016/j.heliyon.2022.e12622

[39] Wanduku, D.: Finite-and multi-dimensional state representations and some fundamental asymptotic properties of a family of nonlinear multi-population models for HIV/AIDS with ART treatment and distributed delays. Discrete Contin. Dyn. Syst., Ser. S 15(2), 283–313 (2022)

[40] Wanduku, D.: A novel complex social network rumor stochastic model: convergence in distribution to a final rumor size. Heliyon 9(4), e15125 (2023)

[41] Wanduku, D.: Statistical analysis of a new Markov chain model for rumor dynamics in heterogeneous complex social networks. In: Wanduku, D., Zheng, S., Zhou, H., Chen, Z., Sills, A., Agyingi, E. (eds.) Applied Mathematical Analysis and Computations I (SGMC 2021). Springer Proceedings in Mathematics & Statistics, vol. 471. Springer, Cham (2024). https://doi.org/10.1007/978-3-031-69706-7_13

[42] Wanduku, D.: Applied Mathematical Analysis and Computations I. Springer (2024). https://doi.org/10.1007/978-3-031-69706-7

[43] Wanduku, D.: Applied Mathematical Analysis and Computations II. Springer (2024). https://doi.org/10.1007/978-3-031-69710-4

[44] Wanduku, D., Hasan, M. M.: Compartmental models driven by renewal processes: survival analysis and applications to SVIS epidemic models. Sci. Rep. 15(1), 1943 (2025). https://doi.org/10.1038/s41598-024-80166-y

[45] Wanduku, D., Ladde, G. S.: A two-scale network dynamic model for human mobility process. Math. Biosci. 229(1), 1–15 (2011)

[46] Wanduku, D., Ladde, G. S.: Fundamental properties of a two-scale network stochastic human epidemic dynamic model. Neural Parallel Sci. Comput. 19(3), 229 (2011 Sep.)

[47] Wanduku, D., Ladde, G. S.: Global stability of two-scale network human epidemic dynamic model. Neural Parallel Sci. Comput. 19(1–2), 65–90 (2011)

[48] Wanduku, D., Ladde, G. S.: Global properties of a two-scale network stochastic delayed human epidemic dynamic model. Nonlinear Anal., Real World Appl. 13(2), 794–816 (2012 Apr 1)

[49] Wanduku, D., Ladde, G. S.: The global analysis of a stochastic two-scale network epidemic dynamic model with varying immunity period. J. Appl. Math. Phys. 5, 1150–1173 (2017)

[50] Wanduku, D., et al.: Modeling the stochastic dynamics of influenza epidemics with vaccination control, and the maximum likelihood estimation of model parameters. In: Mathematical Modelling in Health, Social and Applied Sciences, pp. 23–72 (2020)

[51] Wanduku, D., Omotomilola, J., Chinmoy, R., Oluyede, B., Farotimi, O.: On a Novel SVEIRS Markov chain epidemic model with multiple discrete delays and infection rates: modeling and sensitivity analysis to determine vaccination effects. In: Advances in Epidemiological Modeling and Control of Viruses, pp. 95–143. Academic Press (2023)

[52] Wasserman, S., Faust, K.: Social Network Analysis. Cambridge University Press, UK (1994)

[53] Xuerong, M.: Stochastic Differential Equations and Applications, 2nd edn. Horwood Publishing Ltd. (2008)

[54] Yaesoubi, R., Cohen, T.: Generalized Markov models of infectious disease spread: a novel framework for developing dynamic health policies. Eur. J. Oper. Res. 215(3), 679–687 (2011). https://doi.org/10.1016/j.ejor.2011.07.016

Index

https://doi.org/10.1515/9783111613215-010

De Gruyter Series in Probability and Stochastics

Volume 5
Nicolas Lanchier
Stochastic Interacting Systems in Life and Social Sciences, 2024
ISBN 978-3-11-079175-4, e-ISBN 978-3-11-079188-4,
e-ISBN (ePUB) 978-3-11-079193-8

Volume 4
Giacomo Ascione, Yuliya Mishura, Enrica Pirozzi
Fractional Deterministic and Stochastic Calculus, 2024
ISBN 978-3-11-077981-3, e-ISBN 978-3-11-078001-7,
e-ISBN (ePUB) 978-3-11-078022-2

Volume 3
Andrey A. Dorogovtsev
Measure-valued Processes and Stochastic Flows, 2023
ISBN 978-3-11-099758-3, e-ISBN 978-3-11-098651-8,
e-ISBN (ePUB) 978-3-11-098655-6

Volume 2
Yuliya Mishura, Kostiantyn Ralchenko
Discrete-Time Approximations and Limit Theorems. In Applications to Financial Markets, 2021
ISBN 978-3-11-065279-6, e-ISBN 978-3-11-065424-0,
e-ISBN (ePUB) 978-3-11-065299-4

Volume 1
Abdelhamid Hassairi
Riesz Probability Distributions, 2021
ISBN 978-3-11-071325-1, e-ISBN 978-3-11-071337-4,
e-ISBN (ePUB) 978-3-11-071345-9